Building Computer Vision Applications Using Artificial Neural Networks

With Examples in OpenCV and TensorFlow with Python

Second Edition

Shamshad Ansari

Apress®

Building Computer Vision Applications Using Artificial Neural Networks: With Examples in OpenCV and TensorFlow with Python, Second Edition

Shamshad Ansari
Centreville, VA, USA

ISBN-13 (pbk): 978-1-4842-9865-7
https://doi.org/10.1007/978-1-4842-9866-4

ISBN-13 (electronic): 978-1-4842-9866-4

Managing Director, Apress Media LLC: Welmoed Spahr
Acquisitions Editor: Celestin Suresh John
Development Editor: Laura Berendson
Coordinating Editor: Mark Powers

Cover designed by eStudioCalamar

Cover image designed by Freepik (www.freepik.com)

Distributed to the book trade worldwide by Apress Media, LLC, 1 New York Plaza, New York, NY 10004, U.S.A. Phone 1-800-SPRINGER, fax (201) 348-4505, e-mail orders-ny@springer-sbm.com, or visit www.springeronline.com. Apress Media, LLC is a California LLC and the sole member (owner) is Springer Science + Business Media Finance Inc (SSBM Finance Inc). SSBM Finance Inc is a **Delaware** corporation.

For information on translations, please e-mail booktranslations@springernature.com; for reprint, paperback, or audio rights, please e-mail bookpermissions@springernature.com.

Apress titles may be purchased in bulk for academic, corporate, or promotional use. eBook versions and licenses are also available for most titles. For more information, reference our Print and eBook Bulk Sales web page at https://www.apress.com/bulk-sales.

Any source code or other supplementary material referenced by the author in this book is available to readers on GitHub (https://github.com/Apress). For more detailed information, please visit https://www.apress.com/gp/services/source-code.

Paper in this product is recyclable

In God we trust.

To my wonderful parents, Abdul Samad and Nazhat Parween, who always corrected my mistakes and raised me to become a good person.

To my lovely wife, Shazia, and our two beautiful daughters, Dua and Erum. Without their love and support, this book would not have been possible.

Table of Contents

About the Author

Shamshad Ansari is an author, inventor, and thought leader in the fields of computer vision, machine learning, artificial intelligence, and cognitive science. He has extensive experience in high-scale, distributed, and parallel computing. Sam currently serves as an adjunct professor at George Mason University, teaching graduate-level programs within the Data Analytics Engineering department of the Volgenau School of Engineering. His areas of instruction encompass machine learning, natural language processing, and computer vision, where he imparts his knowledge and expertise to aspiring professionals.

Having authored multiple publications on topics such as machine learning, RFID, and high-scale enterprise computing, Sam's contributions extend beyond academia. He holds four US patents related to healthcare AI, showcasing his innovative mindset and practical application of technology.

Throughout his extensive 20+ years of experience in enterprise software development, Sam has been involved with several tech startups and early-stage companies. He has played pivotal roles in building and expanding tech teams from the ground up, contributing to their eventual acquisition by larger organizations. At the beginning of his career, he worked with esteemed institutions such as the US Department of Defense (DOD) and IBM, honing his skills and knowledge in the industry.

Currently, Sam serves as the president and CEO of Accure, Inc., an AI company that he founded. He is the creator and architect of, and continues to make significant contributions to, Momentum AI, a no-code platform that encompasses data engineering, machine learning, AI, MLOps, data warehousing, and business intelligence. Throughout his career, Sam has made notable contributions in various domains including healthcare, retail, supply chain, banking and finance, and manufacturing.

ABOUT THE AUTHOR

Demonstrating his leadership skills, he has successfully managed teams of software engineers, data scientists, and DevSecOps professionals, leading them to deliver exceptional results. Sam earned his bachelor's degree in engineering from Birsa Institute of Technology (BIT) Sindri and subsequently a master's degree from the prestigious Indian Institute of Information Technology and Management Kerala (IIITM-K), now known as the Digital University of Kerala.

About the Technical Reviewers

Bharath Kumar Bolla has over 12 years of experience and is a senior data scientist at Salesforce, Hyderabad. Bharath obtained a master's degree in data science from the University of Arizona and a master's degree in life sciences from Mississippi State University. Bharath worked as a research scientist for seven years at the University of Georgia, Emory University, and Eurofins LLC. At Verizon, Bharath led a team to build a "Smart Pricing" solution, and at Happiest Minds, he worked on AI-based digital marketing products. Along with his day-to-day responsibilities, Bharath is a mentor and an active researcher with more than 20 publications in conferences and journals. Bharath received the "40 Under 40 Data Scientists 2021" award from *Analytics India Magazine* for his accomplishments.

F. Brett Berlin is currently a faculty member of the George Mason University College of Engineering and Computing Data Analytics Engineering master's program. He brings five decades of professional and life experience to his passion for investing in people with purpose. He has served as professor, executive consultant, mentor, applied global infosystem strategist, high-performance computing executive and policy leader, and early Internet policy partner. Strategically, he seeks engagements with public/private/academic collaborations for value creation with impact. His current eighth decade passion is investing forward, by grace, serving ethically vision-driven digital/AI innovator aspirants seeking to engage the challenges of the times.

Mr. Berlin is a graduate of the University of Texas Austin Graduate School of Arts and Sciences (Computer Science, "Time-Extended Petri Nets"); the USAF Academy (Computer Science, Mathematics & Engineering), and USAF Officer leadership schools. He is married to Kathleen Berlin, with whom he shares children and grandchildren.

Acknowledgments

I wrote this book with two primary objectives in mind. First, I aimed to comprehensively construct the foundations of computer vision concepts, starting from the basics and progressing to an advanced level. Secondly, I aspired to furnish a practical guide for the application of these concepts in real-world computer vision systems by including practical use cases and code examples.

Achieving these objectives required meticulous organization of topics, integration of content with meaningful and practical use cases, and thorough testing of the code, all of which demanded my undivided attention throughout the writing process. This accomplishment wouldn't have been possible without the unwavering support of my family. My gratitude is immeasurable towards my wife, Shazia, who managed the care of our two daughters (Erum and Dua), ensuring they remained engaged while I dedicated myself to the writing of this book. She ingeniously transformed this experience into a positive experience for both them and myself. Remarkably, my children began tracking my progress, celebrating each completion of a section, subsection, or chapter. Their enthusiasm became a wellspring of energy and motivation that significantly enriched my writing experience. I still do not know what magic my wife used to do this.

My life is indebted to Anumati Bhagi and Ashok Bhagi, who are no less than parents to me; their love and support always motivate me.

This book is a collection of my lifetime experiences that I gained by working with some of the greatest engineers, data scientists, and business professionals. I would like to thank all my colleagues at Accure and all the past companies I have worked at. I sincerely thank all my teachers, professors, and mentors who enlightened me with their knowledge and wisdom.

Working with the Apress editorial team, which includes remarkable individuals like Mark Powers and Laura Berendson, has been an incredible experience. I want to extend a special thank you to Shonmirin PA, the project coordinator, who has been instrumental in coordinating with the editorial team and technical reviewers and providing prompt and helpful responses to my questions along the way.

ACKNOWLEDGMENTS

I would like to express my appreciation to John Celestin, Executive Editor at Apress. His thoughtfulness and quick decision-making have left a profound impression on me. Thank you, John, for having faith in my work. I also extend my gratitude to Apress for making the publication of this book possible.

I wish to extend my gratitude to Bharat Bolla for his dedicated technical review of this book. Additionally, I would like to express special appreciation to Professor Brett Berlin for his meticulous review, invaluable insights, and constructive suggestions that have significantly enhanced the quality of this book. Thank you, Bharat and Professor Berlin.

Finally, I extend my thanks to the readers of this book. I am eagerly interested in receiving your feedback. Kindly share your comments, suggestions, and questions with me at ansarisam@gmail.com. Also, given the ever-evolving nature of technology, I may need to update some of this book's code examples via the book's GitHub repository. I am committed to maintaining the code updates, so check the book's GitHub repository for updates. I anticipate your valuable input and look forward to engaging with you.

Introduction

For over 20 years, I have had the privilege of working with distinguished data scientists and computer vision experts, some of the finest in their field. Throughout this journey, I've gained invaluable insights, particularly in the realm of building scalable computer vision systems following the best practices. This book encapsulates the wisdom acquired from my work experience, as well as the collective knowledge amassed from the exceptional individuals with whom I've had the fortune to work or collaborate. This book also presents knowledge I've gained from the works of some of the greatest contributors and thought leaders of computer vision. I have provided references to their work at appropriate places throughout the book.

When I hire new engineers and scientists, one of my main challenges has been giving them effective training so that they can quickly start contributing to the development of computer vision systems. With numerous online resources and books available on various computer vision topics, it's easy to feel overwhelmed by the sheer amount of information. This is especially true since the field of computer vision is extensive and intricate.

In this book, my aim is to offer a structured and methodical approach to learning computer visions systems. You will learn essential concepts first and then build on those concepts by working through practical code examples that pertain to real-world computer vision systems. This approach will help you connect the dots as you progress through the chapters. I've structured this book to be as hands-on and practical as possible to assist you effectively.

This book begins by explaining the core concepts of computer vision. It presents code examples to help you understand these ideas better. In the beginning, the examples mainly use OpenCV with the Python language.

This book also covers the basic concepts of machine learning and then gradually moves into more complex concepts like artificial neural networks and deep learning. Every concept is reinforced with one or more practical code examples that show how the concept is applied in practice. Machine learning–related concepts are illustrated via code examples in TensorFlow with Python.

This book presents 11 examples, including working code, that demonstrate how computer vision is used in real life. These examples are from different domains like healthcare, surveillance, security, and manufacturing. To help you understand the code, I explain each line step by step.

Chapters 7, 8, and 9 are dedicated to building practical computer vision–based systems. These chapters show you all stages of how to create vision systems, from obtaining pictures or videos, to building a data pipeline, to training a model, and finally deploying the model for real situations.

Training state-of-the-art computer vision models requires a lot of hardware resources. It is desirable and economically beneficial to train computer vision models on a cloud infrastructure to leverage the latest hardware resources, such as GPUs, and pay-as-you-go cost models. The final chapter, Chapter 10, provides step-by-step instructions for building machine learning–based computer vision applications on three popular cloud infrastructures: Google Cloud Platform, Amazon AWS, and Microsoft Azure.

Even though this book starts from explaining a single pixel of an image and goes up to teaching how to use cloud computers for neural networks training, there are certain prerequisites to better understand the concepts presented in this book. You should already know how to use the Python programming language.

The purpose of this book is to assist working professionals, programmers, data scientists, and even students (undergraduate and graduate) acquire hands-on skills for building computer vision applications using artificial neural networks.

CHAPTER 1

Prerequisites and Software Installation

This book is a practical guide that explores the process of building computer vision applications using the Python programming language. Throughout this book, you will gain a comprehensive understanding of leveraging OpenCV for image manipulation and harnessing the power of TensorFlow to build machine learning models.

OpenCV, originally developed by Intel and written in C++, is an open source computer vision and machine learning library consisting of more than 2,500 optimized algorithms for working with images and videos. TensorFlow is an open source framework for high-performance numerical computation and large-scale machine learning. It is written in C++ and provides native support for both CPUs and GPUs. Python is the most widely used programming language for developing machine learning applications. Both TensorFlow and OpenCV provide Python interfaces to access their low-level functionality. While both TensorFlow and OpenCV also offer interfaces for other programming languages like Java, C++, and MATLAB, we will focus on the use of Python in this book. Python's vast community support and intuitive syntax make it an accessible language for learners and practitioners alike.

The prerequisites for this book are practical knowledge of Python and familiarity with NumPy and pandas. The book assumes that you are familiar with built-in data containers in Python, such as dictionaries, lists, sets, and tuples. If you need to brush up on these prerequisites, the following resources can be helpful:

- Python: `https://www.w3schools.com/python/`

- pandas: `https://pandas.pydata.org/docs/getting_started/index.html`

- NumPy: `https://numpy.org/devdocs/user/quickstart.html`

© Shamshad Ansari 2023
S. Ansari, *Building Computer Vision Applications Using Artificial Neural Networks*,
https://doi.org/10.1007/978-1-4842-9866-4_1

The first order of business is to prepare your working environment for the exercises presented throughout this book. You will begin by downloading and installing the necessary software libraries and packages.

Python and PIP

As mentioned, Python is the primary programming language that is used in this book. For the installation and management of Python packages, we'll rely on PIP, which is a widely accepted package installer and a standard tool in the Python ecosystem.

To set up your working environment, you will begin by installing Python and PIP on your computer. The installation steps depend on the operating system (OS) you are using. The following sections provide, in order, the installation steps for Ubuntu, macOS, Red Hat Enterprise Linux, and Windows. Make sure you follow the instructions for your OS. If you already have Python and PIP installed, ensure that you are using Python version 3.8 or greater and PIP version 19 or greater. To check the version number of Python, execute the following command on your terminal:

```
$ python3 --version
```

The output of this command should be something like this: Python 3.8.10.

To check the version number of PIP, execute the following command on your terminal:

```
$ pip3 --version
```

This command should show a version number of PIP 3, for example, PIP 20.0.2.

Installing Python and PIP on Ubuntu

Run the following commands in your Ubuntu terminal:

```
sudo apt update
sudo apt install python3 python3-dev python3-pip
```

Installing Python and PIP on macOS

Run the following commands on macOS:

```
brew update
brew install python
```

This will install both Python and PIP.

Installing Python and PIP on Red Hat Linux

Run the following commands on Red Hat Linux:

```
sudo yum update
sudo yum install python3
sudo yum groupinstall 'Development Tools'
```

Installing Python and PIP on Windows

Install the Microsoft Visual C++ 2017 Redistributable Update 3. This comes with Visual Studio 2017 but can be installed separately by following these steps:

1. Go to the Visual Studio downloads at `https://visualstudio.microsoft.com/vs/older-downloads/`.

2. Select Other Tools, Frameworks, and Redistributables.

3. Download and install the Microsoft Visual C++ 2017 Redistributable Update 3.

Make sure long paths are enabled on Windows. Here are the instructions to do that: `https://www.thewindowsclub.com/how-toenable-or-disable-win32-long-paths-in-windows-11-10`.

Install the 64-bit Python 3 release for Windows from `https://www.python.org/downloads/windows/` (select PIP as an optional feature).

If these installation instructions do not work in your situation, refer to the official Python documentation at `https://www.python.org/`.

virtualenv

virtualenv is a tool designed to create isolated Python environments. When you use virtualenv, it generates a directory that includes all the necessary executables for utilizing the packages required by a Python project. virtualenv provides the following advantages:

- It enables the coexistence of two versions of the same library, allowing two programs running different versions of the library to function properly. For example, if you have one program that relies on version 1 of a Python library and another program that relies on version 2 of the same library, virtualenv allows you to run both programs simultaneously.

- It establishes a self-contained and independent environment for your development work, which can be utilized in a production setting without the need to install dependencies separately.

Next, we will install virtualenv and set up the virtual environment with all the necessary software. The rest of the book assumes that the dependencies for our reference program will be contained in this virtualenv.

Installing and Activating virtualenv

To install virtualenv system-wide, execute the following PIP command, which works the same for all operating systems:

```
$ sudo pip3 install -U virtualenv
```

After the installation is complete, create a directory of your choice where you intend to set up the virtualenv. For purposes of following the examples in this book, name this directory cv (short for "computer vision"):

```
$ mkdir cv
```

Then create the virtualenv in this cv directory:

```
$ virtualenv --system-site-packages -p python3 ./cv
```

The following is a sample output from running this command (on my MacBook):

```
Running virtualenv with interpreter /anaconda3/bin/python3
Already using interpreter /anaconda3/bin/python3
Using base prefix '/anaconda3'
New python executable in /Users/sansari/cv/bin/python3
Also creating executable in /Users/sansari/cv/bin/python
Installing setuptools, pip, wheel...
done.
```

Activate the virtual environment using a shell-specific command:

```
$ source ./cv/bin/activate  # for sh, bash, ksh, or zsh
```

When virtualenv is active, your shell prompt is prefixed with (cv). Here's an example:

```
(cv) Shamshads-MacBook-Air:~ sansari$
```

Installing packages within a virtual environment does not affect the host system setup. Start by upgrading PIP as follows. Please note that you should not execute any command as root or sudo while inside the virtual environment.

```
$ pip install --upgrade pip
$ pip list  # show packages installed within the virtual environment
```

When you have finished your programming activities and you want to exit from virtualenv, run the following:

```
$ deactivate
```

TensorFlow

TensorFlow is a widely used open source library for numerical computation and large-scale machine learning. In the upcoming chapters, you will delve deeper into TensorFlow and explore its capabilities. However, before we proceed, let's begin by installing TensorFlow and preparing it for our deep learning exercises. If you have a GPU in your computer and you want to utilize it for deep learning tasks, you typically don't need to install both CPU and GPU versions of TensorFlow. You can install the GPU version, which includes support for both GPU and CPU computation. TensorFlow with GPU support automatically falls back to CPU execution when GPU resources are not available or when the operations are not suitable for GPU acceleration. So, installing TensorFlow with GPU support (tensorflow-gpu) should be sufficient. It will work on both GPU and CPU, making it a versatile choice for most deep learning tasks.

Installing TensorFlow on Mac with M1 Chip

The steps for installing TensorFlow on an M1 chip are as follows:

1. Download the Conda environment from the following link: `https://repo.anaconda.com/miniconda/Miniconda3-latest-MacOSX-arm64.sh`.

2. Assuming that the downloaded file `Miniconda3-latest-MacOSX-arm64.sh` is in the Downloads folder, execute the following command to install Miniconda3:

   `$ sh Downloads/Miniconda3-latest-MacOSX-arm64.sh`

3. Follow the onscreen instructions to complete the installation of Miniconda3 on your computer. If you proceed with the default options, Miniconda3 will be installed in your home directory. For example, if your home directory is `/Users/username`, the path for Miniconda3 after installation will be `/Users/username/miniconda3`.

4. Once the installation is finished, restart the terminal.

5. Create a virtual environment using Miniconda3 by running the following command:

   `$ conda create --name cv python=3.8`

6. To activate or deactivate this virtual environment, use the following commands:

   `$ conda activate cv`
   `$ conda deactivate`

7. Activate the virtual environment by running the command `conda activate cv`, and then install TensorFlow by executing the following commands:

```
$ conda install -c apple tensorflow-deps
$ pip install tensorflow-macos
$ pip install tensorflow-metal
$ conda install tensorflow
```

8. Test the TensorFlow installation by running the command:

```
$ python -c "import tensorflow as tf"
```

9. If this command executes without errors, it means that TensorFlow has been successfully installed on your M1 chip.

10. If you encounter any errors, resolve incompatibility issues by, for example, upgrading the Mac OS.

Installing TensorFlow for CPUs

To install the most recent version of TensorFlow from PyPI (`https://pypi.org/project/tensorflow/`), specifically for CPUs, ensure that you are within the virtual environment and execute the following command:

```
(cv) $ pip install --upgrade tensorflow
```

Test your TensorFlow installation by running this command:

```
(cv) $ python -c "import tensorflow as tf"
```

If TensorFlow is successfully installed, the output should not show any errors.

Installing TensorFlow for GPUs

To install TensorFlow for GPUs, you need to follow a slightly different installation process. Here are the steps:

1. Ensure that you have the appropriate NVIDIA GPU and drivers installed on your system. You can check the TensorFlow documentation for the specific GPU requirements and compatible driver versions.

2. Activate the virtual environment that we created before.

3. Install the necessary GPU drivers and dependencies. Refer to the TensorFlow documentation for the specific instructions based on your operating system.

4. Install the GPU version of TensorFlow using the following command:

    ```
    (cv) $ pip install --upgrade tensorflow-gpu
    ```

    This command installs the latest version of TensorFlow that supports GPU acceleration.

5. After the installation, you can test your TensorFlow-GPU setup by running the following command:

    ```
    (cv) $ python -c "import tensorflow as tf; tf.config.
    list_physical_devices('GPU')"
    ```

If TensorFlow-GPU is installed correctly and GPU support is enabled, it will display information about the available GPUs on your system.

Please note that installing TensorFlow for GPUs requires additional setup and can be more complex compared to the CPU version. It is recommended to refer to the official TensorFlow documentation for detailed instructions and troubleshooting guidance specific to your system configuration.

PyCharm IDE

For consistency, the integrated development environment (IDE) used throughout this book for writing and managing Python code is the community version of PyCharm, a popular Python IDE. You can choose to use your preferred IDE, but if you do so, you'll need to adapt relevant examples and exercises to your IDE.

Installing PyCharm

To acquire PyCharm from its official website, follow these steps:

1. Visit the PyCharm website at `https://www.jetbrains.com/pycharm/download/`.

2. Select the appropriate operating system that corresponds to your machine.

3. Scroll down to the PyCharm Community Edition section and click the Download button.

4. Once the download is complete, locate the downloaded package and execute it.

5. Follow the instructions provided onscreen to complete the installation.

Here are the direct download links for different operating systems:

- Linux: `https://www.jetbrains.com/pycharm/download/download-thanks.html?platform=linux&code=PCC`

- Mac: `https://www.jetbrains.com/pycharm/download/download-thanks.html?platform=mac&code=PCC`

- Windows: `https://www.jetbrains.com/pycharm/download/download-thanks.html?platform=windows&code=PCC`

Please note that these links may change over time, so it's recommended to visit the official PyCharm website to ensure you access the most up-to-date download links.

Configuring PyCharm to Use virtualenv

To use the virtualenv named cv that was created earlier, follow these steps in the PyCharm IDE:

1. Launch PyCharm and open your project. Then choose File ➤ Settings on Windows and Linux or choose PyCharm ➤ Preferences on macOS.

2. In the Settings/Preferences dialog, navigate to Project <project name> ➤ Project Interpreter.

3. Click the icon (usually represented by a gear or a plus sign) and select Add from the drop-down menu.

4. In the left pane of the Add Python Interpreter dialog, select Existing Environment.

5. Expand the Interpreter list and choose one of the existing interpreters displayed, or click the ⬚ icon to specify the path to the Python executable in your file system, for example, /Users/ ansarisam/cv/bin/python3.8 (see Figure 1-1). This path should correspond to the virtual environment cv that you created earlier.

Optionally, you can select the checkbox "Make available to all projects" if you want this interpreter to be accessible for all your projects in PyCharm.

By following these steps, you will configure PyCharm to use the virtual environment cv as the Python interpreter for your project.

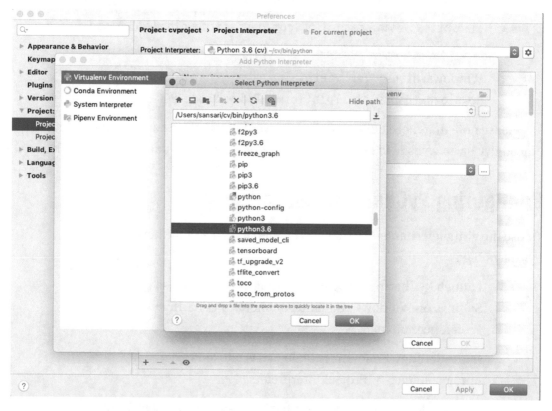

Figure 1-1. *Selecting an interpreter*

OpenCV

OpenCV is one of the most popular and widely used libraries for image processing. All code examples in this book are based on OpenCV 4. Therefore, the installation steps presented in this section are for version 4 of OpenCV.

Working with OpenCV

OpenCV is primarily written in C/C++, and its installation instructions vary depending on which operating system you are using. Since OpenCV is platform-dependent, it needs to be built specifically for your OS to ensure smooth operation. In this book, we will utilize Python bindings to interface with OpenCV for any image processing requirements.

OpenCV is continuously evolving, so if the following installation instructions do not work for your specific case, refer to the official OpenCV website for the most up-to-date and accurate installation process.

To simplify the installation process, we will use PIP to install OpenCV 4 and its Python 3 bindings. If you need additional OpenCV modules and features, you can install the opencv-contrib-python package.

Now, let's proceed with the installation process!

Installing OpenCV 4 with Python Bindings

Make sure you are in your virtual environment. Simply change directory to your virtualenv directory (the cv directory) and type the following command:

```
$ source cv/bin/activate
```

Install OpenCV in a snap using the following command:

```
$ pip install opencv-python
```

To install the opencv-contrib-python package, which includes extra modules and features that are not part of the main OpenCV package, use the following command:

```
$ pip install opencv-contrib-python
```

Once the installation is complete, you can verify it by importing OpenCV in your Python command:

```
$ python -c "import cv2"
```

If there are no errors, it indicates that OpenCV has been successfully installed.

> **Note**   Package names for OpenCV may vary slightly based on your specific operating system or Python version. If you encounter any issues during the installation, refer to the official OpenCV documentation or the PyPI page for the most accurate and up-to-date installation instructions.

Additional Libraries

SciPy and Matplotlib are two additional libraries that we will need as we work on some of the examples. Let's install and keep them in our virtualenv.

Installing SciPy

Install SciPy with the following:

```
$ pip install scipy
```

Installing Matplotlib

Install Matplotlib with the following:

```
$ pip install matplotlib
```

> **Note**   The libraries installed in this chapter are frequently updated. It is strongly advised to check the official websites for updates, new versions of these libraries, and the latest installation instructions.

Summary

The following table summarizes the installation commands for the required libraries within your virtualenv. You can copy and paste these commands into your terminal to install the libraries.

Table 1-1. *Installation Commands for Required Libraries in virtual environment*

Library	Command
TensorFlow	`pip install tensorflow`
TensorFlow GPU	`pip install tensorflow-gpu`
OpenCV	`pip install opencv-python`
OpenCV Contrib	`pip install opencv-contrib-python`
SciPy	`pip install scipy`
NumPy	`pip install numpy`
Matplotlib	`pip install matplotlib`

Ensure that your virtualenv is activated before running these commands. Once you've installed these libraries, you'll have all the necessary dependencies within your virtualenv to work with the examples in the book.

## CHAPTER 2

Core Concepts of Image and Video Processing

This chapter introduces the building blocks of an image and describes various methods to manipulate them. Our learning objectives in this chapter are as follows:

- To understand the smallest unit of an image (a pixel) and how colors are represented

- To learn how pixels are organized in an image and how to access and manipulate them

- To write code in Python and use OpenCV to work with examples to access and manipulate images

- To draw different shapes, such as lines, rectangles, and circles, on an image

Image Processing

Image processing is a method used to manipulate a digital image to obtain an improved image or extract valuable information from it. It involves taking an image as input and producing as output either another image or specific characteristics and features associated with that image. Videos, being composed of a series of images or frames, can also be processed using image processing techniques. In this chapter, we will delve into the fundamental concepts of digital image processing. Additionally, you will acquire the basic skills to work with images and write code to manipulate them.

© Shamshad Ansari 2023
S. Ansari, *Building Computer Vision Applications Using Artificial Neural Networks*,
https://doi.org/10.1007/978-1-4842-9866-4_2

Image Basics

A *digital image* is an electronic representation of an object/scene or scanned document. The digitalization of an image means converting it into a series of numbers and storing these numbers in a computer storage system. Understanding how these numbers are arranged and how to manipulate them is the primary objective of this chapter. This chapter explains the fundamental components that make up an image and guides you through the process of manipulating images using OpenCV and Python.

Pixels

An image can be visualized as a collection of dots arranged in rows and columns, where each dot represents a pixel with a specific color. These pixels are assigned numerical values, which determine their respective colors. You can imagine an image as a grid of square cells, where each cell corresponds to a single pixel with a specific color. For instance, if we have a 300×400-pixel image, it implies that the image is structured as a grid with 300 rows and 400 columns, resulting in a total of 120,000 pixels.

Pixel Color

A pixel can be represented in two different ways: grayscale and color.

Grayscale

In a grayscale image, each pixel takes a value between 0 and 255. The value 0 represents black, and 255 represents white. The values in between are varying shades of gray. The values close to 0 are darker shades of gray, and values closer to 255 are brighter shades of gray.

Color

The RGB (which stands for Red, Blue, and Green) color model is one of the most popular color representations of a pixel. Other color models are available, but this book primarily focuses on the RGB color model.

In the RGB model, each pixel is represented as a tuple of three values, generally represented as follows: (value for red component, value for green component, value for blue component). Each of the three colors is represented by integers ranging from 0 to 255. Here are some examples:

(0,0,0) is a black color.

(255,0,0) is a pure red color.

(0,255,0) is a pure green color.

The W3Schools website is a great place to play with different combinations of RGB tuples to explore more patterns (`https://www.w3schools.com/colors/colors_rgb.asp`). Use the RGB Calculator to explore what color is represented by each of the following tuples:

(0,0,255)

(255,255,255)

(0,0,128)

(128,0,128)

(128,128,0)

Now try to make the color yellow. Here is a clue: red and green make yellow. That means a pure red (255), a pure green (255), and no blue (0) will make yellow. So, the RGB tuple for yellow is (255,255,0).

Now that you have a good understanding of pixels and their color, let's explore how pixels are arranged in an image and how to access them. The following section discusses the concept of coordinate systems in image processing.

Coordinate Systems

As previously mentioned, pixels in an image are arranged in the form of a grid that is made of rows and columns. Imagine a square grid of eight rows and eight columns. This forms an 8×8 or 64-pixel image. Think of it as a 2D coordinate system in which the origin (0,0) is the top-left corner. Figure 2-1 shows our example 8×8-pixel image.

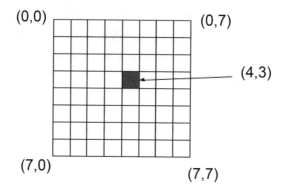

Figure 2-1. *Pixel coordinate system*

With the top-left corner as the start or origin of the image coordinate system, the pixel at the top-right corner is represented by (0,7), the bottom-left pixel is (7,0), and the bottom-right pixel is (7,7). This may be generalized as (x,y), where x is the position of the cell from the left edge of the image and y is the vertical position down from the top edge of the image. In Figure 2-1, the red pixel is in the fifth position from the left and fourth from the top. Since the coordinate system begins at 0, the coordinate of the red pixel shown in Figure 2-1 is (4,3).

To make this concept a little clearer, consider Figure 2-2, an image that is 8×8 pixels, with the letter *H* written on it. Also, for simplicity, assume this is a grayscale image, with the letter *H* written in black and the rest of the area of the image in white.

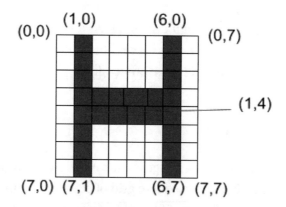

Figure 2-2. *Pixel coordinate system example*

Remember, in the grayscale model, a black pixel is represented by 0 and a white pixel is represented by 255. Figure 2-3 shows the values of each of the pixels within the 8×8 grid.

x

255	0	255	255	255	255	0	255
255	0	255	255	255	255	0	255
255	0	255	255	255	255	0	255
255	0	0	0	0	0	0	255
255	0	0	0	0	0	0	255
255	0	255	255	255	255	0	255
255	0	255	255	255	255	0	255
255	0	255	255	255	255	0	255

y

Figure 2-3. *Pixel matrix and values*

So, what's the value of the pixel at position (1,4)? And at position (2,2)?

Now, you should have a clear understanding of how images are represented as numbers arranged in a grid. These numerical values are serialized and stored in a computer's storage system. When the image is displayed on the screen, these numbers are interpreted and rendered as an actual image.

At this point, you have learned how to access individual pixels using the coordinate system, which allows you to pinpoint specific locations within the image grid. Additionally, you have learned how to assign colors to these pixels by manipulating their numerical values. By grasping these concepts, you now understand how images are structured and how to interact with them programmatically.

With a solid foundation and understanding of image representation, you are now ready to gain practical experience through Python and OpenCV coding. In the following section, I will guide you step-by-step through the process of writing code that loads images from the computer's disk, accesses individual pixels, performs manipulations, and saves the modified images back to the disk.

Let's delve into the exciting world of image processing and begin our hands-on journey!

Using Python and OpenCV Code to Manipulate Images

In OpenCV, the pixel values of an image are represented using a NumPy array. (Not familiar with NumPy? You can find a "getting started" tutorial at `https://numpy.org/devdocs/user/quickstart.html`.) This NumPy array serves as a multidimensional representation of the image grid and its color channels. It allows us to efficiently store, access, and manipulate the pixel values of the image. By leveraging the capabilities of NumPy arrays, OpenCV provides a powerful framework for image processing and analysis tasks. In other words, when you read an image, OpenCV creates a NumPy array. The pixel values can be obtained from NumPy by simply supplying the (x,y) coordinates.

When you give the (x,y) coordinates, NumPy returns the values of colors of the pixel at those coordinates as follows:

- *For a grayscale image*, the returned value from NumPy will be a single value between 0 and 255.

- *For a color image*, the returned value from NumPy will be a tuple for red, green, and blue. Note that OpenCV maintains the RGB sequence in the reverse order. In other words, OpenCV stores the colors in BGR sequence, *not* in RGB sequence. Remember this important feature of OpenCV to avoid any confusion while working with OpenCV.

Before we write any code, let's make sure we always use our virtual environment, in the ~/cv directory, that we already set up with PyCharm in Chapter 1.

Launch your PyCharm IDE and make a new project (I named my project cviz, short for "computer vision"). Refer to Figure 2-4 and ensure that you have selected Existing Interpreter and have our virtualenv Python 3.8 (cv) selected.

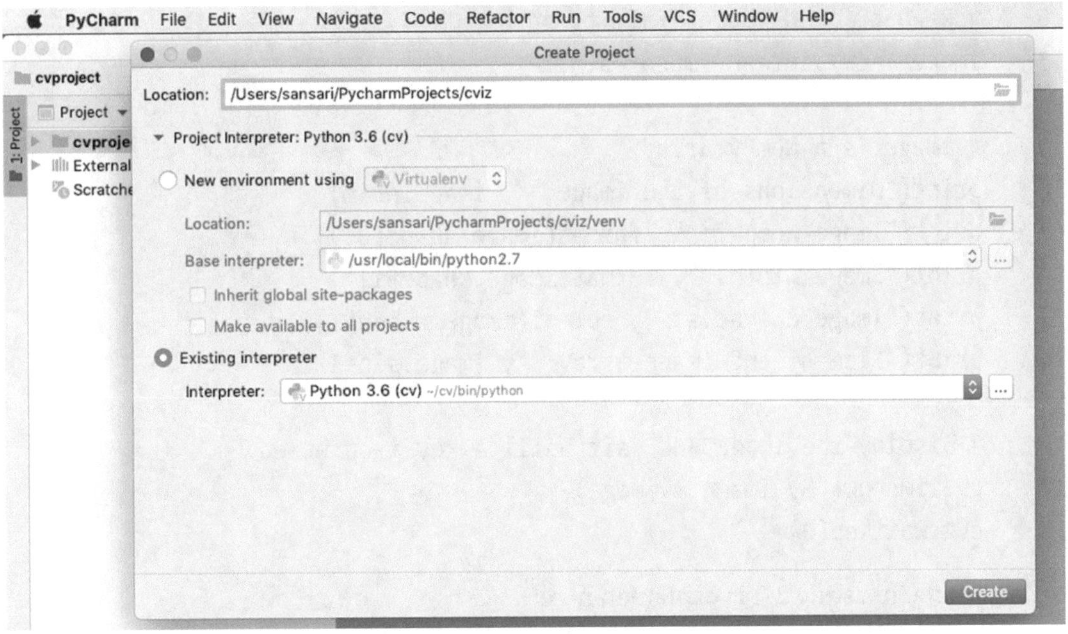

Figure 2-4. *PyCharm IDE, showing the setup of the project with virtualenv*

Program: Loading, Exploring, and Showing an Image

In this section, we will write code to read an image from a file and explore its properties. We will utilize the OpenCV library, which provides powerful tools for image processing and analysis. By loading the image into our program, we will be able to access information such as its dimensions, size, and channel information. Additionally, we will display the image on the screen and wait for a keypress to continue execution. This exploration will give us a better understanding of the image and set the stage for further image processing tasks. Let's dive into the code and start exploring the image!

Listing 2-1 shows the Python code to load, explore, and display an image.

Listing 2-1. Python Code to Load, Explore, and Display an Image

Filename: Listing_2_1.py

```
1     import cv2
2
3     # image path
4     image_path = "images/marsrover.png"
5
```

```
6     # Read or load image from its path
7     image = cv2.imread(image_path)
8
9     # image is a NumPy array
10    print("Dimensions of the image: ", image.ndim)
11    print("Image height: ", format(image.shape[0]))
12    print("Image width: ", format(image.shape[1]))
13    print("Image channels: ", format(image.shape[2]))
14    print("Size of the image array: ", image.size)
15
16    # Display the image and wait until a key is pressed
17    cv2.imshow("My Image", image)
18    cv2.waitKey(0)
```

The code in Listing 2-1 is explained here:

- Line 1: We import the OpenCV library, enabling us to utilize its functions and capabilities.

- Line 4: We assign to a variable the path of the image that we are going to load from the disk. If your input path is in a different directory, you should give either the full or relative path to the image path.

- Line 7: Using the imread() function of OpenCV, we are reading the image into a NumPy array and assigning that to a variable called image (you can name this variable anything you like).

- Lines 10 through 14: Using NumPy features, we are printing the dimensions of the image array, height, width, number of channels, and size of the array (which is the number of pixels).

- Line 17: This line displays the image as is using OpenCV's imshow() function, which takes two arguments: the name of the image window and the Numpy array representing the image.

- Line 18: The waitKey() function allows the program to not terminate immediately and wait for the user to press any key. When you see the image window that will display in line 17, press any key to terminate the program, else the program will block.

> **Note**   The function `waitKey()` waits for a key event infinitely or for a certain delay in milliseconds. Since the OS has a minimum time between switching threads, the `waitKey()` function will not wait, after a key is pressed, for exactly the delay time passed as an argument to the `waitKey()` function. The actual wait time depends on other programs that your computer might be running at the time when a key is pressed and `waitKey()` function is called.

Figure 2-5 shows the output of Listing 2-1.

Dimension of the image: 3
Image height: 400
Image width: 640
Image channels: 3
Size of the image array: 768000

Figure 2-5. *Output and image display*

The NumPy array representing the image has three dimensions: height × width × channel. The first element of the array corresponds to the height, indicating the number of rows in the image pixel grid. Similarly, the second element represents the width, indicating the number of columns in the grid. The three channels represent the color components in the order of blue, green, and red (BGR), rather than the conventional RGB. The overall size of the array is 400×640×3 = 768,000. This means that the image contains a total of 400×640 = 256,000 pixels, with each pixel consisting of three color values.

Program: OpenCV Code to Access and Manipulate Pixels

In the following program, we will explore how to access and modify pixel values using the coordinate system described earlier. Listing 2-2 demonstrates this process and is followed by a line-by-line explanation to help you understand each step of the implementation.

Listing 2-2. Code Example to Access and Manipulate Image Pixels

```
Filename: Listing_2_2.py
1    import cv2
2
3    # image path
4    image_path = "images/marsrover.png"
5    # Read or load image from its path
6    image = cv2.imread(image_path)
7
8    # Access pixel at (0,0) location
9    (b, g, r) = image[0, 0]
10   print("Blue, Green and Red values at (0,0): ", format((b, g, r)))
11
12   # Manipulate pixels and show modified image
13   image[0:100, 0:100] = (255, 255, 0)
14   cv2.imshow("Modified Image", image)
15   cv2.waitKey(0)
```

Listing 2-2 is explained here:

- Lines 1 through 6: We import OpenCV's cv2 package and read the image from a directory path (as explained when discussing Listing 2-1).

- Line 9: We are getting the BGR (not RBG) values of the pixel at coordinates (0,0) and assigning them to the (b,g,r) tuple using the NumPy syntax.

- Line 10: This line prints the BGR values.

- Line 13: We are taking a range of pixels from 0 to 100 along the y-axis and from 0 to 100 along the x-axis to form a 100×100 square and assigning the values (255,255,0)—pure blue, pure green, and no red— to all the pixels within this square.

- Line 14: This line displays the modified image.

- Line 15: The program waits for the user to press any key for the program to exit.

Figure 2-6 shows a sample output of Listing 2-2.

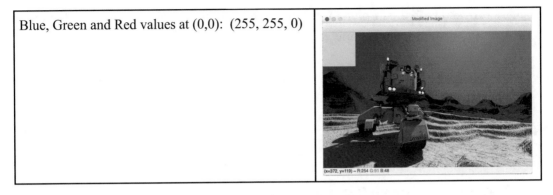

| Blue, Green and Red values at (0,0): (255, 255, 0) | |

Figure 2-6. *Output and modified image display*

As shown in Figure 2-6, the modified image has a 100×100-pixel square at the top-left corner in aqua, represented by (255,255,0) of the BGR scheme.

Drawing

OpenCV provides convenient methods to draw various shapes on an image. In the following sections, we will explore how to draw a line, a rectangle, and a circle using the corresponding methods provided by OpenCV:

- Line: `line()`

- Rectangle: `rectangle()`

- Circle: `circle()`

These methods allow us to easily add these shapes to an image, enhancing our ability to annotate and visualize data.

Drawing a Line on an Image

To draw a line on an image using OpenCV, follow these steps:

1. Load the image into a NumPy array.

2. Determine the coordinates of the starting position of the line.

3. Determine the coordinates of the end position of the line.

4. Specify the color of the line.

5. Optionally, specify the thickness of the line.

Let's proceed with the implementation and explore how each step is carried out in code. Listing 2-3 demonstrates how to draw a line on an image.

Listing 2-3. Drawing a Line on an Image

```
Filename: Listing_2_3.py
1    import cv2
2
3     # image path
4    image_path = "images/marsrover.png"
5    # Read or load image from its path
6    image = cv2.imread(image_path)
7
8    # set start and end coordinates
9    start = (0, 0)
10   end = (image.shape[1], image.shape[0])
11   # set the color in BGR
12   color = (255,0,0)
13    # set thickness in pixel
14   thickness = 4
15   cv2.line(image, start, end, color, thickness)
16
17   # display the modified image
18   cv2.imshow("Modified Image", image)
19   cv2.waitKey(0)
```

Here is the line-by-line explanation of the code:

- Lines 1 through 6: As in the previous listings, these lines import the OpenCV package and load the image from an input directory to a NumPy array.

- Line 9: We set the starting coordinates of the point from where the line will be drawn. Recall that the location (0,0) is the top-left corner of the image.

- Line 10: We set the coordinates of the endpoint of the image. Notice that the expression (image.shape[1], image.shape[0]) represents the coordinates of the bottom-right corner of the image.

You have probably guessed by now that we are drawing a diagonal line from the top-left corner to the bottom-right corner of the image.

- Line 12: We set the color of the line in BGR sequence.

- Line 14: We specify the line thickness.

- Line 15: The actual line is drawn by using OpenCV's line() function, which takes the following arguments:

  - Image NumPy. This is the image on which we are drawing the line.

  - The start coordinates of the line.

  - The end coordinates of the line.

  - The color tuple.

  - The thickness of the line. (This is optional. If you do not pass this argument, the line will have a default thickness of 1.)

- Line 18: The modified image with a line drawn on it is displayed using the OpenCV's imshow() function.

- Line 19: The program waits for the user to press any key to terminate the program.

Figure 2-7 shows the sample output of the image we just drew a line on.

Figure 2-7. *Image with a diagonal line in blue*

Drawing a Rectangle on an Image

Drawing a rectangle is easy with OpenCV. Let's dive into the code directly, shown in Listing 2-4. We first load an image and draw a rectangle on it. We then save the modified image to the disk and display it.

Listing 2-4. Loading an Image, Drawing a Rectangle on It, Saving It, and Displaying the Modified Image

```
Filename: Listing_2_4.py
1    import cv2
2
3    # image path
4    image_path = "images/marsrover.png"
5    # Read or load image from its path
6    image = cv2.imread(image_path)
7    # set the start and end coordinates
8    # of the top-left and bottom-right corners of the rectangle
9    start = (100,70)
10   end = (350,380)
11   # Set the color and thickness of the outline
12   color = (0,255,0)
13   thickness = 5
14   # Draw the rectangle
15   cv2.rectangle(image, start, end, color, thickness)
16   # Save the modified image with the rectangle drawn to it.
17   cv2.imwrite("rectangle.jpg", image)
18   # Display the modified image
19   cv2.imshow("Rectangle", image)
20   cv2.waitKey(0)
```

Here is a line-by-line explanation of Listing 2-4:

- Line 1 is our usual import of OpenCV package.

- Line 4 assigns the image path.

- Line 5 reads the image from its path.

- Line 9 sets the starting point of the rectangle we want to draw on the image. The starting point consists of the coordinates of the top-left corner of the rectangle.

- Line 10 sets the endpoint of the rectangle. This represents the coordinates of the bottom-right corner of the rectangle.

- Line 12 sets the color.

- Line 13 sets the thickness of the outline of the rectangle.

- Line 15 draws the rectangle. We are using OpenCV's `rectangle()` function, which takes the following parameters:

  - NumPy array that holds the pixel values of the image

  - The start coordinates (top-left corner of the rectangle)

  - The end coordinates (bottom-right of the rectangle)

  - The color of the outline

  - The thickness of the outline

  Notice that line 15 does not have any assignment operator. In other words, we did not assign the return value from the `cv2.rectangle()` function to any variable. The NumPy array, `image`, that is passed as an argument to the `cv2.rectangle()` function is modified.

- Line 17 saves the modified image, with the rectangle drawn on it, to a file on the disk.

- Line 19 displays the modified image.

- Line 20 calls the `waitKey()` function to allow the image to remain displayed on the screen until a key is pressed.

Figure 2-8 shows the output of the image with the rectangle drawn on it.

Figure 2-8. *Image with rectangle drawn*

In the previous example, we first read an image from the disk and drew a rectangle on it. We will now slightly modify this example and draw the rectangle on a blank canvas. We will first create a canvas (as opposed to loading an existing image) and draw a rectangle on it. We will then save and display the resultant image. See Listing 2-5.

Listing 2-5. Drawing a Rectangle on a New Canvas and Saving the Image

```
Filename: Listing 2_5.py
1    import cv2
2    import numpy as np
3
4    # create a new canvas
5    canvas = np.zeros((200, 200, 3), dtype = "uint8")
6    start = (10,10)
7    end = (100,100)
8    color = (0,0,255)
9    thickness = 5
10   cv2.rectangle(canvas, start, end, color, thickness)
11   cv2.imwrite("rectangle.jpg", canvas)
12   cv2.imshow("Rectangle", canvas)
13   cv2.waitKey(0)
```

Because you are already familiar with the purpose of most of the lines in Listing 2-5, I'll explain only the lines that are new.

Line 2 imports the NumPy library that we will use to create the canvas.

Line 5 is where we are creating an image (called the *canvas*). Our canvas is 200×200 pixels, with each pixel holding three channels (to hold BGR values). The variable name, canvas, is a NumPy array that, in this case, holds a zero value for each pixel. Notice that the data type of each pixel value of the canvas is an 8-bit unsigned integer (as explained in Chapter 1).

How would you draw a solid rectangle (meaning, a rectangle filled with a particular color)?

Clue: set the thickness to -1.

Figure 2-9 shows the output of Listing 2-5. Figure 2-10 shows a canvas with a solid rectangle drawn on it.

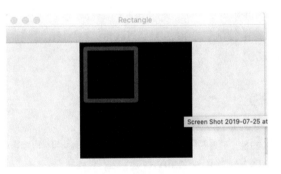

Figure 2-9. *Rectangle with border thickness 5*

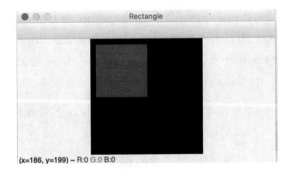

Figure 2-10. *Solid rectangle with a thickness of -1*

Drawing a Circle on an Image

Drawing a circle on an image is equally easy. You create your own canvas or load an existing image and then set the coordinates of the center, radius, color, and thickness of the outline of the circle.

Listing 2-6 shows a working piece of code that draws a circle on a blank canvas. Figure 2-11 shows the output of this code listing.

Listing 2-6. Drawing a Circle on a Canvas

```
Filename: Listing_2_6.py
1    import cv2
2    import numpy as np
3
4    # create a new canvas
5    canvas = np.zeros((200, 200, 3), dtype = "uint8")
6    center = (100,100)
7    radius = 50
8    color = (0,0,255)
9    thickness = 5
10   cv2.circle(canvas, center, radius, color, thickness)
11   cv2.imwrite("circle.jpg", canvas)
12   cv2.imshow("My Circle", canvas)
13   cv2.waitKey(0)
```

The code in Listing 2-6 is not very different from that of Listing 2-5 except that line 6 defines the center of the circle, line 7 sets the radius, line 8 defines the color, and line 9 sets the thickness of the circle. Finally, line 10 draws the circle using OpenCV's `circle()` function, which accepts the following parameters:

- The image on which to draw the circle. This is our NumPy array containing the image pixels.

- The coordinates of the center of the circle.

- The radius of the circle.

- The color of the outline of the circle.

- The thickness of the outline.

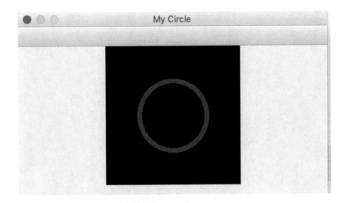

Figure 2-11. *A circle drawn at the center of a black canvas*

Here's an exercise for you:

1. Draw a solid circle at the center of the canvas.

2. Draw two concentric circles and make the radius of the outermost circle 1.5 times the radius of the inner circle.

Summary

In this chapter, we delved into the fundamentals of working with images. We began by describing pixels and their representation in various color schemes, including grayscale and color. We examined how the coordinate system enables us to locate specific pixels and modify their values. We also explored the process of drawing basic shapes such as lines, rectangles, and circles on images. While these concepts may seem simple, they form the foundation for various image processing tasks.

In the next chapter, we will explore image processing techniques and algorithms in depth, including topics such as image filtering, edge detection, image enhancement, and more. By building upon the concepts covered in this chapter, you will expand your understanding and capabilities in the field of image processing.

CHAPTER 3

Techniques of Image Processing

In a computer vision application, images are normally ingested from their source, such as cameras, files stored on a computer disk, or streams from another application. In most cases, these input images are converted from one form into another. For instance, they may need to be resized or rotated or their colors may need to be altered. In some cases, background pixels may need to be removed or two images may need to be merged. Additionally, finding boundaries around specific objects within an image may be a necessary task.

This chapter delves into different techniques of image transformation, providing examples implemented in Python and OpenCV. The learning objectives for this chapter are as follows:

- To explore commonly used transformation techniques in image processing

- To learn the arithmetic operations used in image processing

- To learn techniques for merging two or more images or splitting channels

- To understand techniques for cleaning images, such as noise reduction

- To acquire the skills to detect and draw contours (boundaries) around objects within an image

- To explore the concepts of morphological transformation and template matching

© Shamshad Ansari 2023
S. Ansari, *Building Computer Vision Applications Using Artificial Neural Networks*,
https://doi.org/10.1007/978-1-4842-9866-4_3

35

Transformation

While working on any computer vision problem, you will often need to transform images into different forms. The following sections demonstrate different techniques you can use to transform images through a set of Python examples.

Resizing

Let's start with our first transformation, resizing. To resize an image, we increase or decrease the height and width of the image. An important concept to remember when resizing an image is *aspect ratio*, which is the proportion of width to height and is calculated by dividing width by height. The formula for calculating the aspect ratio is as follows:

aspect ratio = width/height

A square image has an aspect ratio of 1:1, and an aspect ratio of 3:1 means the width is three times larger than the height. If an image's height is 300px and the width is 600px, its aspect ratio is 2:1.

When resizing an image, maintaining the original aspect ratio ensures that the resized image does not look stretched or compressed.

Listing 3-1 shows two distinct techniques of image resizing:

- Resize an image to a desired size in pixels while maintaining the aspect ratio. In other words, if you know the desired height of the image, you can compute the corresponding width using the aspect ratio from the original size of the image.

- Resize an image by a factor. For example, enlarge the image width by a factor of 1.5 or the height by a factor of 2.5.

OpenCV provides a single function, `resize()`, to perform these two techniques of resizing.

Listing 3-1. Code to Calculate Aspect Ratio and Resize the Image

```
Filename: Listing_3_1.py
1    # Image transformation using the resize() function
2    import cv2
3    import numpy as np
4
5    # Load image
6    imagePath = "images/zebra.png"
7    image = cv2.imread(imagePath)
8
9    # Get image shape which returns height, width, and channels as a
     tuple. Calculate the aspect ratio
10   (h, w) = image.shape[:2]
11   aspect = w / h
12
13   # let's resize the image to decrease height by half of the
     original image.
14   # Remember, pixel values must be integers.
15   height = int(0.5 * h)
16   width =  int(height * aspect)
17
18   # New image dimension as a tuple
19   dimension = (height, width)
20   resizedImage = cv2.resize(image, dimension, interpolation=cv2.
     INTER_AREA)
21   cv2.imshow("Resized Image", resizedImage)
22
23   # Resize using x and y factors
24   resizedWithFactors = cv2.resize(image, None, fx=1.2, fy=1.2,
     interpolation=cv2.INTER_LANCZOS4)
25   cv2.imshow("Resized with factors", resizedWithFactors)
26   cv2.waitKey(0)
```

Listing 3-1 shows how to resize an image using OpenCV's `resize()` function, which takes the following arguments as parameters:

- The first argument is the original image represented by a NumPy array.

- The second argument is the dimension of the intended resizing. This is a tuple of integers representing the height and width of the resized image. Pass this argument as None if you want to resize using horizontal or vertical factors, as explained in a moment.

- The third and fourth arguments, fx and fy, are the resize factors in the horizontal (widthwise) and vertical (heightwise) directions. These two arguments are optional.

- The last argument is the interpolation. This is the algorithm name that OpenCV internally uses to resize the image. *Interpolation* is the process of calculating the pixel values when the image is resized. The following five interpolation algorithms are supported in OpenCV:

  - INTER_LINEAR: This is a bilinear interpolation in which the four nearest neighbors (2×2 = 4) are determined and their weighted average is calculated to determine the value of the next pixel.

  - INTER_NEAREST: This uses the nearest-neighbor interpolation method of approximating the value of a function for a nongiven point in some space when given the value of that function in points around (neighboring) that point. In other words, to calculate the value of a pixel, its nearest neighbor is considered to approximate the interpolation function.

  - INTER_CUBIC: This uses a bicubic interpolation algorithm to calculate the pixel value. Like bilinear interpolation, it uses 4×4 = 16 nearest neighbors to determine the value of the next pixel. When speed is not a concern, bicubic interpolation gives a better resized image compared to bilinear.

  - INTER_LANCZOS4: This uses the 8×8 nearest neighbor interpolation.

  - INTER_AREA: The calculation of the pixel value is performed by using the pixel area relation (as described by the OpenCV official documentation). Use this algorithm to create a moiré-free resized image. When the image size is enlarged, INTER_AREA is like the INTER_NEAREST method.

Let's examine the code in Listing 3-1.

Lines 2 and 3 are the library imports.

Line 6 assigns the image path, and line 7 reads the image as a NumPy array and assigns to a variable named image.

NumPy's shape function returns the dimensions of the objects within the array. Calling the shape function for the image returns the height, width, and number of channels as a tuple. Line 10 retrieves only the height and width by specifying the index length 2 (image.shape[,:2]). The height and width are stored in variables h and w.

If we do not specify the index length, it will return the tuple with the height, width, and channels, like the following one:

```
(h, w, c) = image.shape[:]
```

In this example, we want to shrink the size of the image by 50 percent, maintaining the original aspect ratio. We can simply multiply the original height and width by 0.5 to obtain the desired height and width. If we know only the desired height, we can calculate the desired width by multiplying the original new height with the aspect ratio. This is demonstrated in lines 16 and 17.

Line 20 sets the desired height and width as a tuple.

Line 21 calls the resize() function of OpenCV and passes the original image NumPy, the desired dimensions, and the interpolation algorithm (INTER_AREA in this example) to the resize() function as arguments.

Line 24 demonstrates the resize operation using the second approach when we know the factors by which the image height or width or both need to increase or decrease. In this example, both the height and width are enlarged by a factor of 1.2.

Figure 3-1 and Figure 3-2 show the sample output of our resizing program.

Figure 3-1. *Original image*

Figure 3-2. *Resized image*

Translation

Image translation means moving the image either left, right, up, or down along the x- and y-axes. The process of moving an image involves two primary steps: defining a translation matrix and invoking OpenCV's warpAffine() function. The translation matrix determines the direction and magnitude of the movement.

The warpAffine() function is the OpenCV function that does the actual movement. The warpAffine() function takes three arguments:

- The image NumPy

- The translation matrix

- The dimension of the image

To help you understand the translation operation, Listing 3-2 provides a code example.

Listing 3-2. Image Translation Along the x- and y-Axes

```
Filename: Listing_3_2.py
1    # Image translation using OpenCV's warpAffine() to move image along
     x- and y-axes
2    import cv2
3    import numpy as np
4
5    # Load image
6    imagePath = "images/soccer-in-green.jpg"
7    image = cv2.imread(imagePath)
8
9    # Define translation matrix
10   translationMatrix = np.float32([[1,0,50],[0,1,20]])
11
12   # Move the image
13   movedImage = cv2.warpAffine(image, translationMatrix, (image.shape[1],
     image.shape[0]))
14
15   cv2.imshow("Moved image", movedImage)
16   cv2.waitKey(0)
```

Line 10 is where the translation matrix is defined, specifying the movement directions and the number of pixels by which the image should be shifted. In this example, the translation matrix is a 2×3 matrix or a 2D array. The first row, as defined by [1,0,50], represents the movement along the x-axis by 50 pixels to the right. If the third element of this array is a negative number, the movement will be to the left. The second row, as defined by [0,1,20], represents the movement along the y-axis by 20 pixels down. If the third element of this array is a negative number, this will move the image up along the y-axis.

In line 13, we are calling OpenCV's `warpAffine()` function. This function takes the following arguments:

- The NumPy representation of the image we intend to move.

- The translation matrix that defines the movement direction and the amount of the movement.

- The last argument is a tuple that has the width and height of the canvas within which we want to move our image. In this example, we are keeping the canvas size the same as the original height and width of the image.

Figure 3-3 and Figure 3-4 show the results.

Figure 3-3. *Original image*

Figure 3-4. *Moved image*

Here's an exercise for you: move an image by 50 pixels to the left and 60 pixels up.

Rotation

To rotate an image by a specific angle, we begin by defining a rotation matrix using the getRotationMatrix2D() function provided by OpenCV. The example in Listing 3-3 demonstrates the process of creating this rotation matrix. To actually perform the rotation, we can utilize the warpAffine() function, similar to how we applied the translation in the previous example (Listing 3-2).

Listing 3-3. Image Rotation Around the Center of the Image

Filename: Listing_3_3.py

```
1    # Example code to demonstrate image rotation using OpenCV's warpAffine
     function
2    import cv2
3    import numpy as np
4
5    # Load image
6    imagePath = "images/zebrasmall.png"
7    image = cv2.imread(imagePath)
8    (h,w) = image.shape[:2]
9
10   # Define translation matrix
11   center = (h//2, w//2)
12   angle = -45
13   scale = 1.0
14
15   rotationMatrix = cv2.getRotationMatrix2D(center, angle, scale)
16
17   # Rotate the image
18   rotatedImage = cv2.warpAffine(image, rotationMatrix, (image.shape[1],
     image.shape[0]))
19
20   cv2.imshow("Rotated image", rotatedImage)
21   cv2.waitKey(0)
```

Listing 3-3 shows how to rotate an image around its center by a 45-degree angle (clockwise).

Line 11 calculates the center of the image. Notice that we divided the height and width by using // to get only the integer part of it.

Line 12 simply assigns a value to the angle by which we want to rotate the image. A negative value will rotate the image clockwise, while the positive angle will rotate it counterclockwise.

Line 13 sets the rotation scale, which is used to resize the image while rotating. A value of 1.0 keeps the original size after rotation. If we set this to 0.5, the rotated image will be reduced in size by half.

In line 15, we define the rotation matrix by using OpenCV's function getRotationMatrix2D(), which takes the following arguments:

- A tuple that represents the point around which the image needs to be rotated

- The angle of rotation in degrees

- Resizing scale

Line 18 does the work of rotating the image as per the definition of a rotation matrix. We use the same warpAffine() function that we used to translate the image. The only difference is that in the case of rotation, we pass the rotation matrix created in line 15.

Line 20 shows the rotated image, and line 21 waits for the keypress before the displayed image is closed.

Figure 3-5 and Figure 3-6 show the sample outputs of our code.

(x=236, y=428) ~ R:247 G:247 B:247

Figure 3-5. *Original image*

(x=113, y=456) ~ R:0 G:0 B:0

Figure 3-6. *Rotated image*

Flipping

Flipping an image horizontally along the x-axis or vertically along the y-axis can be accomplished effortlessly by utilizing OpenCV's convenient function `flip()`. The `flip()` function takes two arguments.

- The original image

- The direction of the flip

  - 0 means flip vertically.

  - 1 means flip horizontally.

  - -1 means first flip horizontally and then vertically.

Let's explore the process of image flipping in various directions using Listing 3-4.

Listing 3-4. Image Flipping Horizontally, Vertically, and Then Horizontally plus Vertically

```
Filename: Listing_3_4.py
1    # Example to demonstrate image flipping in various directions
2    import cv2
3    import numpy as np
4
5    # Load image
6    imagePath = "images/zebrasmall.png"
7    image = cv2.imread(imagePath)
8
9    # Flip horizontally
10   flippedHorizontally = cv2.flip(image, 1)
11   cv2.imshow("Flipped Horizontally", flippedHorizontally)
12   cv2.waitKey(-1)
13
14   # Flip vertically
15   flippedVertically = cv2.flip(image, 0)
16   cv2.imshow("Flipped Vertically", flippedVertically)
17   cv2.waitKey(-1)
18   # Flip horizontally and then vertically
19   flippedHV = cv2.flip(image, -1)
20   cv2.imshow("Flipped H and V", flippedHV)
21   cv2.waitKey(0)
```

Listing 3-4 is fairly self-explanatory, demonstrating the image flipping process in different directions, but let's examine the lines in Listing 3-4 that perform the image flips.

Line 10 calls the `flip()` function and passes the original image and a value of 0 for the horizontal flip. Line 15 flips the image vertically. Line 19 has an argument of -1 to make the flip first horizontally and then vertically.

Figures 3-7 to 3-10 show how these flips look.

Figure 3-7. *Original image*

Figure 3-8. *Flipped horizontally*

Figure 3-9. *Flipped vertically*

Figure 3-10. *Flipped horizontally and then vertically*

Cropping

Image cropping involves removing unwanted outer areas of an image. As previously described, in OpenCV an image is represented as a NumPy array. An image can be cropped by slicing the image NumPy array. OpenCV does not provide a dedicated function for image cropping; instead, we utilize the slicing capabilities of NumPy arrays.

Let's examine how to crop an image with the help of Listing 3-5.

Listing 3-5. Image Cropping

```
Filename: Listing_3_5.py
1    # Example to demonstrate image cropping
2    import cv2
3    import numpy as np
4
5    # Load image
6    imagePath = "images/zebrasmall.png"
7    image = cv2.imread(imagePath)
8    cv2.imshow("Original Image", image)
9    cv2.waitKey(0)
10
11   # Crop the image to get only the face of the zebra
12   croppedImage = image[0:150, 0:250]
13   cv2.imshow("Cropped Image", croppedImage)
14   cv2.waitKey(0)
```

Line 12 shows how to slice the NumPy array. In this example, we are using a 150-pixel height and a 250-pixel width to crop the image to extract only the face portion of the zebra.

Figure 3-11 shows the original image, and Figure 3-12 shows the cropped images.

Figure 3-11. *Original image*

Figure 3-12. *The cropped image*

Image Arithmetic and Bitwise Operations

When developing computer vision applications, it is often necessary to enhance the properties of input images. This involves performing various arithmetic operations such as addition, subtraction, and bitwise operations like OR, AND, NOT, and XOR.

In the context of image processing, each pixel in an image can have an integer value ranging from 0 to 255 (as described in Chapter 2). However, what happens when we add a constant to a pixel, resulting in a value greater than 255, or subtract a constant from it, resulting in a value less than 0?

For instance, let's consider a scenario where one of the pixels in an image has a value of 230, and we add 30 to it. Clearly, the resulting pixel value cannot be 260. So, what should we do in such cases? Should we truncate the value to maintain a maximum value of 255 for the pixel, or should we wrap it around to make it 4 (which means that after reaching 255, the value wraps back to 0 and any remainder is added)?

There are two methods to handle situations where the pixel value falls outside the range [0,255]:

- *Saturated operation (or trimming)*: In this operation, 230 + 30 ⇒ 255.

- *Modulo operation*: Here it performs a modulo like this: (230+30) % 256 ⇒ 4.

You can perform arithmetic operations by using both OpenCV and NumPy's built-in functions. However, they handle the operations differently.

OpenCV's addition is a saturated operation. On the other hand, NumPy performs a modulo operation.

Note the difference between NumPy and OpenCV, as these two techniques yield different results, so where you use each depends on your situation and needs.

Addition

OpenCV provides two convenient methods to add two images:

- add(): Takes the two equal-sized images as arguments and adds their pixel values to produce the result.

- addWeighted(): Generally used for blending two images. More details about this function are provided in a moment.

Note that to add two images, they must be of the same depth and type.

To better understand the differences between these two approaches of adding images, Listing 3-6 shows both approaches.

Listing 3-6. Addition of Two Images

```
Filename: Listing_3_6.py
1    # Example program to demonstrate two approaches of adding images
2    import cv2
3    import numpy as np
4
5    image1Path = "images/zebra.png"
6    image2Path = "images/nature.jpg"
7
8    image1 = cv2.imread(image1Path)
9    image2 = cv2.imread(image2Path)
10
11   # resize the two images to make them of the same dimension. This is a
     must to add two images
12   resizedImage1 = cv2.resize(image1,(300,300),interpolation=cv2.
     INTER_AREA)
13   resizedImage2 = cv2.resize(image2,(300,300),interpolation=cv2.
     INTER_AREA)
14
15   # This is a simple addition of two images
16   resultant = cv2.add(resizedImage1, resizedImage2)
17
18   # Display these images to see the difference
```

```
19   cv2.imshow("Resized 1", resizedImage1)
20   cv2.waitKey(0)

21

22   cv2.imshow("Resized 2", resizedImage2)
23   cv2.waitKey(0)

24

25   cv2.imshow("Resultant Image", resultant)
26   cv2.waitKey(0)

27

28   # This is weighted addition of the two images
29   weightedImage = cv2.addWeighted(resizedImage1,0.7,
     resizedImage2, 0.3, 0)
30   cv2.imshow("Weighted Image", weightedImage)
31   cv2.waitKey(0)

32

33   imageEnhanced = 255*resizedImage1
34   cv2.imshow("Enhanced Image", imageEnhanced)
35   cv2.waitKey(0)

36

37   arrayImage = resizedImage1+resizedImage2
38   cv2.imshow("Array Image", arrayImage)
39   cv2.waitKey(0)
```

Lines 8 and 9 load two different images from disk. As previously mentioned, to add two images together, they must be of the same size and depth. You may have already guessed the purpose of lines 12 and 13. In the code, the images are resized to dimensions of 300×300 pixels.

Line 16 is where these two images are being added, using OpenCV's simple addition function, add(), that takes the two images as arguments. Refer to the output image in Figure 3-15 to see the result of simply adding two images shown in Figures 3-13 and 3-14.

In line 29, we are doing weighted addition by using OpenCV's addWeighted() function, which works as follows:

$$\text{ResultantImage} = \alpha \times \text{image1} + \beta \times \text{image2} + \gamma \qquad (1)$$

where α is the weight of image 1, β is the weight of image 2, and γ is a constant. By varying the values of these weights, we create the desired effects of additions.

By looking at the previous equation, you can easily guess the arguments you need to pass to the function addWeighted(). Here is the argument list:

- NumPy array of image 1

- The weight, α, of image 1 (we passed a value 0.7 in Listing 3-6)

- NumPy array of image 2

- The weight, β, of image 2 (we passed the value 0.3 in Listing 3-6)

- The last argument, γ (we passed a value 0 in Listing 3-6)

Let's examine the inputs and outputs of Listing 3-6. Figure 3-13 and Figure 3-14 are the original images, resized to 300×300 to make them of equal dimensions.

Figure 3-15 is the output when these two images are added together using the function add(). Figure 3-16 is the resultant image when the inputs are added using the function addWeighted().

Figure 3-13. *Original image*

Figure 3-14. *Original image that is added*

Notice the difference between the simple add() and addWeighted() functions by referring to the outputs shown in Figure 3-15 and Figure 3-16.

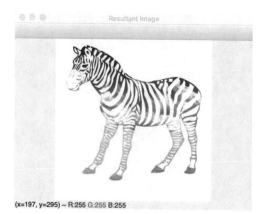

Figure 3-15. *Result of OpenCV's add() function*

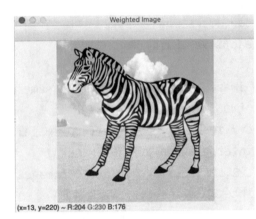

Figure 3-16. *Result of OpenCV's* `addWeighted()` *function*

Subtraction

Image subtraction means subtracting the pixel values of one image from the corresponding pixel values of another image. We can also subtract a constant from the image pixels. When we subtract two images, it is important to note that the two images must be of the same size and depth.

What happens when we subtract an image from itself? All the pixel values of the resultant image will be zeros, producing a completely black image. This property is useful in detecting any change/alteration in an image. If there is no change, the result of subtracting two images will be a completely black image.

Another reason for subtracting images is to level any uneven sections or shadows.

We will see some interesting results of image subtraction through the code examples presented in Listing 3-7.

Listing 3-7. Image Subtraction

Filename: Listing_3_7.py

```
1    import cv2
2    import numpy as np
3
4
5    image1Path = "images/cat1.png"
6    image2Path = "images/cat2.png"
7
```

```
8    image1 = cv2.imread(image1Path)
9    image2 = cv2.imread(image2Path)
10
11   # resize the two images to make them of the same dimensions. This is a
     must to subtract two images
12   resizedImage1 = cv2.resize(image1,(int(500*image1.shape[1]/image1.
     shape[0]), 500),interpolation=cv2.INTER_AREA)
13   resizedImage2 = cv2.resize(image2,(int(500*image2.shape[1]/image2.
     shape[0]), 500),interpolation=cv2.INTER_AREA)
14
15   cv2.imshow("Cat 1", resizedImage1)
16   cv2.imshow("Cat 2", resizedImage2)
17
18   # Subtract image 1 from 2
19   cv2.imshow("Diff Cat1 and Cat2",cv2.subtract(resizedImage2,
     resizedImage1))
20   cv2.waitKey(0)
21
22
23   # subtract images 2 from 1
24   subtractedImage = cv2.subtract(resizedImage1, resizedImage2)
25   cv2.imshow("Cat2 subtracted from Cat1", subtractedImage)
26   cv2.waitKey(0)
27
28   # Numpy Subtraction Cat2 from Cat1
29   subtractedImage2 = resizedImage2 - resizedImage1
30   cv2.imshow("Numpy Subtracted Images", subtractedImage2)
31   cv2.waitKey(0)
32
33   # A constant subtraction
34   subtractedImage3 = resizedImage1 - 50
35   cv2.imshow("Constant Subtracted from the image", subtractedImage3)
36   cv2.waitKey(0)
```

Listing 3-7 demonstrates several interesting behaviors associated with image subtraction. Here is an overview of what is presented in this listing.

Lines 5 through 9 load images from disk (from the directory paths). We are loading two images of cats, and we are trying to determine if there is any difference in these two look-alike cats. The images shown in Figures 3-17 and 3-18 are the input images used in this example.

Lines 12 and 13 are to resize images to ensure that their dimensions are the same. Remember, this is a must for subtracting two image arrays.

In line 19, we are displaying the result of subtracting cat1 from cat2. To determine the difference, we are using OpenCV's subtract() function and passing the NumPy representations of the two images (resized ones). In this case, we want to subtract cat1 from cat2; hence, we pass the resizedImage2 variable first and resizedImage1 as the second argument in the function. The order does matter as is evident from the outputs shown in Figure 3-19 and Figure 3-20.

To demonstrate the effect of the order, line 24 has resizedImage1 first and resizedImage2 as the second argument in the subtract() function.

Line 29 does not use OpenCV's subtraction function. This is a simple NumPy array subtraction. Notice the difference in the output shown in Figure 3-21.

Line 34 subtracts a constant from the image. The output is shown in Figure 3-22.

Figure 3-17. *Cat1 image*

Figure 3-18. *Cat2 image*

Figure 3-19. *Image1 subtracted from Image2*

Figure 3-20. *Image2 subtracted from Image1*

Figure 3-21. *NumPy subtraction*

Figure 3-22. *A constant subtracted from an image*

So far, we have covered two powerful techniques for image arithmetic: addition and subtraction. Now, let's delve into the realm of performing bitwise logical operations on image pixels, which further expands our image processing capabilities.

Bitwise Operations

Some of the most useful operations in computer vision involve the bitwise operations AND, OR, NOT, and XOR. These operations, as you may recall from Boolean algebra, are binary operations that operate on two states of pixels: on and off. In grayscale images, a pixel can have a value ranging from 0 to 255. So, how do we define "on" and "off" in this context? In image processing, for grayscale binary images, a pixel value of 0 represents off and a pixel value greater than 0 represents on. Utilizing this notion of pixels being either on or off, we will explore the following bitwise operations.

AND

The bitwise AND operation performed on the two operands "a" and "b" yields a result of 1 only if both "a" and "b" are 1; otherwise, the result is 0.

In the realm of image processing, the bitwise AND operation performed on two image arrays calculates the element-wise conjunction. It is essential to ensure that both arrays have equal dimensions to carry out the bitwise AND operation successfully. Additionally, it is also possible to perform bitwise AND with an array and a scalar value.

To facilitate the bitwise AND operation, OpenCV provides a convenient function called bitwise_and(imageArray1, imageArray2). This function takes the two image arrays as arguments and performs the bitwise AND operation.

To explore the implementation of the bitwise AND operation, refer to Listing 3-8, which follows the discussion of OR, NOT, and XOR.

Listing 3-8. Bitwise Operations

```
Filename: Listing_3_8.py
1    import cv2
2    import numpy as np
3
4    # create a circle
5    circle = cv2.circle(np.zeros((200, 200, 3), dtype = "uint8"),
     (100,100), 90, (255,255,255), -1)
6    cv2.imshow("A white circle", circle)
7    cv2.waitKey(0)
8
9    # create a square
10   square = cv2.rectangle(np.zeros((200,200,3), dtype= "uint8"), (30,30),
     (170,170),(255,255,255), -1)
11   cv2.imshow("A white square", square)
12   cv2.waitKey(0)
13
14   # bitwise AND
15   bitwiseAnd = cv2.bitwise_and(square, circle)
16   cv2.imshow("AND Operation", bitwiseAnd)
17   cv2.waitKey(0)
18
19   # bitwise OR
20   bitwiseOr = cv2.bitwise_or(square, circle)
21   cv2.imshow("OR Operation", bitwiseOr)
```

```
22    cv2.waitKey(0)
23
24    # bitwise XOR
25    bitwiseXor = cv2.bitwise_xor(square, circle)
26    cv2.imshow("XOR Operation", bitwiseXor)
27    cv2.waitKey(0)
28
29    # bitwise NOT
30    bitwiseNot = cv2.bitwise_not(square)
31    cv2.imshow("NOT Operation", bitwiseNot)
32    cv2.waitKey(0)
```

OR

A bitwise OR of the two operands "a" and "b" results in 1 if either or both of "a" and "b" are 1; otherwise, the result is 0. The bitwise OR operation calculates element-wise disjunction of two arrays or an array and a scalar. In OpenCV, the function bitwise_or(imageArray1, imageArray2) calculates the bitwise OR of the two input arrays. Listing 3-8 shows a working example of the OR operation.

NOT

Bitwise NOT inverts the bit values of its operand. OpenCV's bitwise_not(imageArray) function takes only one image array as an argument to perform the bitwise NOT operation on that image. See Listing 3-8 for an example.

XOR

A bitwise XOR of the two operands "a" and "b" results in 1 if either (but *not* both) "a" or "b" is 1; otherwise, the result is 0. OpenCV provides a convenient function called bitwise_xor(imageArray1, imageArray2) to perform a bitwise XOR. Again, both of the image arrays must have equal dimensions. Listing 3-8 shows a working example of a bitwise XOR.

The following table summarizes the bitwise operations commonly used in image processing, including masking and other related tasks:

Operator	Usage	Description
Bitwise AND	a AND b	Returns a 1 in each bit position for which the corresponding bits of both operands are 1s
Bitwise OR	a OR b	Returns a 1 in each bit position for which the corresponding bits of either or both operands are 1s
Bitwise XOR	a XOR b	Returns a 1 in each bit position for which the corresponding bits of either but not both operands are 1s
Bitwise NOT	NOT a	Inverts the bits of its operand

To gain a better understanding of these bitwise operations, let's examine the program in Listing 3-8, which first creates two images, a circle and a square, and then applies various bitwise operations on the images. We will then observe the effects of these operations in subsequent figures.

The following events are taking place in Listing 3-8.

Line 5 creates a white color circle at the center of a 200×200 canvas. (See Listing 2-5 for an example of how to draw a circle on a canvas.)

Similarly, line 10 draws a white square on a 200×200 canvas. (See Listing 2-4 for an example of how to draw a rectangle on a canvas.)

Line 15 shows the implementation of OpenCV's `bitwise_and()` function. The arguments to this function are the circle and square images (represented by NumPy arrays).

Similarly, lines 20 and 25 show the `bitwise_or()` and `bitwise_xor()` operations, respectively.

All these three functions for AND, OR, and XOR take two arrays to operate on.

Line 30 shows the `bitwise_not()` function that takes only one argument to calculate the bitwise NOT.

Figures 3-23 through 3-28 show the outputs of Listing 3-8.

Figure 3-23.   *White circle*

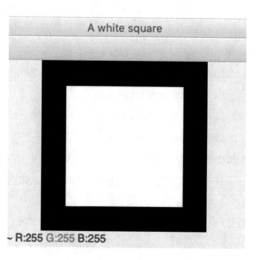

Figure 3-24.   *White square*

Figure 3-25. *Bitwise AND*

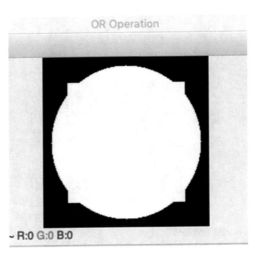

Figure 3-26. *Bitwise OR*

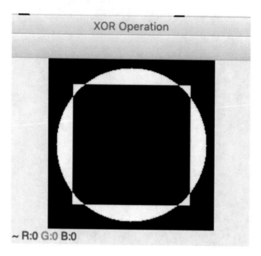

Figure 3-27. *Bitwise XOR*

Figure 3-28. *Bitwise NOT*

Masking

Masking is one of the most powerful techniques in computer vision. Masking refers to the "hiding" or "filtering" of an image.

When we mask an image, we hide a portion of the image with some other image. In other words, we put our focus on a portion of the image by applying a mask on the remaining portion of the image. For example, Figure 3-29 has the digits 1, 2, and 3 in it, while Figure 3-30 is a black image with a white cut-out. When we blend these two images, digits 1 and 3 will get hidden, and the only digit that will be visible is digit 2. The result of masking is shown in Figure 3-31.

The technique of masking is applied in the smoothing or blurring of an image and in detecting the edges and contours within the image. The masking technique is also used in object detection, which we will explore in Chapter 6.

Listing 3-9 shows how to perform masking using OpenCV.

Figure 3-29. *Original image*

Figure 3-30. *A mask image*

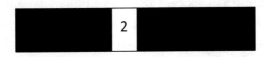

Figure 3-31. *Masking effect*

Listing 3-9. Masking Using Bitwise AND Operation

```
Filename: Listing_3_9.py
1    import cv2
2    import numpy as np
3
4    # Load an image
5    natureImage = cv2.imread("images/nature.jpg")
6    cv2.imshow("Original Nature Image", natureImage)
7
8    # Create a rectangular mask
9    maskImage = cv2.rectangle(np.zeros(natureImage.shape[:2],
     dtype="uint8"),
```

```
10                         (50, 50), (int(natureImage.shape[1])-50,
                           int(natureImage.shape[0] / 2)-50), (255, 255,
                           255), -1)
11   cv2.imshow("Mask Image", maskImage)
12   cv2.waitKey(0)
13
14   # Using bitwise_and operation to perform masking. Notice the
     mask=maskImage argument
15   masked = cv2.bitwise_and(natureImage, natureImage, mask=maskImage)
16   cv2.imshow("Masked image", masked)
17   cv2.waitKey(0)
```

In OpenCV, the image masking is performed by using a bitwise AND operation. Listing 3-9 shows a simple example of how to mask an area of an image. For this example, our goal is to extract a rectangular section of the cloud shown in Figure 3-32.

Line 5 of Listing 3-9 should be familiar to you by now. All we are doing here is loading the image (Figure 3-32).

In line 9, we are creating a black canvas with a white rectangular section at the top (with some margin). The size of the canvas is the same as the size of the original image. Notice in Figure 3-33 that the bigger rectangle has another rectangular white section at the top and the rest of the area of this rectangle is black.

Line 15 is where the masking is performed. Notice that we are using the bitwise_and() function, which takes two mandatory arguments, which in this case are the original image itself and an optional masking argument (mask=maskImage). This function performs the AND operation of the image with itself and applies a mask as instructed by the argument mask=maskImage. When OpenCV sees this mask argument, it will examine only those pixels that are turned on in the mask (maskImage) array. The output of this masking operation is shown in Figure 3-34.

Figure 3-32. *Original image to be masked*

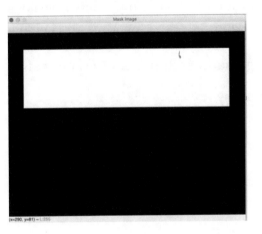

Figure 3-33. *A mask that will be applied to extract the cloud from Figure 3-32*

Figure 3-34. *Masked image*

Masking is one of the most commonly used image processing techniques for computer vision. You will learn more about its practical applications in subsequent chapters on machine learning and neural networks.

Splitting and Merging Channels

Recall from Chapter 2 that a color image consists of multiple channels (R,G,B). You have already learned how to access these channels and represent them as NumPy arrays. In this section, you will learn how to split these channels and store them as separate images. OpenCV provides a convenient function, split(), to do that. Using this split() function, we can split images into respective color components. Listing 3-10 is a working code example to illustrate this. For this example, we will again take our "nature" image (as shown previously in Figure 3-32) and split it into its component colors.

Listing 3-10. Splitting Channels into Color Components

```
Filename: Listing_3_10.py
1    import cv2
2    import numpy as np
3
4    # Load the image
5    natureImage = cv2.imread("images/nature.jpg")
6
7    # Split the image into component colors
8    (b,g,r) = cv2.split(natureImage)
9
10   # show the blue image
11   cv2.imshow("Blue Image", b)
12
13   # Show the green image
14   cv2.imshow("Green image", g)
15
16   # Show the red image
17   cv2.imshow("Red image", r)
18   cv2.waitKey(0)
```

In Listing 3-10, line 5 loads the image. Line 8 splits the image into three components and stores them in separate NumPy variables (b, g, r). Recall that NumPy stores colors in blue, green, and red (BGR) sequences and not as RGB sequences. Lines 11, 14, and 17 show these split images. The outputs are shown in Figures 3-35, 3-36, and 3-37.

Figure 3-35. *Red channel*

Figure 3-36. *Green channel*

Figure 3-37. *Blue channel*

We can merge channels by using OpenCV's merge() function, which takes arrays in BGR sequence. Listing 3-11 shows the use of the merge() function.

Listing 3-11. Split and Merge Functions

```
Filename: Listing_3_11.py
1     import cv2
2     import numpy as np
3
4     # Load the image
5     natureImage = cv2.imread("images/nature.jpg")
6
7     # Split the image into component colors
8     (b,g,r) = cv2.split(natureImage)
9
10    # show the blue image
11    cv2.imshow("Blue Image", b)
12
13    # Show the green image
14    cv2.imshow("Green image", g)
15
16    # Show the red image
17    cv2.imshow("Red image", r)
18
19    merged = cv2.merge([b,g,r])
20    cv2.imshow("Merged Image", merged)
21    cv2.waitKey(0)
```

Line 5 loads the image. Lines 8 through 17 are related to our previous split functions. We did the split so that we have three components to demonstrate the merge() function.

Line 19 is where we are merging the channels. We simply pass the individual channels as the argument to the merge() function. Notice that the channels are in BGR sequence. Execute the program and observe the output. Did you get the original image back?

Splitting and merging are helpful image processing techniques to perform feature engineering for machine learning. We will apply some of these concepts in the upcoming chapters.

Noise Reduction Using Smoothing and Blurring

Smoothing and *blurring*, are important image processing technique to reduce noise present in an image. Smoothing and blurring both involve reducing the high-frequency components or sharp details in an image, resulting in a more uniform appearance. Smoothing is a general term used to describe the process of reducing noise or sharp transitions in an image. Blurring is a specific type of smoothing technique. In image processing, we commonly encounter the following types of noise:

- *Salt and pepper noise*: Random occurrences of black and white pixels

- *Impulse noise*: Random occurrences of white pixels

- *Gaussian noise*: Intensity variation that follows a Gaussian normal distribution

In this section, we will delve into various techniques of blurring and smoothing that can effectively reduce noise in an image.

Mean Filtering or Averaging

In an *averaging* technique, we take a small portion of the image, say *k*×*k* pixels. This small portion of the image is called the *sliding window*. We move this sliding window from left to right and from top to bottom of the image. The pixel at the center of this *k*×*k* matrix is replaced by the average of all the pixels surrounding it. This *k*×*k* matrix is also called a *convolution kernel* or simply a *kernel*. Typically, this kernel is taken as an odd number so that a definite center can be calculated. The larger the kernel size, the blurrier the image will become. For example, a 5×5 kernel will produce a blurrier image compared to a 3×3 kernel.

OpenCV provides a convenient function to blur an image. The function `blur()` is used to blur an image by using mean filtering or averaging technique. This function takes two arguments.

- The NumPy representation of the original image that needs to be blurred

- The *k*×*k* kernel matrix

Listing 3-12 shows a blurring of an image using different kernel sizes.

Listing 3-12. Smoothing/Blurring by Mean Filtering or Averaging

```
Filename: Listing_3_12.py
1    import cv2
2    import numpy as np
3
4    # Load the image
5    park = cv2.imread("images/nature.jpg")
6    cv2.imshow("Original Park Image", park)
7
8    # Define the kernel
9    kernel = (3,3)
10   blurred3x3 = cv2.blur(park,kernel)
11   cv2.imshow("3x3 Blurred Image", blurred3x3)
12
13   blurred5x5 = cv2.blur(park,(5,5))
14   cv2.imshow("5x5 Blurred Image", blurred5x5)
15
16   blurred7x7 = cv2.blur(park, (7,7))
17   cv2.imshow("7x7 Blurred Image", blurred7x7)
18   cv2.waitKey(0)
```

As usual, we start with loading the image and assigning it to an array variable (the park variable in line 5 in Listing 3-12).

Line 9 defines a 3×3 kernel.

In line 10 we are using the blur() function and passing the park image and kernel as arguments. This will produce a blurred image using a 3×3 kernel.

To compare the effects of kernel size, lines 13 and 16 use kernel sizes 5×5 and 7×7. Notice the increasing order of blurriness as the kernel size increases in Figures 3-38 through 3-41.

Figure 3-38. *Original image*

Figure 3-39. *Blurring using a 3×3 kernel*

Figure 3-40. *Blurring using 5×5 kernel*

Figure 3-41. *Blurring using 7×7 kernel*

Gaussian Filtering

Gaussian filtering is one of the most effective blurring techniques in image processing. This is used to reduce Gaussian noise. This blurring technique gives a more natural smoothing result compared to the averaging technique. In this filtering, we supply a Gaussian kernel instead of a boxed fixed kernel.

A Gaussian kernel consists of the height, width, and standard deviations in the X and Y directions.

OpenCV provides a convenient function, GaussianBlur(), to perform the Gaussian filtering. GaussianBlur() takes the following arguments:

- The image represented by the NumPy array

- The *k*×*k* matrix as the kernel height and width

- sigmaX and sigmaY are a standard deviations in the X and Y directions

Here are a few notes about standard deviation:

- If only sigmaX is specified, sigmaY is taken the same as sigmaX.

- If both are taken as zero, the standard deviations are calculated from the kernel size.

- OpenCV provides a function, getGaussianKernel(), to auto-calculate the standard deviations.

For those who are interested in knowing the formula that is used in the Gaussian filtering, here is the Gaussian equation:

$$G_0\left(x,y\right) = Ae^{\frac{-\left(x-\mu_x\right)^2}{2\sigma_x^2} + \frac{-\left(y-\mu_y\right)^2}{2\sigma_y^2}}$$

where μ is the mean (the peak) and σ^2 is the variance (for each of the variables x and y).

Listing 3-13 is a working example to demonstrate Gaussian blurring.

Listing 3-13. Smoothing Using the Gaussian Technique

Filename: Listing_3_13.py

```
1    import cv2
2    import numpy as np
3
4    # Load the park image
5    parkImage = cv2.imread("images/park.jpg")
6    cv2.imshow("Original Image", parkImage)
7
8    # Gaussian blurring with 3x3 kernel height and 0 for standard
     deviation to calculate from the kernel
9    GaussianFiltered = cv2.GaussianBlur(parkImage, (5,5), 0)
10   cv2.imshow("Gaussian Blurred Image", GaussianFiltered)
11
12   cv2.waitKey(0)
```

Here again we are starting with loading our park image (line 5 of Listing 3-13). Line 9 shows the use of OpenCV's GaussianBlur() function. We supplied a 5×5 kernel and a 0 to tell OpenCV to calculate the standard deviations from the kernel size.

Figure 3-42 shows the original image, and Figure 3-43 shows the effect of Gaussian blurring.

Figure 3-42. *Original image*

Figure 3-43. *Gaussian blurred image with a 5×5 kernel*

Median Blurring

Median blurring is an effective technique for reducing salt-and-pepper type of noise. Median blurring is similar to mean blurring except that the central value of the kernel is replaced by the median of the surrounding pixels. We use the `medianBlur()` function of OpenCV to reduce the salt-and-pepper noise (see Listing 3-14). This function takes the following two arguments:

- The original image that needs to be blurred

- The kernel size k (which is similar to the $k×k$ matrix in the case of mean blurring)

Listing 3-14. Salt-and-Pepper Noise Reduction Using Median Blurring

Filename: Listing_3_14.py

```
1    import cv2
2
3    # Load a noisy image
4    saltpepperImage = cv2.imread("images/salt-pepper.jpg")
5    cv2.imshow("Original noisy image", saltpepperImage)
6
7    # Median filtering for noise reduction
8    blurredImage3 = cv2.medianBlur(saltpepperImage, 3)
9    cv2.imshow("Blurred image 3", blurredImage3)
10
11   # Median filtering for noise reduction
12   blurredImage5 = cv2.medianBlur(saltpepperImage, 5)
13   cv2.imshow("Blurred image 5", blurredImage5)
14   cv2.waitKey(0)
```

Listing 3-14 shows the use of the medianBlur() function. Lines 8 and 12 are creating the blurred images from the original image loaded in line 4. Notice the kernel parameter to the function is a scalar and not a tuple or matrix.

Figure 3-44 shows the image with salt-and-pepper noise. Notice the different levels of noise reduction as we apply different kernel sizes. Figure 3-45 shows the output image when the kernel size 3 is applied. Notice that Figure 3-45 still has some noise. Figure 3-46 shows a cleaner output with almost no noise when the kernel size 5 is applied with median blur.

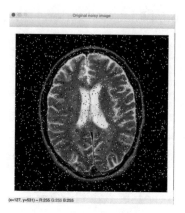

Figure 3-44. *A salt-and-pepper noisy image*

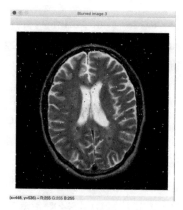

Figure 3-45. *Median blur with kernel size 3 (has some noise)*

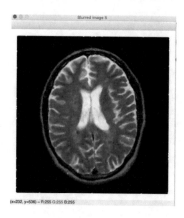

Figure 3-46. *Median blur with kernel size 5 (noise is almost removed)*

Figure 3-44 shows a noisy image with a salt-and-pepper type of noise. You will notice that median blur did a reasonably good job of reducing the noise. Figure 3-45 shows a blurred image by using a kernel size of 3. A good result is achieved by kernel size 5, as shown in Figure 3-46.

Bilateral Blurring

The previous three blurring techniques yield blurred images with the side effect that we lose the edges in the image. To blur an image while preserving the edges, we use *bilateral blurring*, which is an enhancement over Gaussian blurring. Bilateral blurring takes two Gaussian distributions to perform the computation.

The first Gaussian function considers the spatial neighbors (pixels in x and y space that are close together). The second Gaussian function considers the pixel intensity of the neighboring pixels. This makes sure that only those pixels that are of similar intensity to the central pixel are considered for blurring, leaving the edges intact as the edges tend to have higher intensity compared to other pixels.

Although this is a superior blurring technique, it is slower compared to other techniques.

We use `bilateralFilter()` to perform this kind of blurring. The arguments to this function are as follows:

- The image that needs to be blurred.

- The diameter of the pixel neighborhood.

- Color value. A larger value of the color means that more colors of the neighborhood pixels will be considered when computing the blur.

- A space or distance. A larger value of the space means that the pixels farther from the central pixel will be considered.

Let's examine Listing 3-15 to understand bilateral filtering.

Listing 3-15. Bilateral Blurring Example

```
Filename: Listing_3_15.py
1    import cv2
2
3    # Load a noisy image
4    noisyImage = cv2.imread("images/nature.jpg")
5    cv2.imshow("Original image", noisyImage)
6
7    # Bilateral Filter with kernel 5
8    fileteredImag5 = cv2.bilateralFilter(noisyImage, 5, 150,50)
9    cv2.imshow("Blurred image 5", fileteredImag5)
10
11   # Bilateral blurring with kernal 7
12   fileteredImag7 = cv2.bilateralFilter(noisyImage, 7, 160,60)
13   cv2.imshow("Blurred image 7", fileteredImag7)
14
15   cv2.waitKey(0)
```

As shown in Listing 3-15, lines 8 and 12 are for blurring the input image using `bilateralFilter()`. The first set of arguments (in line 8) includes the NumPy-represented image pixels, the kernel or diameter, the color threshold, and the distance from the center.

Figures 3-47 through 3-49 show the outputs of Listing 3-15.

Figure 3-47. *Original image*

Figure 3-48. *Bilateral blurring with diameter 5*

Figure 3-49. *Bilateral blurring with diameter 7*

The knowledge that you have acquired about various techniques for blurring or smoothing images will be applied extensively in the subsequent chapters.

Moving forward, our focus will shift to the process of converting a grayscale image into a binary image using a technique known as thresholding.

Binarization with Thresholding

Image binarization is the process of converting a grayscale or color image into a binary image, where each pixel is represented by only one of two possible values: black or white. Binarization is typically achieved by applying a threshold value to the pixels in the image. Pixels with intensities below the threshold are assigned the value of black (0), while pixels with intensities above the threshold are assigned the value of white (255). The result is a binary image that simplifies the image by emphasizing the object edges and eliminating the details and variations in intensity. Binarization is commonly used in various image processing applications such as object detection, segmentation, and character recognition.

OpenCV supports three types of thresholding techniques: simple thresholding, adaptive thresholding, and Otsu's binarization.

Simple Thresholding

In simple thresholding, we manually select a threshold value, *T*. All pixels greater than this *T* value are set to 255, and all pixels less than or equal to *T* are set to 0.

Sometimes it is helpful to do an inverse of binarization, in which case the pixels greater than the threshold are set to 0, and the pixels less than the threshold are set to 255.

Let's see an example of how to binarize an image using OpenCV's threshold() function. This function takes the following arguments:

- The original grayscale image that needs to be binarized

- The threshold value *T*

- The max value that will be set if the pixel value is greater than the threshold

- A thresholding method such as THRESH_BINARY or THRESH_BINARY_INV

The threshold() function returns a tuple containing the threshold value and the binarized image.

Listing 3-16 converts a grayscale image into a binary image.

Listing 3-16. Binarization Using Simple Thresholding

```
Filename: Listing_3_16.py
1    import cv2
2    import numpy as np
3
4    # Load an image
5    image = cv2.imread("images/scanned_doc.png")
6    # convert the image to grayscale
7    image = cv2.cvtColor(image, cv2.COLOR_BGR2GRAY)
8    cv2.imshow("Original Grayscale Receipt", image)
9
10   # Binarize the image using thresholding
11   (T, binarizedImage) = cv2.threshold(image, 60, 255, cv2.THRESH_BINARY)
12   cv2.imshow("Binarized Receipt", binarizedImage)
13
```

```
14   # Binarization with inverse thresholding
15   (Ti, inverseBinarizedImage) = cv2.threshold(image, 60, 255, cv2.
     THRESH_BINARY_INV)
16   cv2.imshow("Inverse Binarized Receipt", inverseBinarizedImage)
17   cv2.waitKey(0)
```

Listing 3-16 shows the two binarization methods: simple binarization and inverse binarization. Line 5 loads an image, and line 8 converts the image to a grayscale image because the input to the threshold function should be a grayscale image.

Line 11 calls OpenCV's threshold() function and passes as arguments the grayscale image, threshold value, maximum pixel value, and thresholding method THRESH_BINARY. The threshold() function returns a tuple containing the same threshold value that we supply in the argument and the binarized image. Because we are using THRESH_BINARY as the thresholding method, the pixel value will be set to a maximum of 255 for all pixels whose value is greater than 60 and will be set to 0 for those pixels whose value is equal or less than 60.

Line 15 is similar to line 11 except that the last argument to the threshold() function is THRESH_BINARY_INV as the thresholding function. By passing THRESH_BINARY_INV, we are instructing the threshold() method to do just the opposite of what the THRESH_BINARY method does: set the pixel value to 255 if the pixel intensity is less than 60; otherwise, set it to 0.

Sample outputs of the two threshold methods, along with the original image, are shown in Figure 3-50 through 3-52.

Figure 3-50. *Original grayscale image with dark background patches/stains*

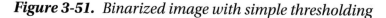

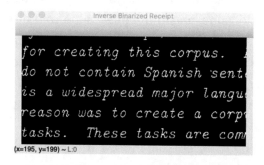

Figure 3-51. *Binarized image with simple thresholding*

Figure 3-52. *Binarized image with simple inverse thresholding*

To demonstrate this example, I took a scanned image of a badly stained document (Figure 3-50) and binarized it using simple thresholding. The method `THRESH_BINARY` generated the output, which contains black text on a white background. The method `THRESH_BINARY_INV` created the image with white text on a black background.

In simple thresholding, one global threshold value is applied to all pixels in the image, and we need to know the threshold value up front. If we are processing a large number of images and want to adjust the threshold values based on the image type and intensity variations, simple thresholding may not be the ideal method. In such cases, alternative thresholding methods such as adaptive thresholding or Otsu binarization can be applied to address this challenge.

Adaptive Thresholding

Adaptive thresholding is a technique that dynamically adjusts the threshold value for each pixel based on its local neighborhood. This allows for better handling of varying lighting conditions and intensity changes within an image.

In adaptive thresholding, the algorithm determines the threshold for a pixel based on a small region around it. As a result, different regions within the same image can have distinct threshold values.

Listing 3-17 shows the usage of adaptive thresholding to binarize a grayscale image.

Listing 3-17. Binarization Using Adaptive Thresholding

Filename: Listing_3_17.py

```
1    import cv2
2    import numpy as np
3
4    # Load an image
5    image = cv2.imread("images/boat.jpg")
6    # convert the image to grayscale
7    image = cv2.cvtColor(image, cv2.COLOR_BGR2GRAY)
8
9    cv2.imshow("Original Grayscale Image", image)
10
11   # Binarization using adaptive thresholding and simple mean
12   binarized = cv2.adaptiveThreshold(image, 255, cv2.ADAPTIVE_THRESH_
     MEAN_C, cv2.THRESH_BINARY, 7, 3)
13   cv2.imshow("Binarized Image with Simple Mean", binarized)
14
15   # Binarization using adaptive thresholding and Gaussian Mean
16   binarized = cv2.adaptiveThreshold(image, 255, cv2.ADAPTIVE_THRESH_
     GAUSSIAN_C, cv2.THRESH_BINARY_INV, 11, 3)
17   cv2.imshow("Binarized Image with Gaussian Mean", binarized)
18
19   cv2.waitKey(0)
```

In Listing 3-17, we have applied adaptive thresholding to an example image that contains varying degrees of shades and color intensity. The objective is to convert the image into a binary image using this technique. Let's explore what is happening in this listing.

Line 5, as usual, loads the image. Line 7 converts the image to a grayscale image as the input to the threshold function is a grayscale image.

Line 12 performs the actual binarization using OpenCV's `adaptiveThreshold()` function. This function takes the following arguments:

- The grayscale image that needs to be binarized

- The maximum value

- The method to calculate the threshold (more information in a moment)

- Binarization method such as `cv2.THRESH_BINARY` or `cv2.THRESH_BINARY_INV`

- Neighborhood size to consider for calculating the thresholds

- A constant value *C* that will be subtracted from the calculated thresholds

In this example, on line 12, we used `cv2.ADAPTIVE_THRESH_MEAN_C` to indicate that we want to calculate the threshold value of a pixel by taking the mean of pixels surrounding it. The size of the neighborhood in our example is 7×7. The last argument, 3, on line 12, is the constant that will be subtracted from the calculated threshold.

Line 16 is similar to line 12 except that we are using `cv2.ADAPTIVE_GAUSSIAN_C` to indicate that we want to calculate the threshold of a pixel by taking the weighted mean of all pixels surrounding it.

Figures 3-53 through 3-55 show sample outputs of Listing 3-17.

Figure 3-53. *Original image*

Figure 3-54. *Binarized image using adaptive thresholding with simple mean*

Figure 3-55. *Binarized image using adaptive thresholding with Gaussian mean*

Otsu's Binarization

Simple thresholding involves selecting a single global threshold value that is applied to the entire image. However, determining the appropriate threshold value can be challenging, as there is no definitive way to know the ideal value beforehand. Finding the right threshold value often requires trial-and-error experimentation. Moreover, even if an optimal value is found for one image, that value may not yield satisfactory results for other images with different pixel intensity characteristics. Hence, the selection of the threshold value in simple thresholding is subjective and may need to be adjusted for different images or scenarios.

Otsu's method is an approach that calculates an optimal global threshold value by analyzing the histogram of an image. We will delve into histograms in detail in the next chapter, so for now, simply consider a histogram as a visual representation of the frequency distribution of pixel values in an image. Otsu's method utilizes this histogram information to automatically determine the threshold value that maximizes the separation between foreground and background pixels. By effectively leveraging the pixel value distribution, Otsu's method helps achieve more accurate and reliable thresholding results.

To perform Otsu's binarization, we pass `cv2.THRESH_OTSU` as an extra flag in the `threshold()` function. For example, we pass `cv2.THRESH_BINARY+cv2.THRESH_OTSU` in the `threshold()` function to indicate the use of Otsu's method. The `threshold()` method requires a threshold value. When using Otsu's method, we pass an arbitrary value (could be 0), and the algorithm automatically calculates the threshold and returns the value of the threshold as one of the outputs.

Listing 3-18 presents a code example that demonstrates the use of Otsu's binarization method.

Listing 3-18. Otsu's Binarization

Filename: Listing_3_18.py

```
1    import cv2
2    import numpy as np
3
4    # Load an image
5    image = cv2.imread("images/scanned_doc.png")
6    # convert the image to grayscale
7    image = cv2.cvtColor(image, cv2.COLOR_BGR2GRAY)
8    cv2.imshow("Original Grayscale Receipt", image)
9
10   # Binarize the image using thresholding
11   (T, binarizedImage) = cv2.threshold(image, 0, 255, cv2.THRESH_
     BINARY+cv2.THRESH_OTSU)
12   print("Threshold value with Otsu binarization", T)
13   cv2.imshow("Binarized Receipt", binarizedImage)
14
15   # Binarization with inverse thresholding
16   (Ti, inverseBinarizedImage) = cv2.threshold(image, 0, 255, cv2.THRESH_
     BINARY_INV+cv2.THRESH_OTSU)
17   cv2.imshow("Inverse Binarized Receipt", inverseBinarizedImage)
18   print("Threshold value with Otsu inverse binarization", Ti)
19   cv2.waitKey(0)
```

The code example in Listing 3-18 and the code example in Listing 3-16 are very similar, with the following exceptions:

- Line 11 uses an additional flag, cv2.THRESH_OTSU, along with cv2. THRESH_BINARY, and the threshold value is passed as 0.

- Line 16 uses the flag cv2.THRESH_OTSU along with cv2.THRESH_ BINARY_INV, and again the threshold value is set to 0.

- We have print statements in lines 12 and 18 to print the calculated threshold values. Figure 3-56 shows the sample output of these print statements.

Figures 3-57 through 3-59 show Otsu's output samples.

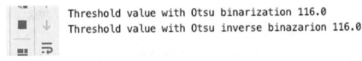

Figure 3-56. *Sample output of threshold values calculated from Otsu's method*

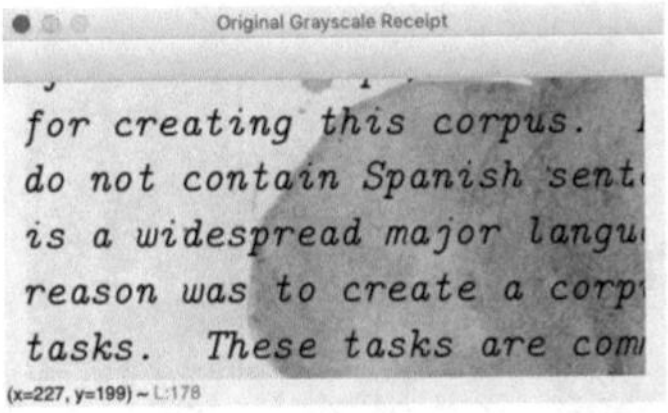

Figure 3-57. *Original image with varying background shades (stains and dark patches)*

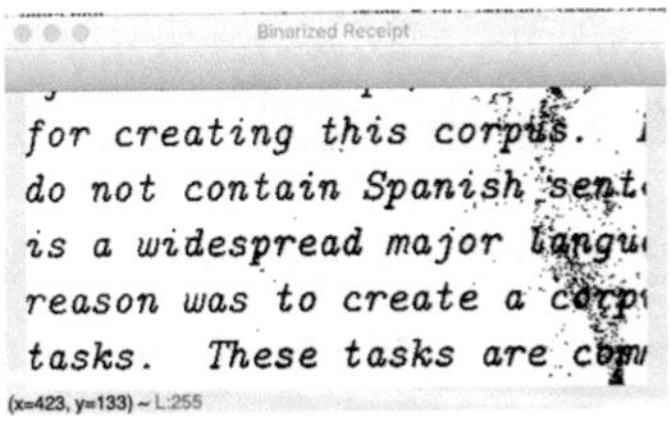

Figure 3-58. *Binarization with Otsu's method*

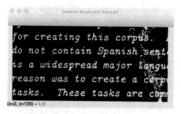

Figure 3-59. *Inverse binarization with Otsu's method*

Binarization is a valuable image processing technique that allows us to extract prominent features from images. Throughout this section, we have explored various binarization techniques and their applications, considering factors such as pixel intensity and intensity variations.

In the next section, we will delve into another powerful image processing technique, edge detection, which enables us to identify and highlight the boundaries and edges of objects within an image, facilitating tasks such as object recognition and image segmentation.

Gradients and Edge Detection

Edge detection encompasses a collection of techniques to identify points in an image where there is a significant change in pixel brightness. In this section, we will explore two methods for detecting edges in an image: gradient-based edge detection and Canny edge detection.

Gradient-Based Edge Detection

Gradient-based edge detection involves computing the gradients of pixel intensities to locate areas of rapid changes. It provides a straightforward approach for identifying edges in an image.

OpenCV provides the two methods for finding gradients: Sobel derivatives and Laplacian derivatives.

Sobel Derivatives (Sobel( ) Function)

The Sobel method is a combination of Gaussian smoothing and Sobel differentiation, which computes an approximation of the gradient of an image intensity function. Because of the Gaussian smoothing, this method is resistant to noise.

We can perform derivatives in either the horizontal direction or vertical direction by passing the arguments xorder and yorder, respectively. The Sobel() function also takes an argument ksize that we use to define the kernel size. If we set ksize to -1, OpenCV will internally apply a 3×3 Scharr filter, which generally gives a better result compared to the 3×3 Sobel filter.

Listing 3-19 shows the Sobel function in action.

Listing 3-19. Sobel and Scharr Gradient Detection

Filename: Listing_3_19.py

```
1    import cv2
2    import numpy as np
3    # Load an image
4    image = cv2.imread("images/sudoku.jpg")
5    cv2.imshow("Original Image", image)
6    image = cv2.cvtColor(image, cv2.COLOR_BGR2GRAY)
7    image = cv2.bilateralFilter(image, 5, 50, 50)
8    cv2.imshow("Blurred image", image)
9
10   # Sobel gradient detection
11   sobelx = cv2.Sobel(image,cv2.CV_64F,1,0,ksize=3)
12   sobelx = np.uint8(np.absolute(sobelx))
13   sobely = cv2.Sobel(image,cv2.CV_64F,0,1,ksize=3)
14   sobely = np.uint8(np.absolute(sobely))
15
16   cv2.imshow("Sobel X", sobelx)
17   cv2.imshow("Sobel Y", sobely)
18
19   # Scharr gradient detection by passing ksize = -1 to Sobel function
20   scharx = cv2.Sobel(image,cv2.CV_64F,1,0,ksize=-1)
21   scharx = np.uint8(np.absolute(scharx))
22   schary = cv2.Sobel(image,cv2.CV_64F,0,1,ksize=-1)
```

```
23    schary = np.uint8(np.absolute(schary))
24    cv2.imshow("Schar X", scharx)
25    cv2.imshow("Schar Y", schary)
26
27    cv2.waitKey(0)
```

A lot of things are happening here, so let's go through the relevant lines of this listing to help clarify the concept of gradients.

Line 4 is simply loading an image from the disk. We applied a bilateral filter to reduce noise in line 7. Figure 3-60 shows the original input image and Figure 3-61 shows the blurred image that is used as an input in the Sobel and Scharr gradient detection functions.

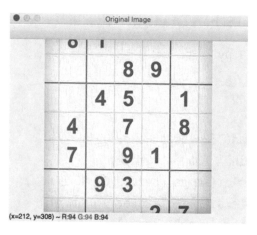

Figure 3-60. *Original image*

Figure 3-61. *Blurred image*

Gradient detection starts from line 11. We use OpenCV's `Sobel()` function, which takes the following parameters:

- The blurred image in which we want to detect gradients.

- A data type, `cv2.CV_64F`, which is a 64-bit float. Why? The transition from black to white is considered a positive slope, while the transition from white to black is a negative slope. An 8-bit unsigned integer cannot hold a negative number. Therefore, we need to use a 64-bit float; otherwise, we will lose gradients when the transition from white to black happens.

- The third argument indicates whether we want to calculate gradients in the X direction. The value 1 means we want to calculate the gradient in the X direction.

- Similarly, the fourth argument indicates whether to calculate gradients in the Y direction. A 1 means yes, and a 0 means no.

- The fifth argument, `ksize`, defines the kernel size. `ksize=5` means the kernel size is 5×5.

Since we want to determine gradients in the X direction on line 11, we set the third parameter in the `Sobel()` function to 1, and we set the fourth parameter to 0.

Line 12 simply takes the absolute value of the gradients and converts them back to 8-bit unsigned integers. Remember, an image is represented as an 8-bit unsigned integer NumPy array.

Line 13 is similar to line 11 except that the third argument is set to 0 and the fourth argument is set to 1 to indicate gradient calculation in the Y direction.

Line 14 converts the 64-bit floats to an 8-bit unsigned integer, as explained earlier.

Figure 3-62 and Figure 3-63 show sample outputs of lines 16 and 17. You will notice that the edge detection in both the X and Y directions is not very sharp. Let's try a simple improvement to see the effect on the sharpness of the edges.

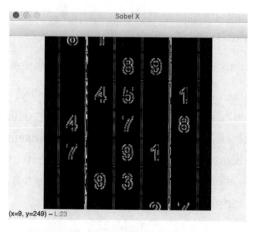

Figure 3-62. *Sobel edge detection in the X direction*

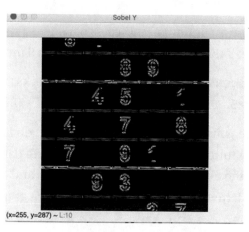

Figure 3-63. *Sobel edge detection in the Y direction*

Line 20 through 23 are similar to lines 11 through 14 of Listing 3-19. The difference is that the value of ksize is -1, which instructs OpenCV to internally call the Scharr function with a kernel size of 3×3. You will notice that the sharpness of the edges is much better compared to the Sobel function. Figure 3-64 and Figure 3-65 are the results of the Scharr filter of the image shown in Figure 3-61.

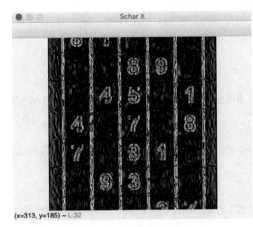

Figure 3-64. *Scharr edge detection in the X direction*

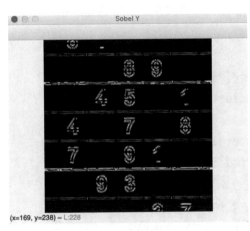

Figure 3-65. *Scharr edge detection in the Y direction*

Sobel and Scharr calculate gradient magnitudes along the X and Y directions, allowing us to determine edges along the horizontal and vertical directions.

Laplacian Derivatives (cv2.Laplacian( ) Function)

The Laplacian operator calculates the second derivative of the pixel intensity function to determine the edges in the image. The Laplacian operator calculates the gradients based on the following equation:

$$\text{Laplace}(f) = \frac{\partial^2 f}{\partial x^2} + \frac{\partial^2 f}{\partial y^2}$$

OpenCV provides a function, Laplacian(), to calculate gradients for edge detection. This function takes the following arguments:

- The image in which edges need to be detected

- The data type, which is normally cv2.CV_64F to hold floating-point values

Listing 3-20 shows a working example of edge detection using the Laplacian function of OpenCV.

Listing 3-20. Edge Detection Using Laplacian Derivatives

Filename: Listing_3_20.py

```
1    import cv2
2    import numpy as np
3
4    # Load an image
5    image = cv2.imread("images/sudoku.jpg")
6    image = cv2.cvtColor(image, cv2.COLOR_BGR2GRAY)
7
8    image = cv2.bilateralFilter(image, 5, 50, 50)
9    cv2.imshow("Blurred image", image)
10
11   # Laplace function for edge detection
12   laplace = cv2.Laplacian(image,cv2.CV_64F)
13   laplace = np.uint8(np.absolute(laplace))
14
15   cv2.imshow("Laplacian Edges", laplace)
16
17   cv2.waitKey(0)
```

As usual, line 5 loads an image, line 6 converts the image to grayscale, and line 8 blurs the image using bilateral filtering.

Line 12 is where the Laplacian() function is called for gradient calculation to detect edges in the image. Again, we passed the CV_64F data type to hold the possible negative values of gradients when the transitions from white to black happen.

Line 13 converts the 64-bit floats to 8-bit unsigned integers.

Figure 3-66 shows a sample display of the `Laplacian()` function.

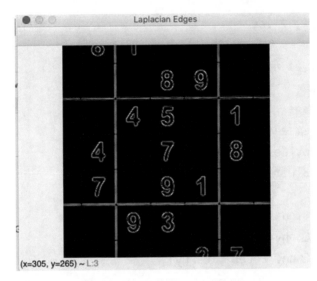

Figure 3-66. *Edge detection using Laplacian derivatives*

Canny Edge Detection

Canny edge detection is one of the most popular methods for detecting edges in image processing. It involves a multistep process that consists of several key operations:

1. The image is blurred to reduce noise.

2. The Sobel gradients are computed in both the X and Y directions to capture edge intensity changes.

3. Non-maximum suppression is applied to suppress nonessential edges. The non-maximum suppression helps refine the detected edges by retaining only the local maximum gradient values in an edge map.

4. A hysteresis thresholding technique is utilized to determine whether a pixel qualifies as an "edge-like" pixel.

OpenCV simplifies the implementation of Canny edge detection by encapsulating all of these steps into a single function called `Canny()`. This function streamlines the process, making it convenient and efficient. Listing 3-21 provides an example of edge detection using the `Canny()` function.

Listing 3-21. Canny Edge Detection

```
Filename: Listing_3_21.py
1    import cv2
2    import numpy as np
3
4    # Load an image
5    image = cv2.imread("images/sudoku.jpg")
6    image = cv2.cvtColor(image, cv2.COLOR_BGR2GRAY)
7    cv2.imshow("Blurred image", image)
8
9    # Canny function for edge detection
10   canny = cv2.Canny(image, 50, 170)
11   cv2.imshow("Canny Edges", canny)
12
13   cv2.waitKey(0)
```

The important line in Listing 3-21 is line 10, where we are calling the Canny() function and passing the minimum and maximum threshold values to the image in which edges need to be detected. Any gradient value larger than the maximum threshold value is considered an edge. Any value below the minimum threshold is not considered an edge. The gradient values in between are considered for edges according to their intensity variations.

Figure 3-67 shows a sample output of the Canny edge detector. Notice that the edges are very crisp in this case.

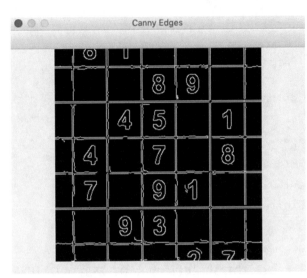

Figure 3-67. *Canny edge detection*

Contours

Contours are curves that connect continuous points of the same intensity. The ability to identify and analyze contours is crucial for tasks such as object identification, face detection, and recognition.

To detect contours in an image, we follow these steps:

1. Convert the image to grayscale to simplify the processing.

2. Binarize the image by using any of the available thresholding methods.

3. Apply the Canny edge detection method to highlight the edges within the image.

4. Utilize the `findContours()` method to locate and extract all the contours present in the image.

5. Optionally, use the `drawContours()` function to visualize and draw the contours on the image.

To give you a practical understanding of how to implement this technique effectively, the example in Listing 3-22 and the following discussion explore the process of contour detection and drawing in action.

Listing 3-22. Contour Detection and Drawing

Filename: Listing_3_22.py

```
1    import cv2
2    import numpy as np
3
4    # Load an image
5    image = cv2.imread("images/sudoku.jpg")
6    image = cv2.cvtColor(image, cv2.COLOR_BGR2GRAY)
7    cv2.imshow("Blurred image", image)
8
9    # Binarize the image
10   (T,binarized) = cv2.threshold(image, 0, 255, cv2.THRESH_BINARY_
     INV+cv2.THRESH_OTSU)
11   cv2.imshow("Binarized image", binarized)
12
13   # Canny function for edge detection
14   canny = cv2.Canny(binarized, 0, 255)
15   cv2.imshow("Canny Edges", canny)
16
17   (contours, hierarchy) = cv2.findContours(canny,cv2.RETR_EXTERNAL,
     cv2.CHAIN_APPROX_SIMPLE)
18   print("Number of contours determined are ", format(len(contours)))
19
20   copiedImage = image.copy()
21   cv2.drawContours(copiedImage, contours, -1, (0,255,0), 2)
22   cv2.imshow("Contours", copiedImage)
23   cv2.waitKey(0)
```

Here is a line-by-line explanation of Listing 3-22.

Line 5 loads the image. Line 6 converts the image to grayscale, and line 10 binarizes the image using Otsu's method. Line 14 calculates gradients for edge detection using Canny's function.

Line 17 calls OpenCV's findContours() function to determine contours. The arguments to this function are as follows:

- The first argument is the image in which we want to detect the edges using Canny's function.

- The second argument, cv2.RET_EXTERNAL, determines the type of contour we are interested in. cv2.RET_EXTERNAL retrieves the outermost contours only. We can also use cv2.RET_LIST to retrieve all contours, cv2.RET_COMP and cv2.RET_TREE, to include hierarchical contours.

- The third argument, cv2.CHAIN_APPROAX_SIMPLE, removes the redundant points and compresses the contour, thereby saving memory. cv2.CHAIN_APPROAX_NONE stores all points of the contour (which require more memory to store them).

The output of the findContours() function is a tuple with the following items in it:

- The first item is a Python list of all the contours in the image. Each individual contour is a NumPy array of (x,y) coordinates of boundary points of the object.

- The second item is the contour hierarchy.

Line 18 is where we are printing the number of contours identified.

We are drawing contours in line 21 of Listing 3-22 by using the drawContours() function. The following are arguments to this function:

- The first argument is the image in which contours are to be drawn.

- The second argument is the list of all contour points.

- The third argument is the index of the contour to be drawn. To draw the first contour, pass a 0. Similarly, pass 1 to draw the second contour, and so on. If you want to draw all contours, pass -1 to this argument.

- The fourth argument is the color of the contour.

- The final argument is the thickness of the contour.

Figures 3-68 through 3-70 show some sample outputs of Listing 3-22.

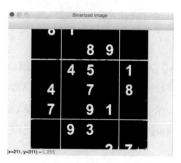

Figure 3-68. *Blurred image*

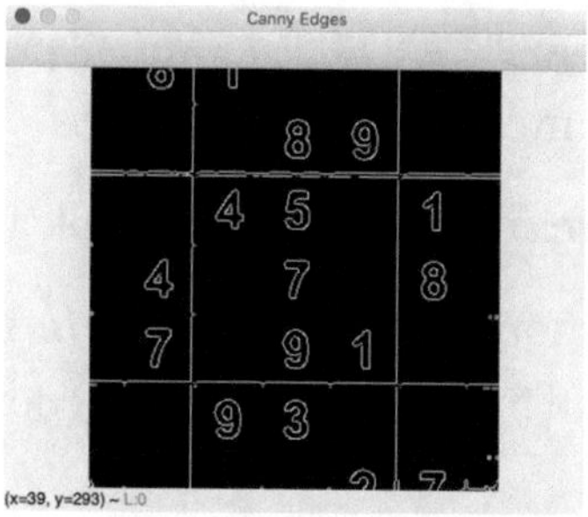

Figure 3-69. *Contours using the Canny function*

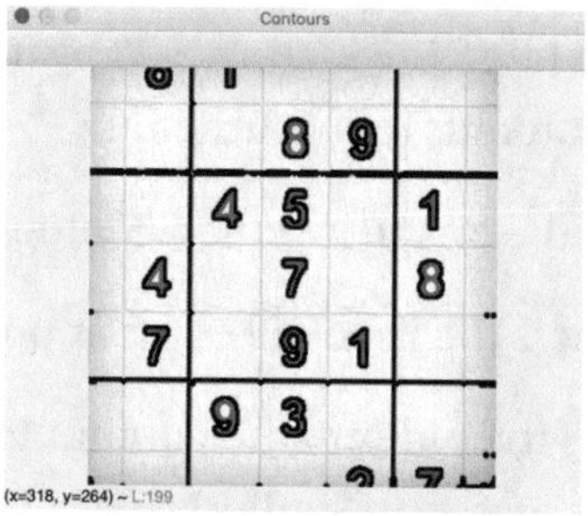

Figure 3-70. *Contours drawn on the original image*

Morphological Transformation

Morphological transformation is a fundamental technique in image processing that involves the manipulation and modification of the geometric structure of objects within an image. It is primarily used for tasks such as noise removal, object extraction, and shape analysis.

Morphological transformations are based on a set of predefined operations known as *morphological operators*. The two most commonly used morphological operators are dilation and erosion. There are several variant forms of morphological transformations that complement the basic dilation and erosion operations. These include opening, closing, gradient, and more. In the following sections, we will explore each of these transformations in detail, using the image shown in Figure 3-71 as an example to illustrate their effects.

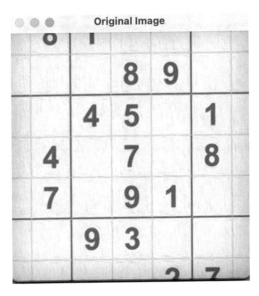

Figure 3-71. *Original image*

OpenCV offers functions dedicated to executing morphological operations, primarily intended for binary images. These operations rely on two inputs: the original image and a structuring element, also known as a kernel, that determines the specific characteristics of the operation to be performed.

Dilation

Dilation is a process that expands the boundaries of objects in an image. It achieves this by convolving the image with a structuring element, which is a small binary matrix. Dilation causes the objects to grow and merge together, making them more prominent and distinct. In the dilation process, a pixel element is assigned the value 1 if there is at least one pixel underneath the kernel that has a value of 1. As a result, the white region within the image expands, causing an increase in the size of the foreground object.

OpenCV provides a function, dilate(), that takes as arguments the original image (as a NumPy array), the structuring element or kernel, and the number of iterations. Refer to Listing 3-23, line 14, for example usage of this function, and Figure 3-72 for an example of its output.

Figure 3-72. *Dilation*

Erosion

Erosion is a process that erodes or shrinks the boundaries of objects in an image. It involves convolving the image with a structuring element. Erosion eliminates small details and breaks up connected regions, resulting in the removal of noise and finer structures. During erosion, the kernel moves across the image in a manner similar to 2D convolution. If all the pixels underneath the kernel in the original image are 1, the corresponding pixel in the output remains 1. However, if any of the pixels beneath the kernel are 0, that pixel is eroded and set to 0.

As a consequence, pixels near the boundary of the object are gradually eliminated based on the kernel's size. This leads to a decrease in the thickness or size of the foreground object, resulting in a reduction of the white region within the image.

OpenCV provides a function, erode(), that takes as arguments the original image (as a NumPy array), the structuring element or kernel, and the number of iterations. Refer to Listing 3-23, line 11, for example usage of this function, and Figure 3-73 for an example of its output.

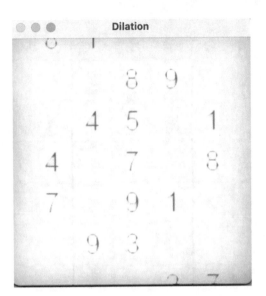

Figure 3-73. *Erosion*

Opening

Opening involves an erosion followed by a dilation and is useful for removing noise and small objects. By first eroding the image, small noisy regions and fine details are removed. Dilation then helps to restore the remaining objects while preserving their overall shape and structure.

To perform opening, OpenCV provides a function, `morphologyEx()`, that takes the following arguments:

- An image as a NumPy array

- `cv2.MORPH_OPEN`, which indicates that the operation being performed is opening

- A kernel that defines the structuring element or convolutional matrix

See Listing 3-23, line 17, for example usage of this function, and Figure 3-74 for an example of its output.

Figure 3-74. *Opening*

Closing

Closing involves a dilation followed by an erosion and is effective for closing small gaps and filling holes in objects. By performing a dilation first, the operation helps to expand the boundaries of the objects and close small gaps between them. The erosion then helps to refine the object boundaries and eliminate any excessive expansion caused by dilation.

To perform closing, we can use the same OpenCV function used in opening, morphologyEx(), and pass the following arguments:

- An image as a NumPy array

- cv2.MORPH_CLOSE, which indicates that the operation being performed is closing

- A kernel that defines the structuring element or convolutional matrix

See Listing 3-23, line 20, for example usage of this function, and Figure 3-75 for an example of its output.

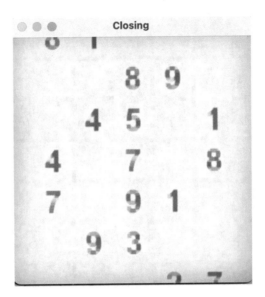

Figure 3-75. *Closing*

Morphological Gradient

Morphological gradient is a process that calculates the difference between a dilation and an erosion of an image using a specific structuring element or kernel. It highlights the boundaries or edges of objects in the image.

The morphological gradient is useful for edge detection, as it emphasizes transitions between regions of different intensities or structures. It provides a representation of the local changes in intensity or shape within an image, allowing for further analysis or processing, such as object segmentation or contour extraction.

Listing 3-23, line 23, shows example usage of morphological gradient, and Figure 3-76 shows an example of the output of this operation.

Figure 3-76. Morphological gradient

Top Hat

Top hat is a morphological operation that calculates the difference between the input image and its opening. It helps in enhancing and extracting small-scale details, such as small objects or fine structures, in an image.

The top hat operation effectively removes the larger-scale structures or background from the original image, leaving behind the smaller-scale details. It is particularly useful in applications such as image enhancement, object detection, and extracting subtle features that may be overshadowed by the overall image structure or background.

The top hat operation is illustrated in Listing 3-23, line 26, and a sample of its output is shown in Figure 3-77.

Figure 3-77. *Top hat*

Black Hat

Black hat is a morphological operation that calculates the difference between the closing of an image and the original image itself. It is used to extract and highlight larger-scale structures or regions in an image.

The black hat operation reveals the larger-scale structures or regions that are present in the image but may be overshadowed or hidden by the surrounding background. It is commonly used in applications such as image analysis, object detection, and text extraction, where the focus is on identifying and extracting significant structures or features that are larger than the background or noise level.

The black hat operation is illustrated in Listing 3-23, line 29, and a sample of its output is shown in Figure 3-78.

Listing 3-23. Morphological Operations: Dilation, Erosion, Closing, Opening, Top Hat, and Black Hat

```
Filename: Listing_3_23.py
1    import cv2
2    import numpy as np
3
4    # Load the image
```

```python
5    image = cv2.imread("images/sudoku.jpg", 0)
6
7    # Define the structuring element
8    kernel = np.ones((5, 5), np.uint8)
9
10   # Perform erosion
11   erosion = cv2.erode(image, kernel, iterations=1)
12
13   # Perform dilation
14   dilation = cv2.dilate(image, kernel, iterations=1)
15
16   # Perform opening
17   opening = cv2.morphologyEx(image, cv2.MORPH_OPEN, kernel)
18
19   # Perform closing
20   closing = cv2.morphologyEx(image, cv2.MORPH_CLOSE, kernel)
21
22   # Perform morphological gradient
23   gradient = cv2.morphologyEx(image, cv2.MORPH_GRADIENT, kernel)
24
25   # Perform top hat
26   tophat = cv2.morphologyEx(image, cv2.MORPH_TOPHAT, kernel)
27
28   # Perform black hat
29   blackhat = cv2.morphologyEx(image, cv2.MORPH_BLACKHAT, kernel)
30
31   # Display the results
32   cv2.imshow("Original Image", image)
33   cv2.imshow("Erosion", erosion)
34   cv2.imshow("Dilation", dilation)
35   cv2.imshow("Opening", opening)
36   cv2.imshow("Closing", closing)
37   cv2.imshow("Morphological Gradient", gradient)
38   cv2.imshow("Top Hat", tophat)
39   cv2.imshow("Black Hat", blackhat)
```

```
40
41    # Wait for key press and exit
42    cv2.waitKey(0)
43    cv2.destroyAllWindows()
```

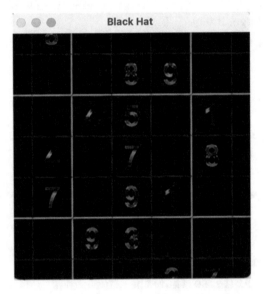

Figure 3-78. *Black hat*

Template Matching

Template matching is a technique used to locate and identify the position of a template image within a larger image. The template matching algorithm compares a template image with the target image at different positions and scales to determine the best match. It works by sliding the template image over the target image and calculating the similarity between the template and the target at each location.

OpenCV implements several template comparison algorithms. It returns a grayscale image, where each pixel represents the similarity between the neighborhood of that pixel and the template image.

If the input image has dimensions (WxH) and the template image has dimensions (wxh), the output image will have dimensions (W-w+1, H-h+1). Once we obtain the result, we can utilize the `cv2.minMaxLoc()` function to identify the location of the maximum/minimum value. This location can be considered as the top-left corner of the rectangle, with the width and height of the rectangle being (w,h). Thus, the identified rectangle represents the region of the template within the larger image.

Listing 3-24 demonstrates the template matching process using all six supported methods.

Listing 3-24. Template Matching with Different Methods

Filename: Listing_3_24.py

```
1    import cv2
2
3    target = cv2.imread('images/obama_with_people.jpg', cv2.IMREAD_
     GRAYSCALE)
4
5    template = cv2.imread('images/obama_face.jpg', cv2.IMREAD_GRAYSCALE)
6    # Resize the template so that the dimensions of template are smaller
     than the target's
7    template = cv2.resize(template, None, fx=0.2, fy=0.2)
8
9    w, h = template.shape[::-1]
10   # All 6 supported comparison methods
11   methods = ['cv2.TM_CCOEFF', 'cv2.TM_CCOEFF_NORMED', 'cv2.TM_CCORR',
12              'cv2.TM_CCORR_NORMED', 'cv2.TM_SQDIFF', 'cv2.TM_SQDIFF_
              NORMED']
13   for m in methods:
14       img = target.copy()
15       method = eval(m)
16       # Apply template Matching
17       result = cv2.matchTemplate(img, template, method)
18       min_val, max_val, min_loc, max_loc = cv2.minMaxLoc(result)
19       # If the method is TM_SQDIFF or TM_SQDIFF_NORMED, take minimum,
         else take maximum
20       if method in [cv2.TM_SQDIFF, cv2.TM_SQDIFF_NORMED]:
```

```
21              top_left = min_loc
22          else:
23              top_left = max_loc
24          bottom_right = (top_left[0] + w, top_left[1] + h)
25          cv2.rectangle(img,top_left, bottom_right, 255, 2)
26          cv2.imshow(m, img)
27          cv2.waitKey(0)
28
29      cv2.destroyAllWindows()
```

The code in Listing 3-24 performs template matching using OpenCV. Here is a breakdown of the code.

Lines 3 and 5 load the target and template images in grayscale format.

Line 7 resizes the template image to ensure that its dimensions are smaller than the dimensions of the target image. This resizing is necessary for accurate template matching.

Lines 11 and 12 define a list of supported template matching methods provided by OpenCV.

Lines 13-28: A for loop is initiated to iterate through the list of matching methods. Inside the loop:

Line 14 creates a copy of the original target image, because the image will be modified when a rectangle is drawn on it (line 25).

In line 17, the function matchTemplate() from OpenCV is used to perform the template matching. It takes three arguments: the target image, the template image, and the matching method.

Lines 18 through 24 calculate the coordinates of the bounding box for the matched portion of the target image.

Line 25 draws a rectangle around the matched section on the target image.

Line 26 displays the resulting image with the drawn rectangle.

Figures 3-79 through 3-86 show the outputs of Listing 3-24.

Figure 3-79. *Template image*

Figure 3-80. *Original image*

Figure 3-81. *Matching using TM_CCOEFF*

Figure 3-82. *Matching using TM_CCOEFF_NORMED*

Figure 3-83. *Matching using TM_CCORR*

Figure 3-84. *Matching using TM_CCORR_NORMED*

Figure 3-85. *Matching using TM_SQDIFF*

Figure 3-86. *Matching using TM_SQDIFF_NORMED*

Template Matching with Multiple Objects

In the previous section, we utilized the matchTemplate() and minMaxLoc() functions to search for Barack Obama's face in an image, assuming it appeared only once. However, if we are searching for an object that appears multiple times within the image, minMaxLoc() will not provide all the locations. In such cases, we need to employ a thresholding technique to filter those bounding boxes that have lower scores compared to the threshold.

Listing 3-25 demonstrates the concept of template matching with multiple objects. It uses a template image of a single soda bottle and performs matching against multiple soda bottles within the target image. Note that the code in Listing 3-25 invokes the non_maximum_suppression() function from Listing 3-26.

Listing 3-25. Template Matching with Multiple Objects

Filename: Listing_3_25.py

```
1    import cv2
2    import numpy as np
3    # load the input image and template image from disk
4    image = cv2.imread("images/pepsico.png")
5    template = cv2.imread("images/pepsi.png")
6    # Resize the template so that the dimensions are less than the image
     dimensions
7    template = cv2.resize(template, None, fx=0.5, fy=0.5)
8    # Get the height and width of the template
9    (tH, tW) = template.shape[:2]
10   # convert both the image and template to grayscale
11   imageGray = cv2.cvtColor(image, cv2.COLOR_BGR2GRAY)
12   templateGray = cv2.cvtColor(template, cv2.COLOR_BGR2GRAY)
13   # perform template matching
14   result = cv2.matchTemplate(imageGray, templateGray, cv2.TM_
     CCOEFF_NORMED)
15   # find all locations in the result where the matched value is
16   # greater than the threshold
17   (yCoords, xCoords) = np.where(result >= 0.5)
18   # clone the original image so we can draw rectangular regions on it.
```

```
19    clone = image.copy()
20    # loop over our starting (x, y)-coordinates
21    for (x, y) in zip(xCoords, yCoords):
22        # draw the bounding box on the image
23        cv2.rectangle(clone, (x, y), (x + tW, y + tH),
24            (0, 255, 255), 2)
25    # show our output image without applying non_max_suppression of
      overlapping bounding boxes
26    cv2.imshow("Without NMS", clone)
27    # Let us now apply NMS
28    # initialize our list of rectangles
29    rects = []
30    # loop over the starting (x, y)-coordinates again
31    for (x, y) in zip(xCoords, yCoords):
32        # update the list of rectangles
33        rects.append((x, y, x + tW, y + tH))
34    # We gave equal scores to all bounding boxes
35    pick = non_max_suppression(np.array(rects),  np.ones(len(rects)), 0.7)
36    # loop over the final bounding boxes
37    for (startX, startY, endX, endY) in pick:
38        # draw the bounding box on the image
39        cv2.rectangle(image, (startX, startY), (endX, endY),
40            (255, 0, 255), 2)
41    # show the output image
42    cv2.imshow("After NMS", image)
43    cv2.waitKey(0)
```

Lines 1 to 14 should be self-explanatory at this point.

In line 14, we execute the template matching operation.

Line 17 applies a threshold to filter out coordinates (x, y) that are below the threshold value, keeping only those that exceed it.

In line 19, we make a copy of the original image and store it in a variable named clone. This clone will serve as the image for drawing rectangles, ensuring the original image remains unaltered.

Lines 21 to 25 comprise a for loop that iterates through all the coordinates, drawing rectangular regions around the matched objects within the image.

Line 26 displays the image, showcasing bounding boxes around the identified objects (refer to Figure 3-89). Multiple overlapping bounding boxes can be observed around the same object. To address this, we need to implement a technique known as *non-maximum suppression (NMS)*, so that only one bounding box is drawn around a matched object. NMS is commonly used in computer vision and object detection algorithms to eliminate overlapping or redundant bounding boxes.

In line 35, non-maximum suppression is applied to the detected bounding boxes to remove duplicate or overlapping detections of the same object. This ensures that each object is represented by a single bounding box, removing redundancy and improving the accuracy of the detection results.

The non-maximum suppression function is implemented in Listing 3-26.

Listing 3-26. Non-Maximum Suppression Function

Filename: Listing_3_26.py

```
1    # The NMS function
2    def non_max_suppression(boxes, scores, overlap_threshold):
3        # If there are no boxes, return an empty list
4        if len(boxes) == 0:
5            return []
6        # Initialize the list of picked indices
7        pick = []
8        # Grab the coordinates of the bounding boxes
9        x1 = boxes[:, 0]
10       y1 = boxes[:, 1]
11       x2 = boxes[:, 2]
12       y2 = boxes[:, 3]
13       # Compute the area of the bounding boxes
14       area = (x2 - x1 + 1) * (y2 - y1 + 1)
15       # Sort the bounding boxes by their scores in descending order
16       idxs = np.argsort(scores)
17       # Keep looping while there are indices to be processed
18       while len(idxs) > 0:
19           # Grab the last index in the sorted list and add it to the list of
20           # picked indices
```

```
21          last = len(idxs) - 1
22          i = idxs[last]
23          pick.append(i)
24          # Find the largest (x, y) coordinates for the start of the
            bounding box
25          # and the smallest (x, y) coordinates for the end of the
            bounding box
26          xx1 = np.maximum(x1[i], x1[idxs[:last]])
27          yy1 = np.maximum(y1[i], y1[idxs[:last]])
28          xx2 = np.minimum(x2[i], x2[idxs[:last]])
29          yy2 = np.minimum(y2[i], y2[idxs[:last]])
30          # Compute the width and height of the bounding box
31          w = np.maximum(0, xx2 - xx1 + 1)
32          h = np.maximum(0, yy2 - yy1 + 1)
33          # Compute the overlap ratio
34          overlap = (w * h) / area[idxs[:last]]
35          # Delete all indices from the index list that have
            overlap greater
36          # than the specified threshold
37          idxs = np.delete(idxs, np.concatenate(([last],
            np.where(overlap > overlap_threshold)[0])))
38      # Return only the bounding boxes that were picked
39      return boxes[pick]
```

The outputs of this code Listings 3-25 and 3-26 are shown in Figures 3-87 through 3-90.

Figure 3-87. *Template image*

Figure 3-88. *Original image*

Figure 3-89. *Multiple object matching without NMS*

Figure 3-90. *Multiple object matching with NMS*

The output after NMS is shown in Figure 3-90. Notice the difference between the outputs before and after the NMS. Before NMS (Figure 3-89), the output contains multiple bounding boxes or detections for the same object, which overlap each other. However, after applying NMS, redundant or overlapping detections are suppressed, resulting in a more refined output with a reduced number of bounding boxes. NMS helps in selecting the most confident and accurate detections by considering their confidence scores and spatial overlap. This leads to a cleaner and more accurate representation of the detected objects.

Summary

In this chapter, we covered a wide range of image processing techniques that are instrumental in constructing computer vision applications. We first explored methods for image transformation, including resizing, rotation, flipping, and cropping, enabling us to manipulate images according to our requirements. Next, we examined how to perform arithmetic and bitwise operations on images, expanding our capabilities for image manipulation.

Furthermore, we delved into powerful image processing functions such as masking, noise reduction, binarization, edge detection, contour detection, and morphological transformation. These techniques provide valuable tools for enhancing image quality, segmenting objects, and extracting important visual features.

It is worth noting that the techniques you learned in this chapter will be used extensively in subsequent chapters, particularly when we delve into feature extraction and engineering for machine learning tasks. Understanding and applying these image processing techniques will be fundamental to harnessing the full potential of computer vision in various applications and domains.

CHAPTER 4

Building a Machine Learning–Based Computer Vision System

Chapter 3 introduced and surveyed diverse image processing approaches and methods. In Chapter 5, we will delve into the process of constructing computer vision systems using machine learning. In this chapter, we will explore the fundamental techniques of feature extraction from images and the art of feature selection. We will demystify how machines learn from these features. This chapter serves as an introduction to Chapter 5, where you will gain insights into different deep learning algorithms and learn how to implement them in Python with TensorFlow.

Image Processing Pipeline

Computer vision (CV) refers to the capacity of computers to capture, analyze, and draw interpretations and conclusions from images. For instance, CV can be applied to image detection and recognition tasks, as well as to identifying patterns or objects within images. When an artificial intelligence (AI) system encounters images, it undergoes a series of processes: image ingestion, processing, feature extraction, and interpretation. This involves the transformation of images between different system components, enabling machines to recognize patterns and detect objects within them.

The *image processing pipeline*, also referred to as the *computer vision pipeline*, encompasses a series of components that process images through various transformations, resulting in a final output. Figure 4-1 provides a high-level overview of this processing pipeline.

© Shamshad Ansari 2023
S. Ansari, *Building Computer Vision Applications Using Artificial Neural Networks*,
https://doi.org/10.1007/978-1-4842-9866-4_4

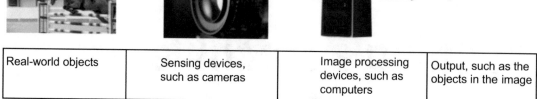

Real-world objects	Sensing devices, such as cameras	Image processing devices, such as computers	Output, such as the objects in the image

Figure 4-1. *Image pipeline*

As depicted in Figure 4-1, real-world objects are captured by sensing devices, such as cameras, and transformed into digital images. These digital images are then processed by computer systems, resulting in final outputs. The outputs can pertain to the image itself, such as image classification, or the identification of patterns and objects within the image. For instance, in the field of healthcare, an image obtained from MRI or X-ray instruments may be fed into an image processing pipeline to detect the presence or absence of a tumor.

This book delves into the inner workings of computer processing units and explores the process of generating outputs. The data flow pipeline for image processing within a computer system is illustrated in Figure 4-2.

Figure 4-2. *Image processing pipeline in computer vision*

Here is a brief description of the computer vision pipeline:

1. **Image ingestion**: Images are captured, converted into digital format, and stored on computer disks. In the case of videos, digital frames are ingested and stored, either from disk or directly from the camera.

2. **Image transformations**: After ingestion, the images undergo various preprocessing transformations to standardize them. These include resizing, color manipulation, translation, rotation, and cropping, and advanced techniques like binarization, thresholding, gradient detection, and edge detection. These transformations ensure consistency in the size, shape, and color schema of the images. For a review of these techniques, please see Chapter 3.

3. **Feature extraction**: Feature extraction is a crucial step in the vision pipeline and involves identifying and extracting relevant features from the images. A good feature set is essential for accurate machine learning outcomes. Further details on feature extraction are covered in the next section, "Feature Extraction."

4. **Machine learning algorithm**: The pipeline utilizes a machine learning algorithm in two stages. In the first stage, a large dataset is used to train the algorithm, resulting in a trained model. In the second stage, the trained model is fed with new datasets to predict outcomes or classes. This stage is known as the *prediction stage*. Chapter 5 will explore popular and effective machine learning models for computer vision. It will introduce Keras and TensorFlow and provide code examples for model training and prediction.

5. **Output generation**: The final component of the vision pipeline is the desired output, which represents the end goal of the vision system.

Feature Extraction

In machine learning, a *feature* is a measurable property of an object or event that is being observed. In computer vision, features provide distinctive information about an image. Feature extraction holds significant importance in machine learning, as the entire process revolves around these features. Identifying and extracting relevant and independent features to achieve reliable machine learning outcomes is crucial to success.

For instance, let's examine the scenario of differentiating between cars and motorcycles solely based on their wheels. A wheel alone is not a distinguishing feature. Additional features, such as the presence or absence of doors, a roof, and other characteristics, are necessary. Moreover, relying on features extracted from a single motorcycle or car may not be sufficient for practical machine learning applications. It is essential to establish patterns by observing repeated occurrences of events or characteristics, because objects in the real world may not always be presented in the exact same way as the features. Hence, repeatability is a crucial characteristic of a good feature.

In the example of the wheel, we had only one feature, but in practical scenarios, there can be numerous features to consider. These may include color, contour, edges, corners, angle, light intensity, and many more. Extracting a larger number of distinctive features enhances the quality and effectiveness of the model.

The performance of a machine learning model heavily relies on the quality of the features used for training. The challenge lies in extracting a robust set of features. While there is no universal solution that applies to all scenarios, the following list provides several practical approaches (among others) to assist you in your feature extraction exercise:

- *Distinctiveness*: Features should possess distinguishing characteristics that enable them to differentiate between different classes or categories.

- *Nonoverlapping*: Features should not have overlapping properties that may lead to confusion or ambiguity in classification or identification.

- *Frequency of occurrence*: Features should be based on properties or events that occur with sufficient frequency, avoiding rare occurrences that may hinder reliable pattern recognition.

- *Robustness*: Features should exhibit consistency across various conditions, such as different lighting conditions, viewing angles, or environmental factors.

- *Identifiability*: Features should be easily identifiable either directly or with the aid of appropriate processing techniques to extract relevant information.

- *Sufficient samples*: Collecting a substantial number of samples is essential to establish reliable patterns and ensure generalizability of the feature extraction process.

By adhering to these principles, we can improve the quality and effectiveness of the features extracted, leading to better machine learning outcomes and more accurate classification or identification results.

How to Represent Features

Features extracted from an image are typically represented as a *feature vector*. As a simple example, consider a grayscale image, in which the features correspond to the pixel values of the image. The pixels are organized in a two-dimensional matrix, where each pixel holds a value ranging from 0 to 255.

To represent these pixel values as features, we convert them into a one-dimensional (1D) row matrix, which can be thought of as a vector or a 1D array. Figure 4-3 provides a visual depiction of this representation.

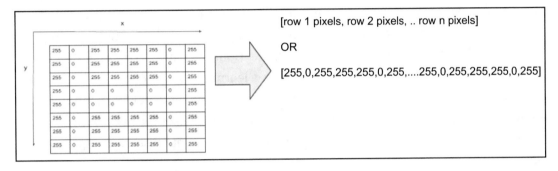

Figure 4-3. *Vector representation of features*

For most machine learning algorithms, it is necessary to extract features from images and input them into the algorithm for model training. However, certain deep learning algorithms, like convolutional neural networks (CNNs), have the capability to automatically extract features during the model training process. (Chapter 5 provides comprehensive information on deep learning algorithms and the techniques required to train computer vision models.)

The following sections explore different methods of feature extraction from images, providing practical examples of feature extraction. Python code using the OpenCV library will demonstrate the concepts and implementation.

Color Histogram

A *histogram* represents the frequency distribution of pixel intensities within an image. A histogram typically is displayed as a graph or chart, with the x-axis representing the pixel values or value ranges, and the y-axis indicating the frequency or count of pixels associated with each value or range. The highest point on the graph indicates the color that appears most frequently in the image.

Given that a pixel's value can range from 0 to 255, the histogram's x-axis consists of 256 values, representing the full spectrum of possible pixel intensities. However, having such a large number of values can be overwhelming for practical purposes. To address this, we commonly divide the pixel values into smaller groups known as *bins*. For instance, we might divide the x-axis into 8 bins, with each bin covering a range of 32 pixel colors. By doing so, we can calculate the y-axis values by summing up the number of pixels falling within each bin.

A histogram provides insights into the distribution of color, contrast, and brightness within an image. Because a color image in the RGB scheme consists of three channels (a grayscale image has only one), a common practice when generating a histogram for a color image is to plot three separate histograms, one for each channel, to gain a comprehensive understanding of the intensity distribution of each color channel. These histograms can be utilized as features in machine learning algorithms.

Another interesting use of histograms is in image enhancement. One technique employed for this purpose is histogram equalization, which uses the histogram to enhance the quality of an image. You'll learn more details about histogram equalization later in this chapter.

How to Calculate a Histogram

To demonstrate how to calculate a histogram, we will utilize Python and OpenCV. To visualize the histogram graph, we will employ the `pyplot` module from the Matplotlib package. If you did not install and configure Matplotlib in Chapter 1, do so now.

OpenCV simplifies the process of calculating a histogram by providing a user-friendly function called `calcHist()`:

```
calcHist(images, channels, mask, histSize, ranges, accumulate)
```

This function takes the following arguments:

- `images`: This is a NumPy array of image pixels. If you have only one image, just wrap the NumPy variable within a pair of square brackets, e.g., `[image]`.

- `channels`: This is an array of indexes of channels we want to calculate the histogram for. This will be `[0]` for grayscale images and `[0,1,2]` for RGB color images.

- mask: This is an optional argument. If you do not supply a mask, the histogram will be calculated for all the pixels in the image or images. If you supply a mask, the histogram will be calculated for the masked pixels only. Remember masks from Chapter 3?

- histSize: This is the number of bins. If we pass this value as [64,64,64], this means that each channel will have 64 bins. The bin size can vary for different channels.

- ranges: This is the range of pixel values, which is normally [0,255] for grayscale and RGB color images. This value may be different in other color schemes, but for now, let's stick to RGB only.

- accumulate: This is the accumulation flag. If it is set, the histogram is not cleared in the beginning when it is allocated. This feature enables you to compute a single histogram from several sets of arrays or to update the histogram in time. The default value is None.

By utilizing the calcHist() function with appropriate parameters, we can efficiently calculate the histogram of an image.

Grayscale Histogram

To gain a practical understanding of how to calculate and visualize the histogram of a grayscale image, check out Listing 4-1 and the discussion that follows it. This code imports pyplot from the Matplotlib package. We will use this library to plot the graph that represents our histogram.

Listing 4-1. Histogram of a Grayscale Image

```
Filename: Listing_4_1.py
1    import cv2
2    import numpy as np
3    from matplotlib import pyplot as plot
4
5    # Read an image and convert it to grayscale
6    image = cv2.imread("images/nature.jpg")
7    image = cv2.cvtColor(image, cv2.COLOR_BGR2GRAY)
8    cv2.imshow("Original Image", image)
```

```
9
10   # calculate histogram
11   hist = cv2.calcHist([image], [0], None, [256], [0,255])
12
13   # Plot histogram graph
14   plot.figure()
15   plot.title("Grayscale Histogram")
16   plot.xlabel("Bins")
17   plot.ylabel("Number of Pixels")
18   plot.plot(hist)
19   plot.show()
20   cv2.waitKey(0)
```

Line 11 of Listing 4-1 calculates the histogram of our grayscale image. Notice that the image variable is wrapped within a pair of brackets because cv2.calcHist() functions take an array of NumPy arrays. Even though we have only one image, we still need to wrap it in an array.

The second argument, [0], denotes that we want to calculate the histogram of the zeroth color channel. Since we have only one channel, we pass only one index value in the array: [0].

The third argument, None, means that we do not want to provide any masking. In other words, we calculate the histogram of all pixels.

The fourth argument, [256], is the bin information. This specifies that we want 256 bins, meaning one bin for each pixel. This may not be useful unless we want to perform a fine-grained analysis of the image pixel distribution. For the majority of practical purposes, we want to pass smaller bin sizes such as [32] or [64].

The last argument, [0,255], tells the function that there are pixel values between 0 and 255.

The hist variable holds the calculation output. If you print this variable, you will see a bunch of numbers that may not be easy to interpret. To make the interpretation easier, we plot the histogram in the form of a graph.

Line 14 configures a blank plot. Line 15 assigns a name to our plot. Lines 16 and 17 set the x-axis and y-axis labels, respectively. Line 18 actually plots the graph. Finally, line 19 displays the plot on the screen. Figure 4-4 shows the original image, and Figure 4-5 shows the histogram output.

Figure 4-4. *Original grayscale image*

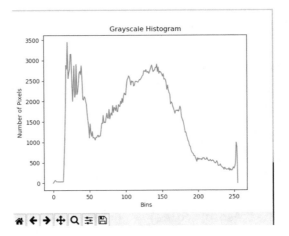

Figure 4-5. *Histogram of the grayscale image in Figure 4-4*

In this histogram, the color value with the highest frequency is 20, which is close to black, and it corresponds to a maximum of 3,450 pixels. Additionally, the majority of pixels fall within the color range of 100 to 150.

Here is another exercise for you: plot a histogram with 32 bins for an image and interpret the resulting graph.

RGB Color Histogram

A previously discussed, a color image has three channels in an RGB scheme. The program in Listing 4-2 demonstrates how to plot histograms of all three channels of an RGB-based color image. It is important to reiterate that OpenCV maintains color information in BGR sequence and not in RGB sequence.

Listing 4-2. Histogram of Three Channels of RGB Color Image

Filename: Listing_4_2.py

```
1    import cv2
2    import numpy as np
3    from matplotlib import pyplot as plot
4
5    # Read a color image
6    image = cv2.imread("images/nature.jpg")
7
8    cv2.imshow("Original Color Image", image)
9    #Remember OpenCV stores color in BGR sequence instead of RBG.
10   colors = ("blue", "green", "red")
11   # calculate histogram
12   for i, color in enumerate(colors):
13       hist = cv2.calcHist([image], [i], None, [32], [0,256])
14       # Plot histogram graph
15       plot.plot(hist, color=color)
16
17   plot.title("RGB Color Histogram")
18   plot.xlabel("Bins")
19   plot.ylabel("Number of Pixels")
20   plot.show()
21   cv2.waitKey(0)
```

In Listing 4-2, line 6 reads a color image from the disk and assigns it to a variable. Line 10 creates a tuple of colors in BGR sequence to hold all our channel colors.

Why do we have a `for` loop in line 12? The second argument of the `calcHist()` function takes an array with a value of 0, 1, or 2. If we pass the value [0], we instruct the `calcHist()` function to calculate the histogram of the color channel in the zeroth index, which is the blue channel. Similarly, a value of [1] instructs the `calcHist()` function to calculate the histogram of the red channel, and a value of [2] says to calculate for the green channel. The first iteration of the `for` loop is for calculating and plotting the histogram of the blue color, the second iteration is for green, and the last iteration is for the red channel.

Notice again that we have passed [32] as the fourth argument to our `calcHist()` function. This is to let the function know that we want to calculate the histogram with 32 bins for each of the channels. The last argument, [0,256], gives the color range.

Within the `for` loop in line 15, the `plot()` function takes the histogram as the first argument and an optional color as the second argument.

Figure 4-6 and Figure 4-7 show the output of Listing 4-2.

Figure 4-6. *Original color image*

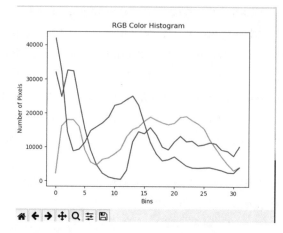

Figure 4-7. *Histogram of three color channels of the image in Figure 4-6*

In Figure 4-7, the x-axis has only up to 32 values because we used only 32 bins for each channel.

Here's an exercise for you: create a histogram of a masked image.

Hint   Create a mask NumPy array and pass this array as the third argument in the `cv2.calcHist()` function. Read Chapter 3 to refresh your memory on how to create a mask.

Histogram Equalizer

Now that you have a clearer understanding of what a histogram represents, we can leverage this concept to enhance the quality of an image. *Histogram equalization* is an image processing technique that is used to adjust the contrast of an image. It involves redistributing the pixel intensities in such a manner that the intensities of underpopulated pixels are equalized with the intensities of overpopulated pixels, as illustrated in Figure 4-8.

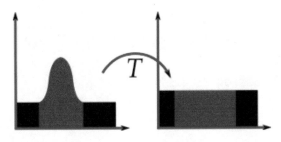

Figure 4-8. *Histogram equalization (source: Wikipedia, https://wikipedia.org/ wiki/Histogram_equalization)*

Listing 4-3 provides an example to observe histogram equalization in action. Although Listing 4-3 contains a substantial amount of code, you can quickly pare it down by first focusing on lines 1 to 19 and observing that they are identical to those in Listing 4-1 for calculating and plotting the histogram of a grayscale image.

In line 21, we utilize OpenCV's equalizeHist() function, which takes the original image as input and adjusts its pixel intensities to enhance the contrast. Lines 22 through 33 calculate and display the histogram of the enhanced (equalized) image.

Figures 4-9 through 4-12 show the outputs of Listing 4-3 and a comparison of the histograms for the original and equalized images.

Listing 4-3. Histogram Equalization

```
Filename: Listing_4_3.py
1    import cv2
2    import numpy as np
3    from matplotlib import pyplot as plot
4
5    # Read an image and convert it into grayscale
6    image = cv2.imread("images/nature.jpg")
7    image = cv2.cvtColor(image, cv2.COLOR_BGR2GRAY)
8    cv2.imshow("Original Image", image)
9
10   # calculate histogram of the original image
11   hist = cv2.calcHist([image], [0], None, [256], [0,255])
12
13   # Plot histogram graph
```

```
14    #plot.figure()
15    plot.title("Grayscale Histogram of Original Image")
16    plot.xlabel("Bins")
17    plot.ylabel("Number of Pixels")
18    plot.plot(hist)
19    plot.show()
20
21    equalizedImage = cv2.equalizeHist(image)
22    cv2.imshow("Equalized Image", equalizedImage)
23
24    # calculate histogram of the original image
25    histEqualized = cv2.calcHist([equalizedImage], [0], None, [256],
      [0,255])
26
27    # Plot histogram graph
28    #plot.figure()
29    plot.title("Grayscale Histogram of Equalized Image")
30    plot.xlabel("Bins")
31    plot.ylabel("Number of Pixels")
32    plot.plot(histEqualized)
33    plot.show()
34    cv2.waitKey(0)
```

Figure 4-9. *Original grayscale image*

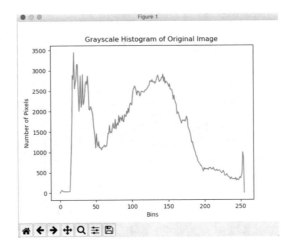

Figure 4-10. *Histogram of the image in Figure 4-9*

Figure 4-11. *Equalized image with enhanced contrast*

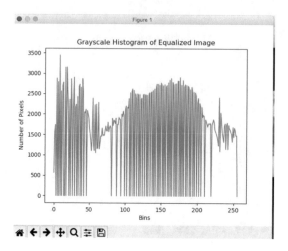

Figure 4-12. *Histogram of equalized image of Figure 4-11*

GLCM

The *gray-level co-occurrence matrix (GLCM)* captures the statistical distribution of pixel values that occur simultaneously within a specified pixel offset. An *offset* represents the relative position and direction of neighboring pixels. As its name indicates, the GLCM is specifically designed for grayscale images.

The GLCM calculates how many times a pixel value *i* co-exists either horizontally, vertically, or diagonally with a pixel value *j*.

For GLCM calculation, we specify an offset distance *d* and an angle Θ (theta). The angle Θ (theta) may be 0° (horizontally), 90° (vertically), 45° (diagonally to the right up), or 135° (diagonally to the left up), as shown in Figure 4-13.

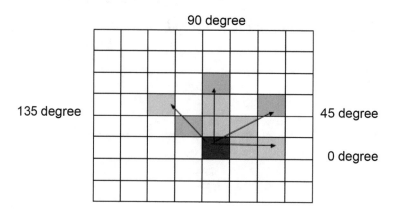

Figure 4-13. *Illustration of adjacent pixel position (distance and angle)*

The importance of the GLCM is that it provides information on spatial relationships over an image. This differs from a histogram because a histogram does not provide any information about the image size, pixel location, or their relationship.

Although the GLCM is such an important matrix, we do not directly use it as a feature vector for machine learning. We calculate certain key statistics about the image using the GLCM, and those statistics are used as features for any machine learning training. You will learn about these statistics and how to calculate them in this section.

Though OpenCV uses the GLCM internally, it does not directly expose any function to calculate it. To calculate the GLCM, we will use another Python library: skimage's `feature` package.

Here is a description of the function we are going to use to compute the GLCM:

```
graycomatrix(image, distances, angles, levels, symmetric,normed)
```

The `graycomatrix()` function takes the following arguments:

- `image`: This is the NumPy representation of a grayscale image. Remember, the image must be grayscale.

- `distances`: This is a list of pixel-pair distance offsets.

- `angles`: This is a list of angles between the pair of pixels. Make sure the angle is a radian and not a degree.

- `levels`: This is an optional parameter and meant for images having 16-bit pixel values. In most cases, we use 8-bit image pixels that can have values ranging from 0 to 255. For an 8-bit image, the max value for this parameter is 256.

- `symmetric`: This is an optional parameter and takes a Boolean. The value `True` means the output matrix will be symmetric. The default is `False`.

- `normed`: This is also an optional parameter that takes a Boolean. The Boolean `True` means that each output matrix is normalized by dividing by the total number of accumulated co-occurrences for the given offset. The default is `False`.

The `graycomatrix()` function returns a 4D ndarray. This is the gray-level cooccurrence histogram. The output value $P[i,j,d,theta]$ represents how many times the gray-level j occurs at a distance d and angle $theta$ from the gray-level j. If the parameter normed is `False` (which is the default), the output is of type `uint32` (a 32-bit unsigned integer); otherwise, it is `float64` (a 64-bit floating point).

Listing 4-4 shows how to calculate the GLCM using the skimage library to compute feature statistics.

Listing 4-4. GLCM Calculation Using the graycomatrix() Function

```
Filename: Listing_4_4.py
1    import cv2
2    import skimage.feature as sk
3    import numpy as np
4
5    #Read an image from the disk and convert it into grayscale
6    image = cv2.imread("images/nature.jpg")
7    image = cv2.cvtColor(image, cv2.COLOR_BGR2GRAY)
8
9    #Calculate GLCM of the grayscale image
10   glcm = sk.graycomatrix(image,[2],[0, np.pi/2])
11   print(glcm)
```

Line 10 calculates the GLCM using graycomatrix() by passing the image NumPy variable and a distance of [2]. The third argument is in radians. np.pi/2 is the radian for a 90-degree angle. The last line, line 11, simply prints the 4D ndarray.

As mentioned, the GLCM is not directly used as a feature, but we use this to calculate some useful statistics, which gives us an idea about the texture of the image. The following table lists the statistics we can derive:

Statistic	Description
Contrast	Measures the local variations in the GLCM.
Correlation	Measures the joint probability occurrence of the specified pixel pairs.
Energy	Provides the sum of squared elements in the GLCM. Also known as uniformity or the angular second moment.
Homogeneity	Measures the closeness of the distribution of elements in the GLCM to the GLCM diagonal.

The following set of high-level formulas is used to compute the aforementioned statistics. A detailed mathematical analysis of these formulas is beyond the scope of this book, but you are encouraged to delve into the mathematical foundations underlying these statistical calculations.

$$\text{Contrast} = \sum_{i,j=0}^{levels-1} P_{i,j} \left(i-j\right)^2$$

$$\text{Dissimilarity} = \sum_{i,j=0}^{levels-1} P_{i,j} \left|i-j\right|$$

$$\text{Homogeneity} = \sum_{i,j=0}^{levels-1} \frac{P_{i,j}}{1+\left(i-j\right)^2}$$

$$\text{ASM} = \sum_{i,j=0}^{levels-1} P_{i,j}^2$$

$$\text{Energy} = \sqrt{ASM}$$

$$\text{Correlation} = \sum_{i,j=0}^{levels-1} P_{i,j} \left[\frac{\left(i-\mu_i\right)\left(j-\mu_j\right)}{\sqrt{\left(\sigma_i^2\right)\left(\sigma_j^2\right)}} \right]$$

In these formulas, P is the GLCM histogram for which to compute the specified property. The value $P[i,j,d,theta]$ is the number of times that gray-level j occurs at the distance d and at the angle *theta* from the gray-level i.

We will use `graycoprops()` from the `skimage` package to compute these statistics from the GLCM. Here is the definition of this function:

```
graycoprops(P, prop='contrast')
```

The first argument is the GLCM histogram (see Listing 4-4, line 10).

The second argument is the property we want to calculate. We can pass any of the following properties for this argument: `contrast`, `dissimilarity`, `homogeneity`, `energy`, `correlation`, and `ASM`. If you do not pass the second argument, it will default to `contrast`.

Listing 4-5 shows how to calculate these statistics.

Listing 4-5. Calculation of Image Statistics from the GLCM

```
Filename: Listing_4_5.py
1     import cv2
2     import skimage.feature as sk
3     import numpy as np
4
5     #Read an image from the disk and convert it into grayscale
6     image = cv2.imread("images/nature.jpg")
7     image = cv2.cvtColor(image, cv2.COLOR_BGR2GRAY)
8
9     #Calculate GLCM of the grayscale image
10    glcm = sk.graycomatrix(image,[2],[0, np.pi/2])
11
12    #Calculate Contrast
13    contrast = sk.graycoprops(glcm)
14    print("Contrast:",contrast)
15
16    #Calculate 'dissimilarity'
17    dissimilarity = sk.graycoprops(glcm, prop='dissimilarity')
18    print("Dissimilarity: ", dissimilarity)
19
20    #Calculate 'homogeneity'
21    homogeneity = sk.graycoprops(glcm, prop='homogeneity')
22    print("Homogeneity: ", homogeneity)
23
24    #Calculate 'ASM'
25    ASM = sk.graycoprops(glcm, prop='ASM')
26    print("ASM: ", ASM)
27
28    #Calculate 'energy'
29    energy = sk.graycoprops(glcm, prop='energy')
30    print("Energy: ", energy)
31
```

```
32    #Calculate 'correlation'
33    correlation = sk.graycoprops(glcm, prop='correlation')
34    print("Correlation: ", correlation)
```

Listing 4-5 shows how to use the graycoprops() function and pass different parameters to prop to calculate respective statistics. Figure 4-14 shows the output of Listing 4-5.

Contrast: [[291.1180688 453.41833488]]

Dissimilarity: [[9.21666213 12.22730486]]

Homogeneity: [[0.32502798 0.23622148]]

ASM: [[0.00099079 0.00055073]]

Energy: [[0.03147683 0.02346761]]

Correlation: [[0.95617083 0.93159765]]

Figure 4-14. *GLCM-based output of various statistics*

HOGs

Histograms of oriented gradients (HOGs) are important feature descriptors used in computer vision and machine learning for object detection. HOGs describe the structural shape and appearance of an object in an image. The HOG algorithm computes the occurrences of gradient orientation in localized portions of the image.

The HOG algorithm works in five stages, as described here:

- *Stage 1: Global image normalization*: This is an optional stage and is needed only to reduce the influence of illumination effects. At this stage, the image is globally normalized by one of the following methods:

    - *Gamma (power law) compression*: Each pixel value, *p*, is changed by applying log(p). This compresses the pixels too much and is not recommended.

    - *Square-root normalization*: Each pixel value, *p*, is changed to [?][?][?]*p* (square root of pixel value). This compresses the pixels less than the gamma compression and is considered a preferred normalization technique.

- *Variance normalization*: For most machine learning work, this technique provides better results compared to the other two methods. In this method, we first compute the mean (μ) and standard deviation (σ) of pixel values. Then, each pixel value, p, is normalized according to the following formula:

$$Tp = (p - \mu)/\sigma$$

- *Stage 2: Compute the gradient image in x and y*: The second stage computes the first-order image gradients to capture contour, silhouette, and some texture information. If you need to capture bar-like features, such as limbs in humans, you will also need to include second-order image derivatives. Listings 3-19 and 3-20 (in Chapter 3) show how to calculate gradients in the X and Y directions. For a refresher, refer to the Chapter 3 section "Gradients and Edge Detection." Assuming the gradients in the X direction are G_x and the gradients in the Y direction are G_y, the gradient magnitude is calculated using the following formula:

$$|G| = \sqrt{G_x^2 + G_y^2}$$

  Finally, the gradient orientation is calculated by using the following formula:

$$\Theta = \arctan\left(G_y / G_x\right)$$

  Once the value of gradient and orientation is calculated, the histogram is then computed.

- *Stage 3: Compute gradient histograms*: The image is divided into small spatial regions, called *cells*. Using the previous formulae for $|G|$ and Θ ♐, we accumulate a local 1D histogram of gradient or edge orientations over all the pixels in each cell. Each orientation histogram divides the gradient angle range into a fixed number of predetermined bins. The gradient magnitudes of the pixels in the cell are used to vote into the orientation histogram. The weight of the vote is simply the gradient magnitude $|G|$ at the given pixel.

- *Stage 4: Normalizing across blocks*: A small number of cells are grouped together to form a square block. The entire image is now divided into blocks (which consists of a group of cells). The formation of blocks is typically done by sharing cells between several blocks. The cell thus appears several times in the final output vector with different normalizations. Then normalization is performed over these localized blocks. It is performed by accumulating a measure of local histogram "energy" within the local blocks. These normalized block descriptors are the HOG. Figure 4-15 shows the block formation.

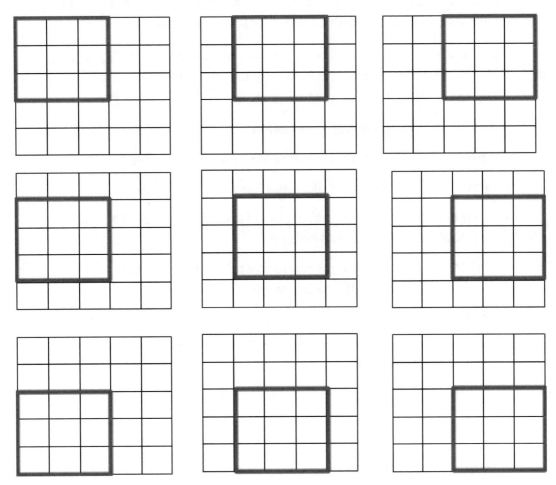

Figure 4-15. *Block formation by grouping cells (block of 3×3 cells)*

- *Stage 5: Flatten into a feature vector*: After all blocks are normalized, we take the resulting histograms and concatenate them to construct our final feature vector.

Don't worry if this explanation of the HOG algorithm seems overwhelming. You don't have to write code from scratch to implement HOG on your own. Numerous libraries are readily available that offer prebuilt functions for effortless calculation of HOG.

We will use the `scikit-image` library to calculate the HOG of an image. The subpackage, `feature`, within the package `skimage` of the `scikit-image` library provides a convenient method to calculate HOG. Here is the function signature:

```
out, hog_image = hog(image, orientations=9, pixels_per_cell=(8, 8), cells_
per_block=(3, 3), block_norm='L2-Hys', visualize=False,
transform_sqrt=False, feature_vector=True, multichannel=None)
```

The description of the parameters is as follows:

- `image`: This is the NumPy representation of the input image.

- `orientation`: The number of orientation bins defaults to 9.

- `pixels_per_cell`: This is the number of pixels in each cell as a tuple; it defaults to (8,8) for an 8×8 cell size.

- `cells_per_block`: This is the number of cells in each block, as a tuple; it defaults to (3,3), which is for 3×3 cells, not pixels.

- `block_norm`: This is the block normalization method as a string with one of these values: L1, L1-sqrt, L2, L2-Hys. These normalization strings are explained here:

  - L1: Normalization using L1-norm using this formula:

  $$- \text{L1-norm} = \sum_{r=1}^{n} |X_r|$$

  - L1-sqrt: Square root of the L1-normalized value. It uses this formula:

  $$- \text{L1-sqrt} = \sqrt{\sum_{r=1}^{n} |X_r|}$$

– L2: Normalization using L2-norm using this formula:

$$- \text{L2-norm} = \sqrt{\sum_{r=1}^{n} |X_r|^2}$$

– L2-Hys: This is the default normalization for the parameter block_norm. L2-Hys is calculated by first taking the L2-normalization, limiting the result to a maximum of 0.2, and then recalculating the L2-normalization.

- visualize: If this is set to True, the function also returns an image of the HOG. Its default value is set to False.

- Transform_sqrt: If set to True, the function will apply power law compression to normalize the image before processing.

- feature_vector: The default value of this argument is set to True, which instructs the function to return the output data as a feature vector.

- multichannel: Set the value of this argument to True to indicate the input image contains multichannels. The dimensions of an image are generally represented as height × width × channel. If the value of this argument is True, the last dimension (channel) is interpreted as the color channel, otherwise as spatial.

What does this hog() function return?

- out: The function returns an ndarray containing (n_blocks_row, n_blocks_col, n_cells_row, n_cells_col, n_orient). This is the HOG descriptor for the image. If the argument feature_vector is True, a 1D (flattened) array is returned.

- hog_image: If the argument visualize is set to True, the function also returns a visualization of the HOG image.

Listing 4-6 shows how to calculate the HOG using the skimage package.

Listing 4-6. HOG Calculation

```
Filename: Listing_4_6.py
1    import cv2
2    import numpy as np
```

```
3    from skimage import feature as sk
4
5    #Load an image from the disk
6    image = cv2.imread("images/obama.jpg")
7    #Resize the image.
8    image = cv2.resize(image,(int(image.shape[0]/5),int(image.
     shape[1]/5)))
9
10   # HOG calculation
11   (HOG, hogImage) = sk.hog(image, orientations=9, pixels_per_
     cell=(8, 8),
12       cells_per_block=(2, 2), visualize=True, transform_sqrt=True,
         block_norm="L2-Hys", feature_vector=True)
13
14   print("Image Dimension",image.shape)
15   print("Feature Vector Dimension:", HOG.shape)
16
17   #showing the original and HOG images
18   cv2.imshow("Original image", image)
19   cv2.imshow("HOG Image", hogImage)
20   cv2.waitKey(0)
```

The concept of HOG is important to understand because we will leverage it in Chapters 6, 7, and 8 to create practical implementations. While we devoted several pages to comprehending the concept, calculating the HOG is remarkably concise, requiring just a single line of code (line 11, Listing 4-6).

We use the hog() function from the feature subpackage of the skimage package. Parameters passed to the hog() function were explained earlier.

How do we know that we are passing the right values of the parameters in the hog() function? Well, there is really no established rule. As a rule of thumb, we should start with all default parameters and tune them as we analyze the result.

It is worth mentioning that the hog() function generates a histogram of very high dimensionality. A 32×32 image with pixel_per_cell=(4,4) and cells_per_block=(2,2) will generate 1,764-dimension results. Similarly, a 128×128 pixel image will generate 34,596-dimension output. It is, therefore, extremely important to pay attention to the parameters, and resize your image appropriately to reduce the output dimensions. This will have a huge impact on the memory, storage requirement, and network transfer time.

Figures 4-16 through 4-18 show the output of Listing 4-6.

Figure 4-16. *Resized image*

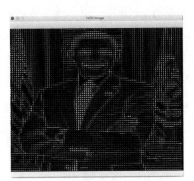

Figure 4-17. *HOG image*

Image Dimension (537, 671, 3)

Feature Vector Dimension:
(194832,)

Figure 4-18. *Dimension output from the* `print()` *statement*

LBP

Local Binary Patterns (LBP) serves as a feature descriptor utilized in the classification of image textures.

The LBP feature extraction works as follows:

1. For every pixel in the image, compare the pixel values of the surrounding pixels. If the value of a surrounding pixel is less than the central pixel, mark it to 0; otherwise 1. In Figure 4-19, the central pixel has the value 20 and is surrounded by eight neighbors. The middle portion of Figure 4-19 shows the pixel value conversion to 0 or 1 based on whether they are smaller or greater than the central pixel (20 in this case).

2. Starting from any of the neighbor's pixels and going in any direction, we assemble the sequence of 0s and 1s to make an 8-bit binary number. In the following example shown in Figure 4-19, we started from the top-right corner and moved clockwise to assemble digits to form 10101000 binary numbers. This binary number is converted into a decimal to get the pixel value of the central pixel, as shown in Figure 4-19.

3. For each pixel in the image, we repeat the previous steps to obtain the pixel values based on the neighbors' pixels. Make sure that for all pixels the starting position and direction remain consistent.

4. When all pixels are done, we arrange the pixel values in an LBP array.

5. Finally, we calculate a histogram over the LBP array. This histogram is taken as an LBP feature vector.

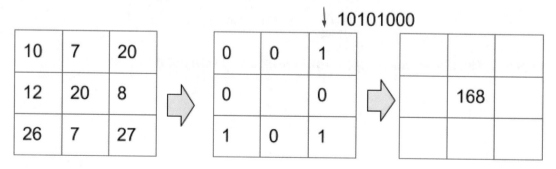

Figure 4-19. LBP pixel value calculation

This approach of calculating an LBP feature vector allows us to capture finer details of the image texture. But for most machine learning classification problems, fine-grained features may not give the desired outcome, especially when the input images are of varying scales of texture.

To overcome this problem, we have an enhanced version of LBP that allows for variable neighborhood sizes. It gives us two additional parameters to work with:

- Instead of a fixed square neighborhood, we can define the number of points, p, in a circularly symmetric neighborhood.

- The radius of the circle, r, allows us to define different neighborhood sizes.

Figure 4-20 shows the green dots as the number of points and the dotted circle with varying radius. The smaller the radius, the finer the texture captured. Increasing the radius enables us to classify textures of varying scales.

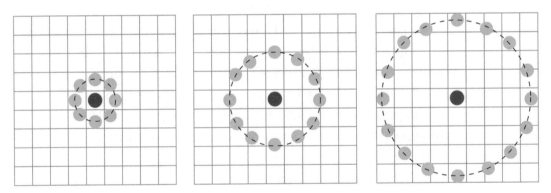

Figure 4-20. *LBP calculation based on neighborhood size and number of points*

With this background knowledge, we are now prepared to delve into the implementation of LBP. Once again, we will rely on `scikit-image`, specifically the `feature` subpackage from the `skimage` package. Here is the designated function signature that we will utilize for calculating LBP:

```
local_binary_pattern(image, P, R, method='default')
```

The parameters are explained here:

- `image`: The NumPy representation of a grayscale image.

- P: The number of neighborhood points along the circle surrounding the point for which LBP is being calculated. This is the number of green dots in Figure 4-20.

- R: This is a floating-point number and defines the radius of the circle.

- method: This parameter takes any of the following string values:

  - default: This instructs the function to calculate original LBP based on grayscale without considering the rotation invariant. The description of a rotationally invariant binary descriptor is beyond the scope of this book. To learn more about this, review the paper "OSRI: A Rotationally Invariant Binary Descriptor" at http://ivg.au.tsinghua.edu.cn/~jfeng/pubs/Xueta1_TIP14_Descriptor.pdf.

  - ror: This method instructs the function to use a rotationally invariant binary descriptor.

  - uniform: This uses an improved rotation invariance with uniform patterns and finer quantization of the angular space, which is grayscale and rotation invariant. A binary pattern is considered uniform if there are at most two 0-1 to 1-0 transitions in the binary sequence of digits. For example, 00100101 is a uniform pattern as it has two transitions (shown in red and blue). Similarly, 00010001 is also a uniform pattern as it has one 0-1 to 1-0 transition. On the other hand, 01010100 is not a uniform pattern. In the computation of the LBP histogram, the histogram has a separate bin for every uniform pattern, and all nonuniform patterns are assigned to a single bin. Using uniform patterns, the length of the feature vector for a single cell reduces from 256 to 59.

  - nri_uniform: Non-rotation-invariant uniform patterns variant, which is only grayscale invariant.

  - var: Rotation invariant variance measures of the contrast of local image texture, which is rotation but not grayscale invariant.

The output of the function local_binary_pattern() is an ndarray representing an LBP image.

We have covered enough background to start implementing LBP and see it in action. Listing 4-7 demonstrates the use of the local_binary_pattern() function.

It starts with loading an image from the disk, resizing it, and converting it to grayscale.

Line 12 calculates the histogram of the original image. Lines 14 through 16 plot the original image histogram.

Listing 4-7. LBP Image and Histogram Calculation and Comparison with Original Image

```
Filename: Listing_4_7.py
1    import cv2
2    import numpy as np
3    from skimage import feature as sk
4    from matplotlib import pyplot as plt
5
6    #Load an image from the disk, resize and convert to grayscale
7    image = cv2.imread("images/obama.jpg")
8    image = cv2.resize(image, (int(image.shape[0]/5), int(image.
     shape[1]/5)))
9    image = cv2.cvtColor(image, cv2.COLOR_BGR2GRAY)
10
11   # calculate Histogram of original image and plot it
12   originalHist = cv2.calcHist(image, [0], None, [256], [0,256])
13
14   plt.figure()
15   plt.title("Histogram of Original Image")
16   plt.plot(originalHist, color='r')
17
18   # Calculate LBP image and histogram over the LBP, then plot the
     histogram
19   radius = 3
20   points = 3*8
21   # LBP calculation
22   lbp = sk.local_binary_pattern(image, points, radius, method='default')
23   lbpHist, _ = np.histogram(lbp, density=True, bins=256, range=(0, 256))
24
25   plt.figure()
```

```
26   plt.title("Histogram of LBP Image")
27   plt.plot(lbpHist, color='g')
28   plt.show()
29
30   #showing the original and LBP images
31   cv2.imshow("Original image", image)
32   cv2.imshow("LBP Image", lbp)
33   cv2.waitKey(0)
```

The calculation of the LBP image is performed in line 22. Notice that we used the default method for LBP calculation, which takes a radius of 3 and the number of points as 24. Line 22 uses the local_binary_pattern() function from the feature subpackage of the skimage package.

Line 23 calculates the histogram over the LBP image. Notice that we use NumPy's histogram() function. If you try to use the cv2.calcHist() function for the LBP image, you will receive an error message saying "-210 Unsupported format or combination of formats." This is because the output format of local_binary_pattern() is different and not supported by OpenCV's calcHist() function. For that reason, we are using NumPy's histogram() function.

Figure 4-21 shows the original image. Let's look at the output of Listing 4-7. Figure 4-22 is the LBP image calculated from an input image (Figure 4-21). Notice how neatly it has captured the texture of the original image. Compare Figure 4-23 with Figure 4-24 for histograms plotted from the original image and from the LBP image, respectively.

Figure 4-21. *Original grayscale image*

Figure 4-22. *LBP image*

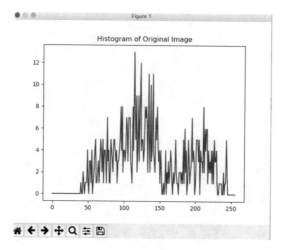

Figure 4-23. *Histogram of the original image*

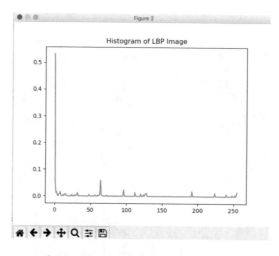

Figure 4-24. *Histogram of the LBP image*

Note   In some cases, LBP is combined with HOG to enhance the accuracy of object detection.

In this section, the primary objective was for you to acquire knowledge about various techniques for performing feature extraction. We concentrated on the fundamental concepts underlying these techniques, as they will prove valuable in the subsequent chapters, where we delve into the realms of machine learning and neural networks. The concepts you have learned here will be applied in practical scenarios throughout Chapters 6 to 9.

The following section explores the strategies involved in feature selection.

Feature Selection

Feature selection in machine learning is the process of choosing variables or attributes that are relevant and useful in model training and eliminating irrelevant or redundant features, resulting in the selection of a subset of influential features that greatly contribute to the learning process of the model. Understanding feature selection is crucial for the following reasons:

- To reduce the complexity of the model and enhance interpretability

- To decrease the training time of machine learning algorithms

- To enhance the model's accuracy by utilizing a well-chosen set of variables

- To mitigate the risk of overfitting, thereby improving the model's generalization capabilities

Feature selection and feature extraction are distinct processes within feature engineering. *Feature extraction* involves generating new features from the existing data, whereas *feature selection* focuses on choosing a subset of relevant features or eliminating redundant ones. Both techniques play vital roles in enhancing the performance of machine learning models. When combined, feature extraction and feature selection collectively form the practice of feature engineering.

It has been statistically proven that there is an optimum number of features beyond which the model performance starts degrading. The question is, how do we know what the optimum number is, and how do we decide which features to use and which not to use? This section attempts to answer this question.

There are many feature selection techniques used for machine learning, so we will focus on three of the most commonly used techniques.

Filter Method

Suppose that you have a feature set and you want to select a subset to feed to your machine learning algorithm. In other words, you want to select the subset of features prior to triggering machine learning. *Filtering* is a process that allows you to do preprocessing to select the feature subset. In this process, you determine a correlation between a feature and the target variable and determine their relationship based on statistical scores. Note that the filtering process is independent of any machine learning algorithm. A feature is selected (or rejected) only on the basis of the relationship between feature variables and the target variable.

Several statistical methods exist to help us score a feature against the target variable. The following table provides a practical guide for selecting methods to determine the feature–target relationship:

Feature Variable Type	Target Variable Type	Statistical Method Name
Continuous	Continuous	Pearson's correlation
Continuous	Categorical	Linear discriminant analysis (LDA)
Categorical	Categorical	Chi-square
Categorical	Continuous	ANOVA

Descriptions of these statistical methods are outside of the scope of this book. A wide variety of books and online resources are available on these age-old methods.

Wrapper Method

In the *wrapper method*, you use a subset of features and train the model. You evaluate the model and, based on the result, either add or remove features and retrain the model. Repeat this process until you get a model with acceptable accuracy. It's more like a trial-anderror approach to finding the right subset of features. This process is computationally expensive because you have to actually train multiple models (and most likely throw away all that you are not happy with).

The following are the practical approaches used to perform feature selection under a wrapper method:

- *Forward selection*: Start with one feature and build and evaluate the model. Iteratively add features that best improve the model.

- *Backward elimination*: Start with all features and build and evaluate the model. Iterate through by eliminating features until you get the best model. Repeat this until no improvement is observed on feature removal.

- *Recursive feature elimination*: Repeatedly create models and set aside the best- or worst-performing feature at each iteration. Rank the features either by their coefficients or by feature importance, and eliminate the least important features. Recursively create new models with the leftover features until all features are exhausted.

Embedded Method

In an *embedded method*, the feature selection is performed by the machine learning algorithm while the model is being trained. LASSO (Least Absolute Shrinkage and Selection Operator) and Ridge regularization methods for regression algorithms are examples of such algorithms where the best suitable features contributing to the model accuracy are evaluated.

Lasso regression uses L1 regularization and adds a penalty equivalent to the absolute value of the magnitude of coefficients. Ridge regression uses L2 regularization and adds a penalty equivalent to the square of the magnitude of coefficients.

Since the model itself evaluates the feature importance, the embedded method is one of the least expensive methods of feature selection.

Note   The primary focus of this book is constructing computer vision applications based on machine learning and deep learning techniques. While feature extraction and selection are crucial components of any machine learning algorithm, this book provides only an introductory level of information regarding this extensive topic. If you want to explore feature extraction and selection in depth, you can fine entire books solely dedicated to this topic.

Model Training

Take a moment to revisit the image processing pipeline depicted in Figure 4-2 earlier in this chapter. Up to this point, you have learned how to acquire images and perform preprocessing techniques to enhance their quality. This preprocessing step ensures that the input image is appropriately formatted for the subsequent stages of the pipeline, namely feature extraction and selection. In the preceding section, you learned about different feature engineering techniques. I hope you have thoroughly grasped the concepts covered thus far and are prepared to delve into the realm of machine learning as it pertains to computer vision.

How to Do Machine Learning

Imagine you have successfully extracted and selected features from a large number of images. What constitutes a "large" number of images? No definitive magic number universally applies. The ideal number should accurately represent or closely approximate the real-world scenario you are attempting to model. It's important to remember that one of the key attributes of a strong feature set is its repeatability. Although there isn't a precise approach to determining a "large" number, the general rule of thumb is that the more images you have, the better the potential outcome of the model.

The mathematical algorithms utilize these sets of features to identify specific patterns, in the form of mathematical equations called models, which we will discuss in more detail in the next chapter. The result generated by the algorithm is known as a *model*, and the process of developing this model is referred to as *model training*. Essentially, as depicted in Figure 4-25, the computer leverages an algorithm to identify patterns from the provided input feature set, a.k.a. the *training set*.

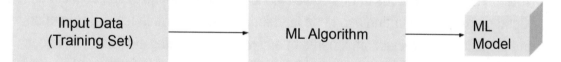

Figure 4-25. *Illustration of ML model training*

In general, there are two categories of training sets, which correspond to two types of machine learning: supervised learning and unsupervised learning. The following sections describe each in turn.

Supervised Learning

Assume you have an 8×8 image and the values of all 64 pixels are your features. Also, assume that you have several of such images and you have extracted pixel values from them to make a feature set. All 64 features of one image are arranged as an array (or vector). The feature set will have as many rows as the number of images in the training set, with each row representing one distinct image. Now, with this dataset, you want to train a model that can classify an input image to a certain class. For example, you want to classify an image based on whether it contains a dog or a cat (let's keep it simple for now).

Assume further that these training images are already labeled, meaning that they are already identified and marked as to which images contain a dog and which images contain a cat. That means we have the correct class identified for each image.

Figure 4-26 shows a sample of a labeled training set. Column 1 of Figure 4-26 is the image ID that uniquely identifies each image. Columns 2 through 65 (truncated) show the pixel values of all 64 columns (because our image dimension is 8×8 in this example). These pixel values together form our feature vector (X). The last column is the label column (y) and has the value 0 for a dog and the value 1 for a cat (labels must be numeric to be fed in machine learning). The labels are also known as *target variables* or *dependent variables*.

Image ID	Feature Vector (X)											Label (y)
image100	159	191	30	161	...	218	137	87	49	193	144	0
image101	103	184	133	125	...	144	85	7	152	247	143	0
image102	15	249	237	200	...	152	107	227	80	207	106	1
image103	217	152	226	122	...	195	95	229	199	36	107	1
..	..	..	..	..	..	..	..	..	..	..	..	..

Figure 4-26. *Example dataset with labeled feature vectors*

When we train a machine learning model by feeding a dataset containing feature vectors and associated labels to a learning algorithm, it is called *supervised learning*.

The supervised learning algorithm (see Figure 4-27) learns by optimizing a function that takes a feature vector as input and generates the label as output. We will explore various optimization functions in Chapter 5.

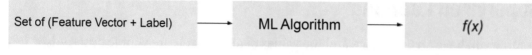

Figure 4-27. *Illustration of supervised learning*

There are several supervised learning algorithms, such as support vector machine (SVM), linear regression, logistic regression, decision tree, random forest, artificial neural network (ANN), and convolutional neural network (CNN).

The focus of this book is the application of deep learning techniques, specifically neural networks such as artificial neural networks (ANN) and convolutional neural networks (CNN), in training models for computer vision tasks. In Chapter 5, you will learn about the intricacies of these deep learning algorithms and discover how to effectively train models for computer vision purposes.

Unsupervised Learning

In the example shown in Figure 4-26, each feature vector has a corresponding label. The primary aim with such a labeled dataset is to establish a relationship between each feature vector and its label. However, what if we do not have labels for the feature vectors? In other words, our model's inputs solely consist of the feature vectors, without any associated outputs or labels. In this scenario, we want our machine learning algorithm to learn from this input dataset alone. The model we train under these circumstances is referred to as *unsupervised learning*.

Figure 4-28 depicts unsupervised learning algorithms, which are designed to handle datasets that solely consist of feature vectors. These algorithms aim to identify structures or patterns within the data, such as data grouping or clustering. In other words, they learn from a training set that lacks labeled data and uncover similarities or commonalities within the dataset.

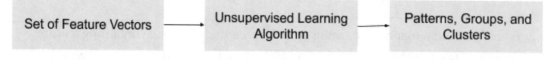

Figure 4-28. *Unsupervised learning*

Unsupervised learning serves multiple purposes, including dataset clustering and grouping. Additionally, it can be utilized to generate labels for supervised learning algorithms. Several widely used unsupervised algorithms include K-means clustering, auto-encoders, deep belief nets, and Hebbian learning.

This book covers only supervised learning techniques used in the field of computer vision.

Model Deployment

After creating a trained machine learning model, the next step is to deploy it. But before we delve into that, let's quickly recap what exactly we do with a trained model:

- In the case of supervised learning, a trained model provides us with a function that takes a feature set as input and gives us an output. The output is generally known as *prediction*. In other words, the model predicts outcomes based on input data. Such predictions may be continuous values or classes.

- In the case of unsupervised learning, a trained model takes a feature set and gives output as the group or cluster that an input feature falls into. The groupings or clustering may be further used to create labels for supervised learning.

Deploying a fully trained model entails making the model available for use in predicting or classifying images, or any other input dataset that may be provided by external business applications. These predictions or classes derived from the model are then utilized in diverse analyses and decision-making processes, tailored to the specific requirements of the business use case.

External applications can generate input images that need to be processed. These images undergo the same ingestion and processing steps as those used during the feature engineering phase for model training. Features are extracted from the ingested images and passed through the model function to obtain predictions or classes.

While the model development is an iterative process, once a model gives acceptable accuracy, we usually version that model and deploy it in production. In practice, models are not changed or retrained with new data until the accuracy starts decreasing or until retraining with the new dataset is expected to increase the accuracy.

The utilization of a model is often more frequent than the retraining of the model itself. In some scenarios, there may be a need to predict or classify hundreds or thousands of input images per second. In other scenarios, there might be a need to classify millions of images in batches over the course of a day or at specific intervals. Consequently, it is crucial to deploy our models in a manner that allows them to scale according to the volume of input data and processing requirements.

Achieving an effective model utilization in production necessitates having the appropriate deployment architecture in place. There are various methods of serving our models in a production environment:

- *Embedded model*: Model artifacts are used as a dependency in the consuming application code. It is built and deployed along with the application that calls the model function as an internal library function. This is a good approach for embedded applications for edge computing devices (such as in the case of IoT) but not suitable for enterprise applications where the data volume is large and the processing needs to scale. Also, deploying new versions of models is harder in this case; you may have to rebuild the entire application code and deploy again.

- *Model deployed as a separate service*: In this approach, the model is wrapped in a service. The service is independently deployed and separated from consuming applications. This allows us to update the models and redeploy them without affecting other applications. The consuming applications make service calls via remote invocation, which may introduce some latencies.

- *Model deployed as a RESTful web service*: This is similar to the approach described earlier. In this case, models are called via RESTful API calls using the TCP/IP protocol. This approach provides scalability and load balancing, but the network latency may be a concern.

- *Model deployed for distributed processing*: This is a highly scalable model deployment. In this approach, the input images (dataset) are stored in a distributed storage location that is accessible by all nodes of a cluster. The models are deployed in all cluster nodes. All participating nodes take input data from the distributed storage, process the data, and store the prediction outcome to distributed storage for applications to consume. Some examples of distributed storage are Hadoop Distributed File System (HDFS), Amazon Simple Storage Service (Amazon S3), Google Cloud Storage, and Azure Blob Storage.

In Chapter 10, we explore the topic of scaling model development and deployment on the cloud. We examine various strategies and techniques to effectively leverage cloud infrastructure for accommodating the needs of model development and deployment at scale.

Summary

This chapter along with the preceding chapters have collectively built a strong foundation for the development of computer vision applications utilizing artificial neural networks. In this chapter, we explored the image processing pipeline, its components, and their roles in constructing machine learning–based computer vision systems. You learned various techniques for feature extraction and selection and gained a high-level understanding of different machine learning algorithms, types of model training, and model deployment methods.

The next chapter, Chapter 5, serves as the central theme of this book. We will examine various machine learning models and implement artificial neural networks (ANN), convolutional neural networks (CNN), Single Shot Multibox Detection (SSD), and the YOLO (You Only Look Once) model specifically tailored for computer vision tasks. Our implementation will involve writing Python code utilizing the Keras deep learning library and executing it on TensorFlow.

Now would be an opportune time to review the concepts covered in the previous chapters. If you have followed along with the provided code examples, your development environment is most likely prepared for the next chapter. However, if you haven't done so, I recommend going back to Chapter 1 to install all the necessary software and set up your development computer. We are about to embark on an exciting journey of serious work and fascinating discoveries. If you are ready, let's proceed!

CHAPTER 5

Deep Learning and Artificial Neural Networks

In this chapter, we delve into the world of applied deep learning and artificial neural networks. Our primary focus will be on their application in computer vision. To facilitate your understanding, working code examples are provided throughout the chapter.

Our learning objectives in this chapter are as follows:

- To gain a comprehensive understanding of neural networks, including their architecture, underlying mathematical functions, and algorithms.

- To acquire proficiency in writing TensorFlow code to process and analyze images, extract essential features, and train various types of neural networks.

- To develop the skills necessary to work with pretrained models and to create custom-trained models for image classification tasks. Additionally, we will explore the process of retraining existing models.

- To learn how to evaluate the performance of a model and fine-tune its parameters to optimize accuracy and overall effectiveness.

Note This chapter includes technical descriptions of underlying mathematical concepts, notations, and equations, but you do not need a formal understanding of them to apply neural network methods. However, if you wish to further explore the mathematical treatment of these equations, this chapter provides references to guide you.

© Shamshad Ansari 2023
S. Ansari, *Building Computer Vision Applications Using Artificial Neural Networks*,
https://doi.org/10.1007/978-1-4842-9866-4_5

Introduction to Artificial Neural Networks

An *artificial neural network (ANN)* is a computational model inspired by the structure and functioning of biological neural networks in the brain. An ANN is a machine learning algorithm that learns from data and is capable of solving complex problems.

To help illustrate the concept, let's consider a simplistic but instructive scenario. Imagine that you see an object that you've never seen before. Someone informs you that it is a car. Initially, your level of confidence in identifying the object is low because you have limited knowledge about cars.

As you continue to encounter various objects and learn to recognize them, your understanding and familiarity with different objects grows. Now, let's say you come across another object, and based on your past experiences, you make a guess about what it is. In this case, you might say something like, "I think I've seen this before" or "I guess it is a car." These statements reflect your uncertainty in identifying the object accurately, because you still lack sufficient information or training to make a definitive conclusion.

However, if you have observed a large number of cars in different shapes, sizes, orientations, and colors during your learning process, you become better trained to identify the "car" object. As a result, when you encounter a new object that clearly resembles a car, your level of confidence significantly increases and you can confidently state, without any hesitation or uncertainty, "It is a car!"

This higher confidence arises from the extensive training you have undergone by observing numerous instances of cars, allowing you to recognize common design characteristics and patterns associated with the "cars" class of object.

In summary, the more exposure and training you have in recognizing a particular object or concept, the higher your confidence becomes in accurately identifying it. Training yourself with a larger and more diverse set of examples enhances your ability to make confident and reliable judgments.

In this context, let's explore conceptually what is happening in mental classification and the learning process. When you encounter a car for the first time or first few times, you begin to develop the ability to recognize similar objects, especially when one appears in a context similar to known cars you've seen before. However, of course, your recognition may not be perfect, and there can be some level of uncertainty.

On the other hand, when exposed to a large number of car samples presented in various ways, your learning becomes more comprehensive, and your recognition of the object reaches a high level of accuracy, often approaching 100 percent.

To understand the underlying mechanism, refer to the simplified diagram in Figure 5-1, which depicts how information is processed in our brains. Although the diagram represents a simplified version of human brain function, it provides insights into the overall process by which the brain processes information in different stages. The initial stage involves receiving sensory inputs, such as visual dimensional and characteristic information about the object. That information then is transmitted to subsequent neural processing areas in the brain.

As the information propagates through neural pathways, various processing layers naturally and continuously analyze and extract meaningful features and observed patterns. These layers progressively build a sophisticated representation of the object, based on the incoming data in context. The deeper layers capture higher-level abstractions and observe complex relationships between different and similar aspects of the object.

Through exposure to a large and diverse set of car samples, the brain's processing pathways become finely tuned to recognize the defining characteristics of a car. This comprehensive training enables the brain to achieve near-perfect accuracy in identifying cars, as it can robustly recognize them across different contexts, orientations, and variations.

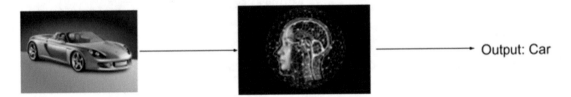

Figure 5-1. *Human eyes as sensing device feeding input to the brain that stores patterns*

The human eye functions as a sensory device, capturing images of objects which then are transmitted to the brain as input signals via neuron tree structure edges, known as *dendrites*. These signals then are further processed by the neurons in our brain, which perform pattern recognition computations and generate input signals for use by the brain system.

In the context of a neural network, Figure 5-2 illustrates the role of dendrites in receiving the input signals (X). These input signals then are combined and processed within the neuron, and computations are performed using neuronal-specific functions or algorithms. The resulting output is subsequently transmitted through axon terminals.

Essential to our brain's continuous recognition processing, an extensive network of neurons, numbering in the billions, with trillions of interconnections between them. As illustrated notionally in Figure 5-2, this intricate web of interconnected neurons is commonly referred to as a *neural network* and is the inspiration of today's artificial neural networks and related computing and listening complex computing systems.

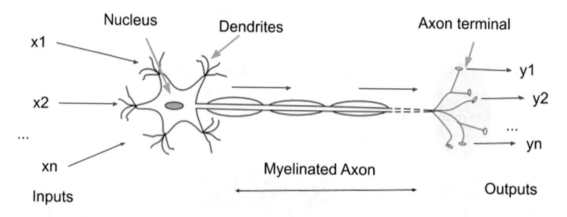

Figure 5-2. *Information processing in human neurons*

Computer scientists, starting (1956) with the first concept of an electromagnetic artificial "neuron" design; and, a decade later were first inspired by the idea of a human vision system and tried to mimic neural networks, by creating a computer system that learns and functions the way our brains do. This learning system today is called an *artificial neural network* (ANN).

Figure 5-3 shows the analogous ANN version of Figure 5-1. A camera works as a sensing device, much like our eyes capture images of objects. The images are transmitted to an interpreter system, such as a computer, where they are processed in a similar way as a neuron processes the input signals. Some examples of other sensing devices are X-ray, CT-scan, and MRI machines; satellite imaging systems; and document scanners.

The interpreting devices, such as computers, provide the processing of the data acquired by the camera. Most of the computer vision–related computations, such as feature extraction and pattern determination, are performed within the computer.

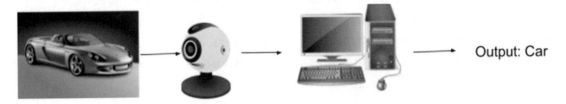

Figure 5-3. *Artificial sensing device (a camera) feeding image input to computers*

Figure 5-4 is analogous to the human neuron shown in Figure 5-2. The variables x1, x2, .. xn are the input signals (e.g., image features), with certain weights w1, w2, .. wn associated with each input signal. These input signals are processed using some mathematical functions to generate outputs. The processing unit that combines these input signals is called a *neuron*, named after the human neuron. The mathematical function that computes the output from the neuron is called an *activation function*. In Figure 5-4, the circle marked with the function symbol *f(x)* is the neuron. The output y is generated from the neuron.

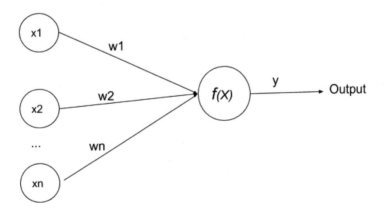

Figure 5-4. *Artificial neuron*

Perceptron

A single neuron of a neural network is called a *perceptron*. A perceptron implements a mathematical function that operates on the input signals and generates outputs. Figure 5-4 is an example of a perceptron. A perceptron is the simplest neural network. We will see a bit later that a typical neural network for machine learning consists of several neurons. The inputs to the neuron come either from the source (camera or sensing devices) or from the outputs of other neurons.

How a Perceptron Learns

The learning objective of a perceptron is to determine the ideal weights for each input signal. The learning algorithm arbitrarily assigns weights to each input signal. The signal value is multiplied by its corresponding weight. The product (weight times signal value) of each signal is added to compute an output. The computation is represented by the following equations:

$$f(x) = w_1 x_1 + w_2 x_2 + w_3 x_3 + \ldots + w_n x_n \qquad \text{(Equation 5-1)}$$

Sometimes a bias, x_0, is also added to the equation, as shown here:

$$f(x) = x_0 + w_1 x_1 + w_2 x_2 + w_3 x_3 + \ldots + w_n x_n \qquad \text{(Equation 5-2)}$$

Equation 5-2 can also be written as follows:

$$f(x) = X_0 + \sum_{i=1}^{i=n} W_i X_i \qquad \text{(Equation 5-3)}$$

The neuron computes using Equation 5-2 over a large number of inputs. An optimization function optimizes the weights by using certain mathematical algorithms, called an *optimizer*, and the computation is repeated with the new weights. This weight optimization and computation and re-optimization are performed in multiple iterations until the weights are fully optimized for the given set of inputs. We will explore more about this optimization function later in this chapter. The fully optimized weights are the actual learning of the neuron.

Multilayer Perceptron

Just as the human brain contains billions of neurons, artificial neural networks may have the capacity for billions or more neurons, yet in most practical scenarios, they typically comprise a significantly smaller number of neurons. Inputs are processed by a group of neurons. Each neuron in the group processes the inputs independently. Outputs from this group of neurons are fed to another neuron or group of neurons for further processing. You can imagine these neurons arranged as layers where the output from one layer is fed as inputs to the next layer. You can have as many layers as needed to train your neural network. This multilayer approach of arranging neurons in a neural network is commonly known as a *multilayer perceptron (MLP)*. Figure 5-5 shows an example MLP.

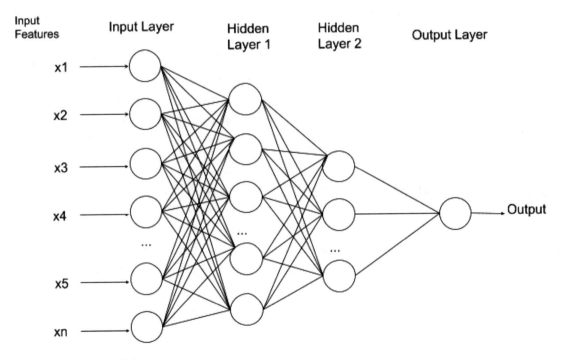

Figure 5-5. *Multilayer perceptron*

What is the advantage of using an MLP over a single perceptron? Let's consider a single neuron with a single input. Equation 5-1 will look like the following:

$$f(x) = x_0 + w_1 x_1$$

This represents the equation of a straight line with an intercept as x_0 and a slope (angle with the horizontal or x-axis) that equals w_1.

Don't worry if you do not understand this math. This is to show you that a single neuron models a linear relationship of input to output. Machine learning algorithms, such as linear regression and logistic regression, model linear relationships. Most real-world problems do not exhibit linear relationships. Multilayer perceptrons model nonlinearity and can model real-world problems more accurately than single neuron–based models.

What Is Deep Learning?

Deep learning is another name for a multilayer artificial neural network or multilayer perceptron. We have different types of deep learning systems depending upon the neural network architecture and its working principles. For example, feed-forward neural networks, convolutional networks, recurrent neural networks, autoencoders, and deep beliefs are different types of deep learning systems.

The following sections start with an explanation of the high-level architecture of the multilayer perceptron. In this book, we will use the terms MLP and deep learning (DL) interchangeably.

Deep Learning or Multilayer Perceptron Architecture

A multilayer perceptron is composed of three essential layer types, as depicted in Figure 5-5: the input layer, one or more hidden layers, and the output layer. Each layer consists of one or more neurons, which process the received inputs and produce outputs. The outputs from the neurons are then forwarded as inputs to the subsequent layer, except from the output layer. The output layer generates the final outputs that can be utilized by various applications.

An MLP architecture consists of the following:

- *Input layer*: This is the first layer of a neural network. This layer takes the input from the external source, such as images from the sensing devices. The inputs to this layer are the *features* (see Chapter 4 for details on features).

    The nodes in the input layer do not do any computation. These nodes simply pass their inputs to the next layer.

The number of neurons in the input layer is the same as the number of features. Sometimes, an additional node is added in each layer. This additional node is called a *bias node*. The bias node is added to have control over the output from the layer. In deep learning, the bias is not required, but it is a common practice to add one.

Figure 5-6 shows a neural network architecture with bias nodes. The nodes shown in orange color are the bias nodes added in each layer.

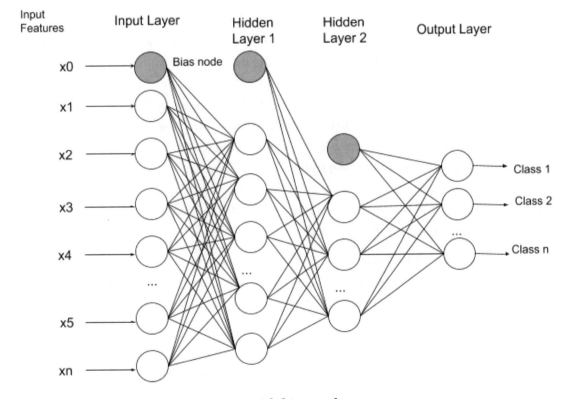

Figure 5-6. *Multilayer perceptron with bias nodes*

Question: What is the total number of neurons in the input layer of a neural network?

Answer: The number of input layer neurons = The number of input features without a bias = (The number of input features + 1 if biased is used).

- *Hidden layer*: The hidden layers are the intermediary layers between the input and output layers within a neural network. At least one hidden layer is necessary for a neural network to facilitate the learning process. The neurons within these layers perform the computations required for learning. While one hidden layer is often sufficient for learning purposes, the number of hidden layers can be increased to better model real-world scenarios. As the number of hidden layers grows, so does the computational complexity, leading to longer computation times.

  Determining the ideal number of neurons for each hidden layer is not an exact science and various practical strategies are available. A commonly used approach is to set the number of neurons in the hidden layer to two-thirds (or approximately 66 percent) of the number of neurons in the previous layer. For instance, if the input layer has 100 neurons, the first hidden layer would have around 66 neurons, and subsequent hidden layers would follow this pattern (e.g., 43 neurons in the next hidden layer). However, it's important to reiterate that there is no universally optimal number of neurons, and the counts should be adjusted based on the desired accuracy and performance of the model.

- *Output layer*: This is the final layer of the neural network. The output layer gets its inputs from the last hidden layer. The number of neurons in the output layer depends on the type of problem you want the neural network to solve:

  - *Regression problems*: When the network has to predict a continuous value, such as the pixel intensity for an image, the output node has only one neuron.

  - *Classification problems*: When the network has to predict one of many classes, the output layer has as many neurons as the number of all possible classes. For example, if the network is trained to predict one of four classes of animals—cat, dog, lion, bull—the output layer will have four neurons, one for each class.

- *Edges or weight connections*: Weights are also referred to as *coefficients* or *input multipliers*. Each input feature to a neuron is multiplied by a weight. Pictorially, each connection from input to a neuron is linked with a weighted line. The weighted line signifies the contribution of the feature in predicting the outcome we are trying to model for. Think of weight as the contribution or significance of an input feature. The higher the weight, the more the contribution of the feature. If weight is negative, the feature has a negative effect. If the weight is zero, the input feature is not important and can be removed from the training set.

The training objective of a neural network is to determine the optimal weights for each connection between input features and neurons across all layers. In this chapter, we will delve deeper into how a neural network learns by iteratively adjusting these weights. It's important to note that if bias is incorporated into the network, the learning process also involves determining the optimal bias values alongside the weights.

Activation Functions

The mathematical function that determines the output of a neuron is called the *activation function.*

Neurons operate on inputs using the following linear equation:

$$z = X_0 + \sum_{i=0}^{i=n} w_i x_i \qquad \text{(Equation 5-4)}$$

But the output of a neuron is not the result of Equation 5-4. It is the activation function that operates on the value of z (calculated from Equation 5-4) and determines the output from the neuron.

The activation function determines whether the neuron it's attached to should be activated (turned on), based on whether the neuron's input is relevant for model prediction. The activation function normalizes the output of each neuron to a range between 0 and 1 or between -1 and 1.

There are various mathematical functions that can be utilized as activation functions, depending on their specific purposes. In the upcoming sections, we will explore some of the activation functions supported by TensorFlow, a framework we will delve into later in this chapter.

Linear Activation Function

The *linear activation function* calculates the neuron output by multiplying weights to inputs as per the equation $f(x) = x_0 + w_1x_1 + w_2x_2 + w_3x_3 + ... + w_nx_n$. The output of the linear activation function varies from $-\infty$ to $+\infty$, as shown in Figure 5-7. That means the linear activation function is as good as having no activation.

Linear Function

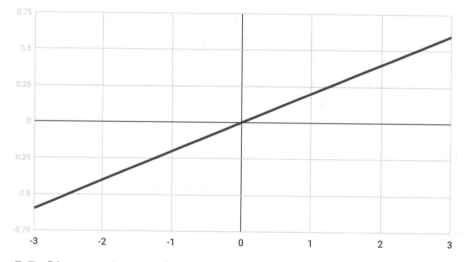

Figure 5-7. *Linear activation function graph*

The linear activation function has the following two main problems and is not used in deep learning:

- Deep learning uses a method called *backpropagation* (more on this later), which uses a technique called *gradient descent*. The gradient descent requires calculating a first-order derivative of the input, which, in the case of linear activation, is a constant. The first derivative of a constant is a zero. That means it has no relationship with the input. Therefore, it is not possible to go back and update the weights of the inputs.

- If you use a linear activation function, the last layer will be the linear function of the first layer, regardless of the number of layers in the neural network. In other words, a linear activation function turns your network into just one layer. That means your network can learn only the linear dependencies of inputs to output, and that is not suitable for solving complex problems such as computer vision.

Sigmoid or Logistic Activation Function

The *sigmoid activation function*, also known as logistic activation function, calculates the neuron output using the sigmoid function, as shown here, where z is calculated using Equation 5-4:

$$\sigma(z) = 1/\left(1 + e^{-z}\right) \qquad\qquad \text{(Equation 5-5)}$$

The sigmoid activation function always yields a value between 0 and 1. This makes the output smooth without many jumps as the input value fluctuates. The other advantage is that this is a nonlinear function and does not generate a constant value from a first-order derivative. This makes it suitable for deep learning with backpropagation that updates weights based on gradient descent. See Figure 5-8.

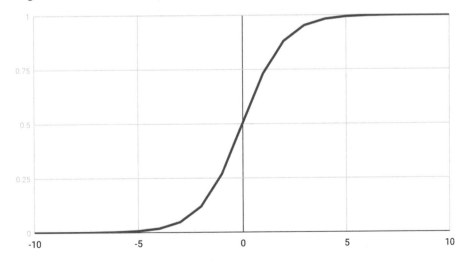

Figure 5-8. Sigmoid activation function graph

The biggest disadvantage of the sigmoid activation function is that the output does not change between large or small input values, which makes it unsuitable for cases where the feature vector contains large or small values. One way to overcome this disadvantage is to normalize your feature vector to have values between -1 and 1 or between 0 and 1.

Another characteristic that you will notice from Figure 5-8 is that the S-shaped curve is not centered at zero.

Hyperbolic Tangent (TanH)

TanH is similar to the sigmoid activation function except that TanH is zero-centered. See Figure 5-9 and notice that the S-shaped curve passes through the origin.

TanH Activation Function Graph

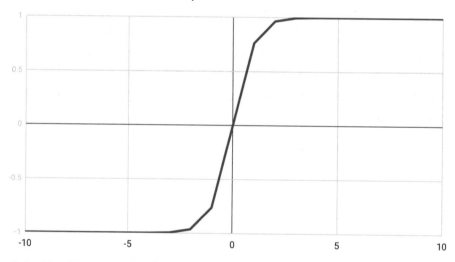

Figure 5-9. *TanH activation function graph (zero-centered)*

The TanH activation function calculates the neuron output using this formula:

$$\tanh(z) = \left(e^z - e^{-z}\right) / \left(e^z + e^{-z}\right)$$ (Equation 5-6)

Because the TanH function is zero-centered, it models with inputs having small, large, and neutral values.

Rectified Linear Unit

The *rectified linear unit (ReLu)* determines the neuron output based on the value of z as computed from Equation 5-4. If the value of z is positive, ReLU takes that value as an output; otherwise, it outputs as zero. The output from ReLU ranges between 0 and +∞. The ReLU function is represented as shown here (see also Figure 5-10):

$$f(z) = max(0, z)$$

(Equation 5-7)

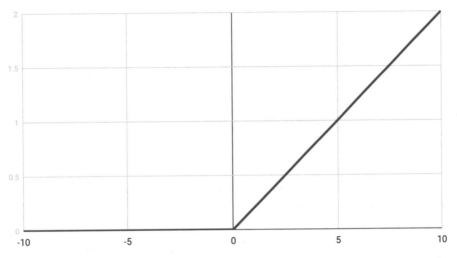

Figure 5-10. *ReLU activation graph (with value ranges between 0 and infinity)*

The advantage of the ReLU activation function is that it is computationally efficient and allows the network to converge quickly. Also, ReLU is nonlinear, and it has a derivative function that makes it suitable for backpropagation for weight adjustment as the neural network learns.

The biggest disadvantage of the ReLU function is that the gradient of the function becomes zero for zero or negative inputs. This makes it not suitable for backpropagation when the input has negative values.

ReLU is widely used for most computer vision model training because the image pixels do not have negative values.

Leaky ReLU

Leaky ReLU provides a slight variation of ReLU. Instead of making the negative value of z (as calculated from Equation 5-3) zero, it multiplies the negative value of z by a small number such as 0.01. Figure 5-11 depicts the Leaky ReLU outputs.

Leaky ReLU Graph

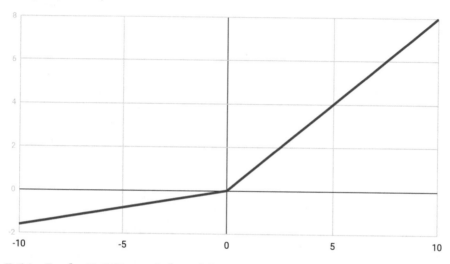

Figure 5-11. *Leaky ReLU graph (modified ReLU by taking negative value multiplied with a small number)*

The leaky ReLU has a small slope in the negative area and allows for backpropagation for negative inputs.

The disadvantage is that the result of the leaky ReLU is not consistent with negative values.

Scaled Exponential Linear Unit

A *scaled exponential linear unit (SELU)* computes neuron outputs using the following equation:

$$f(\alpha, x) = \lambda \begin{cases} \alpha\left(e^x - 1\right) & \text{for } x < 0 \\ x & \text{for } x \geq 0 \end{cases}$$ (Equation 5-8)

where the value of Lambda = 1.05070098 and the value of α = 1.67326324. These values are fixed and do not change during backpropagation.

The graph in Figure 5-12 shows the SELU characteristics.

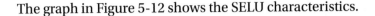

SELU Activation Graph

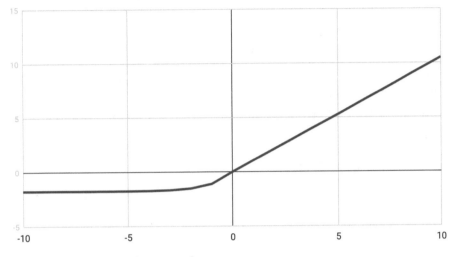

Figure 5-12. *SELU activation graph*

SELU has the "self-normalizing" properties (refer to the original paper of SELU https://arxiv.org/pdf/1706.02515.pdf). The inventors of SELU have proven mathematically that SELU generates output that is normalized with mean 0 and standard deviation 1.

In TensorFlow or Keras, if you use the weight initialization method as truncated normal distribution centered around zero by using the method tf.keras. initializers.lecun_normal, you will get the normalized output of all network components, such as weights, biases, and activations, at each layer.

So, why do we care about the network generating normalized outputs? The initialization function lecun_normal initializes the parameters of the network as a normal distribution or Gaussian distribution. SELU also generates normalized outputs. That means the entire network exhibits normalized behavior. Therefore, the output in the last layer is also normalized.

With SELU, the learning is highly robust and allows training networks that have many layers.

Since with SELU the entire network is self-normalizing, it is efficient in terms of computation and tends to converge faster. Another advantage is that it overcomes the problems of exploding or vanishing gradients when the input features are too high or too low.

Softplus Activation Function

The *softplus activation function* applies smoothing to the activation function value z (as calculated by Equation 5-4). It uses the log of exponent as follows:

$$f(x) = \ln(1 + e^z)$$
(Equation 5-9)

Softplus is also called the *SmoothReLU function*. The first derivation of the softplus function is $1/(1+e^{-z})$, which is the same as the sigmoid activation function. See Figure 5-13.

Softplus Activation Graph

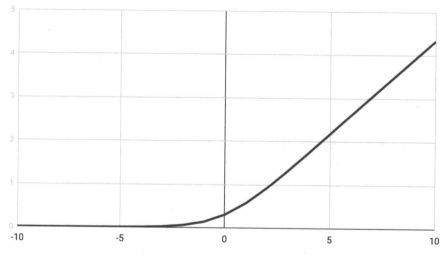

Figure 5-13. *Softplus activation graph*

Softmax

Softmax is a function that takes an input vector of real numbers, normalizes it into a probability distribution, and generates outputs in the range (0,1) with the sum of output values equal to 1.

Softmax is most often used as the activation for the last layer (output layer) of a classification neural network. The result is interpreted as the prediction probability of each class.

The softmax transformation is calculated using the following formula:

$$\sigma(z)_i = \frac{e^{z_i}}{\sum_{j=1}^{K} e^{z_j}} \text{ for } i = 1, \ldots, K \text{ and } z = (z_1, \cdots, z_K) \in \mathbb{R}^K \quad \text{(Equation 5-10)}$$

The normalized output of the previous equation is always between 0 and 1. When you add these outputs, the result will be 1.

Feedforward

A *feedforward neural network* is a type of artificial neural network in which information flows in only one direction: forward from the input layer to the output layer, through the hidden layers. It is one of the simplest and most common types of neural networks used for various machine learning tasks. In this network, there is no loopback or feedback mechanism. In other words, connections between neurons don't form any cycling connections.

The example networks shown earlier in Figures 5-2 and 5-3 are feedforward artificial neural networks.

Throughout this book, our primary focus will be on utilizing feedforward networks.

Error Function

What is an error? An *error*, in the context of machine learning, is the difference between the expected outcome and the predicted outcome. The equation of error may be written in a simplified form as follows:

Error = expected outcome – predicted outcome

As previously discussed, the learning objective of a neural network is to calculate optimized values of weights. The weights are considered optimized for a given dataset when the errors are at a minimum (ideally, zero). We have seen that when the network starts the learning process, it initializes weights and calculates the output from each neuron by using one of the activation functions. It then calculates the error, adjusts the weights, calculates outputs, and recalculates the errors and compares them with previously calculated errors, until it finds the minimum error. The weights that give the minimum errors are taken as the final weights. The network is considered "learned" at this stage.

From calculus, if the first derivative of a function is zero, the function at that point is either minimum or maximum. Finding this minimum point where the first derivative is zero is the goal of the neural network training process. Therefore, a neural network must have an error function that will calculate the first derivative and find the points (weights and biases) where the error function is minimum. What this error function should be depends on the type of model we want to train. Error functions are also known as *loss function*, or simply *loss*.

The mathematics that computes the derivatives and finds the optimum values of weights is beyond the scope of this book. We will explore a few commonly used error functions and where they should be applied. We will not get deep into the mathematics behind these error functions to keep this book focused on our learning objectives: building computer vision applications. If you do not have any background in calculus, do not worry about it. Just make sure that you understand which error functions should be used in solving computer vision problems.

The error functions are broadly divided into the following three categories:

- *Regression loss functions*: Used when we want to train models to predict continuous value outcomes, such as stock prices and housing prices.

- *Binary classification loss functions*: Used when we want to train models to predict a maximum of two classes, such as cat versus dog or cancer versus no cancer.

- *Multiclass classification loss functions*: Used when our models need to predict more than two classes, such as object detection.

The following sections provide an overview of these three categories of error functions, their usages, and the types of activation functions they are compatible with. Use this information as a guide to determine the appropriate error functions for your particular modeling work.

Regression Loss Function

Error function name: Mean squared error (MSE) loss

- *Brief description*: This is the default error function for regression problems. This is the preferred loss function if the distribution of the target variable is normal or Gaussian.

- *Where to use*: When the distribution of target variables is normally distributed.

- *Applicable activation function*: `model.add(Dense(1, activation='linear'))`

- *TensorFlow example:* `model.compile(loss='mean_squared_error')` or `model.compile(loss='mse')`

Error function name: Mean squared logarithmic error (MSLE) loss

- *Brief description*: This function first calculates the logarithm of predicted values and calculates the mean squared error.

- *Where to use*: When the target variable has a spread of values or when predicting a large value. In either case, you may not want to punish a model as heavily as the mean squared error. This is normally used when your model is predicting unscaled values.

- *Applicable activation functions*: `model.add(Dense(1, activation='linear'))`

- *TensorFlow example*: `model.compile(loss='mean_squared_logarithmic_error')`

Error function name: Mean absolute error loss

- *Brief description*: This is calculated as the average of the absolute difference between the expected and predicted values.

- *Where to use*: When the target variable is normally distributed and has some outliers.

- *Applicable activation functions*: `model.add(Dense(1, activation='linear'))`

- *TensorFlow example*: `model.compile(loss='mean_absolute_error')`

Binary Classification Loss Function

Error function name: Binary cross-entropy

- *Brief description*: This is the default loss function for binary classification problems and is preferred over other functions. Cross-entropy calculates a score that summarizes the average difference between actual and predicted probability distributions for predicting class 1. The score is minimized, and a perfect cross-entropy value is set to 0.

- *Where to use*: When the target value is in the range (0, 1).

- *Applicable activation functions*: `model.add(Dense(1, activation='sigmoid'))`

- *TensorFlow example*: `model.compile(loss='binary_crossentropy', metrics=['accuracy'])`

Error function name: Hinge loss

- *Brief description*: This function is used largely in support of vector machine–based binary classification.

- *Where to use*: When the target variable is in the range (-1, 1).

- *Applicable activation functions*: `model.add(Dense(1, activation='tanh'))`

- *TensorFlow example*: `model.compile(loss='hinge', metrics=['accuracy'])`

Error function name: Squared hinge loss

- *Brief description*: This function calculates the square of the score hinge loss. It smooths the surface of the error function and makes it numerically easier to work with.

- *Where to use*: When the target variable is in the range (-1, 1).

- *Applicable activation functions*: `model.add(Dense(1, activation='tanh'))`

- *TensorFlow example*: `model.compile(loss='squared_hinge', metrics=['accuracy'])`

Multiclass Classification Loss Function

Error function name: Multiclass cross-entropy loss

- *Brief description*: This is the default loss function for multiclass classification problems and is preferred over other functions. Cross-entropy calculates a score that summarizes the average difference between the actual and predicted probability distributions for predicting class 1. The score is minimized, and a perfect cross-entropy value is set to 0.

- *Where to use*: When the target values are in the set {0, 1, 3, 4,..., n}, where each class is assigned a unique integer value.

- *Applicable activation functions*: `model.add(Dense(4, activation='softmax'))`

- *TensorFlow example*: `model.compile(loss='categorical_ crossentropy', metrics=['accuracy'])`

Error function name: Sparse multiclass cross-entropy loss

- *Brief description*: This function performs the same cross-entropy calculation of error, without requiring the target variable to be one hot-encoded prior to training.

- *Where to use*: When you have a large number of classes in the target; for example, predicting dictionary words.

- *Applicable activation functions*: `model.add(Dense(100, activation='softmax'))`

- *TensorFlow example*: `model.compile(loss='sparse_categorical_ crossentropy', metrics=['accuracy'])`

Error function name: Kullback-Leibler divergence (KLD) loss

- *Brief description*: This function measures how a probability distribution differs from a baseline distribution. A KLD loss of 0 means distributions are identical. It determines information loss (in terms of bits), if the predicted probability distribution is used to approximate the desired target probability distribution.

- *Where to use*: When you need to solve complex problems such as auto-encoders for learning dense features. If this is used for multiclass classification, it works as multiclass cross-entropy.

- *Applicable activation functions*: `model.add(Dense(100, activation='softmax'))`

- *TensorFlow example*: `model.compile(loss='kullback_leibler_ divergence', metrics=['accuracy'])`

Optimization Algorithms

The learning objective of a neural network is to determine the most optimized weights (and biases) at which the loss is minimum. When the network starts learning, it assigns weights to each input connection. Initially, these weights are rarely optimized. How much the weights are off from optimization is determined by measuring the loss (or error). To determine the ideal weights, the learning algorithm optimizes the loss function so that it finds weights that make the loss function have the minimum value. The weights (and biases) are updated, and the process is repeated until there is no more scope for optimization. The mathematical function that optimizes the loss function is called the *optimization algorithm* or *optimizer*.

There are several optimization algorithms that offer different degrees of accuracy, speed, and parallelism. We will explore some of the most popular optimizers in this section. I will provide introductory-level information, without going deep into the mathematics used in these algorithms, to give you a basic understanding of where to use which optimization algorithms.

Gradient Descent

Gradient descent is an optimization algorithm that finds weights where the loss function (also known as *cost function*) is zero or minimum. Gradient descent is a technique to find the minimum cost function. This section describes how it works.

The cost function or error function is represented by the following equation:

$$f(w) = \frac{1}{N}\Sigma(y_i - w_i x_i) \qquad \text{(Equation 5-11)}$$

where y_i is the actual/known value and w_i is the weight corresponding to the feature vector x_i of ith sample. $w_i x_i$ is the predicted value that is subtracted from the actual value y_i to calculate the error or loss.

From calculus, we know that the first derivative of a function at a point gives the slope or gradient of the function at that point. If you plot the cost function f(w), you will see a multidimensional curve (as shown in Figure 5-14). The derivative is calculated to get the gradient to determine which direction along the curve to move to get the new set of weights. Since the goal is to minimize the cost, the algorithm moves to the direction of the negative gradient.

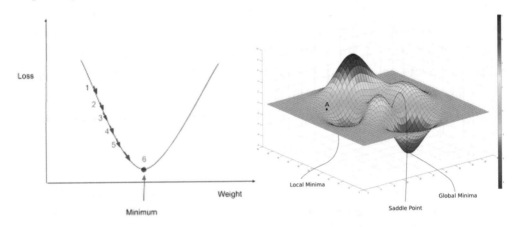

Figure 5-14. *Cost function with gradient movement toward minimum*

For example, let's assume there is only one feature, and hence we need to compute only one weight (w). The cost function will look like the left image in Figure 5-14.

1. The algorithm first calculates the cost or loss for the initial weights, assuming this loss is f(w) and assuming the loss is calculated at point 1 in Figure 5-14 (left).

2. The algorithm then computes the gradient (delta) and moves down the curve; the direction is decided by the negative gradient.

3. As it descends, the algorithm computes the new weights using the following formula:

$$weight = weight + alpha^* \left(-delta \right) = weight - alpha^* delta \qquad \text{(Equation 5-12)}$$

Here, alpha is called the *learning rate*. The learning rate determines the size of the steps through which the gradient descends the curve to reach the minimum point.

4. The error is again computed using the new value of the weight, and the process is repeated until the algorithm finds the ultimate minimum cost.

Local and Global Minima

For simplicity in the previous example, we considered only one feature and hence only one weight. But in practice, there may be tens or even hundreds of features for which weights need to be learned. The image on the right of Figure 5-14 shows the error curve when more than one weight needs to be optimized. In this case, the curve may have multiple points that would appear as minimums, called *local minima*. The objective of the gradient descent algorithm is to find the global minimum to optimize the weights.

Learning Rate

In Equation 5-12, the parameter alpha is referred to as the *learning rate*, which plays a crucial role in the gradient descent algorithm. The learning rate determines the size of the steps taken by the algorithm as it descends along the curve in search of the global minimum.

Choosing an appropriate value for the learning rate is important. If the learning rate is set to a large value, it can cause the algorithm to overshoot the minimum point, leading to oscillations and failure to converge. Conversely, selecting a small learning rate will necessitate a large number of steps to reach the minimum point, resulting in slow convergence.

It is worth noting that a small learning rate can significantly slow down the learning process. Figure 5-15 provides a visual representation of the impact of both large and small learning rates.

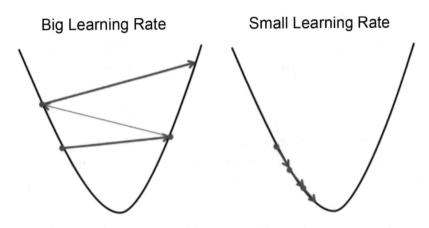

Figure 5-15. *Effect of big and small learning rates*

To ensure effective learning, it is crucial to set the learning rate appropriately. Typically, a practical range for the learning rate falls between 0.01 and 0.1. It is common practice to start with a learning rate within this range and make adjustments as necessary based on the specific problem at hand.

Adaptive Learning Rate

In addition to manually setting the learning rate within the practical range, another approach to consider is using an *adaptive learning rate*. Adaptive learning rate algorithms dynamically adjust the learning rate during the training process based on the behavior of the optimization algorithm.

One popular adaptive learning rate algorithm is called *Adaptive Gradient (AdaGrad)*. AdaGrad adjusts the learning rate for each parameter individually based on its historical gradients. It gives larger updates for parameters with smaller historical gradients and smaller updates for parameters with larger historical gradients. This adaptive adjustment helps to alleviate the need for manual tuning of the learning rate.

Another widely used adaptive learning rate algorithm is *Adaptive Moment Estimation (Adam)*. Adam combines adaptive learning rates with momentum to achieve faster convergence. It computes adaptive learning rates for each parameter and maintains exponentially decaying averages of both past gradients and squared gradients. This helps to adaptively scale the learning rates based on the magnitudes of the gradients.

By using adaptive learning rate algorithms like AdaGrad or Adam, the learning rate is automatically adjusted during the training process, potentially leading to improved convergence and performance without the need for manual fine-tuning.

Regularization

Regularization is a technique used in machine learning and statistical modeling to prevent overfitting and improve the generalization capability of a model. *Overfitting* occurs when a model becomes too complex and starts to memorize the training data, leading to poor performance on unseen data.

The basic idea behind regularization is to introduce additional constraints or penalties on the model's weights during the training process. These constraints encourage the model to learn simpler patterns and avoid overly complex or intricate relationships in the data.

The following different types of regularization techniques are commonly employed:

- *L1 regularization (LASSO)*: L1 regularization adds a penalty term to the loss function, proportional to the absolute values of the model's weights. This penalty encourages sparsity, meaning it drives some weights to become exactly zero, effectively performing feature selection and reducing the model's complexity.

- *L2 regularization (Ridge)*: L2 regularization adds a penalty term to the loss function, proportional to the squared magnitudes of the model's weights. This penalty encourages small weights but does not drive them to zero as aggressively as L1 regularization. L2 regularization is also known as *ridge regression* and can help with reducing the impact of multicollinearity in linear models.

- *Elastic net regularization*: Elastic net regularization combines both L1 and L2 regularization. It adds a penalty term to the loss function that is a combination of the absolute and squared magnitudes of the weights. The combination of L1 and L2 penalties provides a balance between feature selection (sparsity) and weight shrinkage.

- *Dropout*: Dropout is a regularization technique commonly used in neural networks. During training, dropout randomly sets a fraction of the units (neurons) in a layer to zero with a predefined probability.

This process helps prevent co-adaptation of neurons and encourages the network to learn more robust representations. Dropout is only applied during training, and during inference, all units are used, but their activations are scaled by the dropout probability.

Regularization techniques help to control the model's complexity and reduce overfitting by adding constraints or penalties to the learning process. By encouraging simpler and more generalized models, regularization can improve performance on unseen data and enhance the overall reliability of machine learning models.

Stochastic Gradient Descent

Gradient descent computes the gradients for all training examples in each step and iteration. This process can be computationally expensive and time-consuming, especially for large datasets. Additionally, it may not be feasible to run the algorithm on a single machine if the data cannot fit into memory (RAM), and parallelized computing is not possible. *Stochastic gradient descent (SGD)* addresses these challenges.

SGD calculates the gradients based on a small subset of the training set that can easily fit into memory. Here is an overview of how SGD works:

1. Randomize the input dataset to eliminate biases and ensure randomness in gradient estimation.

2. Calculate the gradients using a randomly selected single data point or a small batch of data.

3. Update the weights using the formula *weight = weight – alpha * delta*, where *alpha* represents the learning rate.

In SGD, it is common to compute weight updates for a few training examples at a time rather than a single example. This helps reduce weight variances and leads to more stable convergence. A mini-batch size of 128 or 256 is often a good starting point. However, the optimal batch size may vary depending on the application, architecture, and hardware capacity of the computer.

By using SGD, the computational and memory requirements are significantly reduced, making it more feasible to train models on large datasets. The use of smaller subsets also introduces randomness and can help the algorithm avoid local minima and explore different parts of the optimization space.

SGD for Distributed and Parallel Computing

When dealing with a large training dataset, it is possible to divide the randomized set into smaller mini-batches and distributed them across multiple computers in a cluster architecture. By using SGD, each computer can independently and in parallel compute the weights based on its assigned mini-batch of data. The results from the individual computers can then be combined on a central computer to obtain the final optimized weights.

Furthermore, SGD can also take advantage of parallel processing within a single computer that has multiple CPUs or GPUs. This allows for simultaneous computation of weights, further enhancing the speed of convergence.

By leveraging distributed and parallel operations to compute optimized weights using the SGD algorithm, the convergence process can be accelerated. This approach is particularly useful for handling large datasets and exploiting the computational power of multiple machines or parallel processing within a single machine.

SGD with Momentum

If your cost function exhibits ravine-shaped curves characterized by steep walls and narrow bottoms, it is advisable to employ *SGD with momentum*. These ravines are particularly noticeable around local minima. In such scenarios, standard SGD may oscillate around the minimum and struggle to reach the desired target. It often experiences a delayed convergence, especially after several iterations. For a visual representation, refer to Figure 5-16.

Figure 5-16. *SGD without momentum*

Momentum is a method that controls the oscillation by controlling the gradient movement. The momentum update is given by the following equation:

$$v = yv + \text{alpha} * \text{delta}$$ (Equation 5-13)

where the delta is gradient calculated using SGD and alpha is the learning rate. v is the velocity vector having the same dimension as the parameters (or weights). The value of γ is in the range (0, 1) and generally taken as 0.9 by default.

Finally, the weights are updated using the following equation:

$$weight = weight + v$$

SGD with momentum is a powerful optimization technique that can accelerate convergence, aid in escaping local minima, provide a smoother optimization trajectory, and improve generalization performance. It is particularly useful in scenarios with ravine-shaped cost functions and complex optimization landscapes.

Adaptive Gradient (AdaGrad) Algorithm

As explained earlier in the section "Adaptive Learning Rate," both gradient descent and stochastic gradient descent (SGD) require manual setting and fine-tuning of the learning rate. If the learning rate is too high, the algorithm may overshoot the minimum point, while a too low learning rate can result in slow convergence. Finding an ideal learning rate typically involves a manual trial-and-error process, which becomes particularly challenging when dealing with multidimensional neural networks. With hundreds or thousands of dimensions, manually selecting an appropriate learning rate becomes nearly impossible.

To address this issue, AdaGrad offers a solution by automatically determining the suitable learning rate for each parameter based on historical information. It assigns a larger learning rate to infrequent features and a smaller learning rate to more frequent features. Consequently, each parameter possesses its own learning rate, enhancing performance when dealing with problems involving sparse gradients. AdaGrad proves highly effective in handling sparse data, such as in computer vision or natural language processing (NLP).

However, one significant drawback of AdaGrad is that its adaptive learning rate tends to decrease significantly over time.

Root Mean Squared Propagation (RMSProp)

As previously described, SGD with momentum introduces a mechanism to regulate the movement of gradients along a steep curve. RMSProp improves upon SGD with momentum by addressing the gradient movement specifically in the vertical direction.

Consider a situation where you encounter a steep curve. Even a small horizontal movement can result in significant vertical displacement. RMSProp tackles this issue by controlling the vertical movement, ensuring that the overall movement in both vertical and horizontal directions is more balanced. As a result, this balanced movement accelerates the process of reaching the minimum point.

Adaptive Moment (Adam)

The Adam optimization algorithm, also introduced in the "Adaptive Learning Rate" section, is widely used in deep learning and is a favored choice among optimizers. It combines the benefits of both SGD with momentum and RMSProp. Adam performs iterative weight updates for the neural network based on the training data.

Unlike RMSProp, which adjusts the learning rates of parameters based on the average first moment (the mean), Adam takes advantage of the average of the second moments of the gradients. By utilizing these second moments, Adam offers improved performance in updating network weights.

The math behind Adam is out of the scope of this book (again, to stay focused on the core theme of the book). See the original paper at `https://arxiv.org/pdf/1412.6980.pdf` for more detailed information about how gradients are calculated and updated.

The paper describes the following benefits of Adam:

- Straightforward to implement

- Computationally efficient

- Little memory requirements

- Invariant to diagonal rescale of the gradients

- Well-suited for problems that are large in terms of data and/or parameters

- Appropriate for nonstationary objectives

- Appropriate for problems with noisy/or sparse gradients

- Hyperparameters that have intuitive interpretation and typically require little tuning

In summary, Adam offers ease of implementation, computational efficiency, and minimal memory usage. It is versatile in handling various types of problems, including those with nonstationary objectives or noisy/sparse gradients. The hyperparameters used in Adam are intuitive and often require little fine-tuning.

Backpropagation

To train a neural network, three essential components are required:

- Input data or input features

- A feedforward multilayer neural network

- An error function

To initiate the training process, the network assigns initial weights to each input feature. An optimization algorithm such as SGD or Adam is utilized to optimize the error function and compute the minimum error. This optimization process involves updating the weights.

As discussed earlier in the chapter, a multilayer perceptron consists of at least three layers: the input layer, one or more hidden layers, and the output layer. The feedforward network calculates the output of each neuron in a forward direction, starting from the first hidden layer, then proceeding to the subsequent hidden layers, and finally reaching the output layer.

The next step involves estimating the error, which in turn leads to the update of weights. The backpropagation method is employed, where the gradients of the weights are first calculated at the last layer, and then the gradients of the preceding layers are computed. This backward flow of error information facilitates efficient gradient computation at each layer. In other words, the gradient calculations are not performed independently for each layer.

Why is the error of the last layer computed first? The simple reason is that the hidden layers do not have target variables. It is the output layer that maps to the target variables present in the labeled dataset. Therefore, it is logical to calculate the errors at the last layer initially.

This section (Introduction to Artificial Neural Networks) provided an overview of how neural networks function and the underlying algorithms involved. It also highlighted the various parameters, such as learning rates and momentum, that can be controlled to fine-tune the training process. These adjustable parameters are known as *hyperparameters* and will be covered in depth later in this chapter.

In the following sections, we will write code to implement the neural network concepts covered so far in this chapter. We will use Python, along with TensorFlow, to work through the examples. We will begin with a high-level introduction to TensorFlow, focusing on features and functions relevant to computer vision, and then discuss specific implementations of neural network concepts.

Introduction to TensorFlow

TensorFlow is a comprehensive machine learning platform that is open source and supports end-to-end model development. It offers a user-friendly and intuitive API for creating machine learning models. TensorFlow serves as the execution engine for Keras, a Python-based high-level neural network API.

In this second edition of the book, we leverage the advanced capabilities of TensorFlow, with particular focus on the update functions for object detection tasks. The introduction of these new functionalities enhances the book's coverage of object detection, allowing you to explore the latest developments in this field using TensorFlow's powerful tools.

TensorFlow Installation

If you followed the TensorFlow installation instructions in Chapter 1, you should have TensorFlow and Keras installed in your working environment. If you skipped those instructions, refer to Chapter 1 now and follow the installation instructions for TensorFlow.

How to Use TensorFlow

To use TensorFlow in your code, you must import it as follows:

```
import tensorflow as tf
```

You can access Keras API by using the following:

```
tf.keras
```

TensorFlow Terminology

Before you delve into the intricacies of implementing neural networks, you should be familiar with the key TensorFlow terminology covered here.

Tensor

A *tensor* is a data structure containing n-dimensional arrays of a base data type.

> If the value of n is 0, it's called a *scalar*, and the rank of the scalar is 0 or 0-dimensional.

> If the value of n is 1, it's called a *vector*, and the rank of the vector is 1 or 1-dimensional.

> If the value of n is 2, it's called a *matrix*, and the rank of the matrix is 2 or 2-dimensional.

> If the value of n is 3 or more, it's called a *tensor*. Depending on the value of n, its rank is 3 or more.

Hence, a tensor serves as an extension of vectors and matrices to higher dimensions. The distinctions between scalar, vector, matrix, and tensor are outlined in Table 5-1 for reference.

Table 5-1. *Definitions of Scalar, Vector, Matrix, and Tensor*

Data Structure	Dimension or Rank (the Value of *n*)	Example
Scalar	0	`scalar_s = 231`
Vector	1	`vector_v = [1,2,3,4,5]`
Matrix	2	`matrix_m =` `[[1,2,3],[4,5,6],[7,8,9]]`
Tensor	3 or more	`tensor_3d = [` `[[1,2,3],    [4,5,6],` `[7,8,9]],` `[[11,12,13], [14,15,16],` `[17,18,19]],` `[[21,22,23], [24,25,26],` `[27,28,29]],` `]`

TensorFlow internally defines, manipulates, and performs computations on tensors. To facilitate this functionality, it offers a `Tensor` class that can be accessed using the following:

`tf.Tensor`

The `Tensor` class has the following properties:

- A data type, e.g., `uint8`, `int32`, `float32`, or `string`. Every element of a tensor must be of the same data type.

- A shape, which is the number of dimensions and size of each dimension.

Tensors can be created using various methods in TensorFlow, such as by explicitly defining the values, by initializing them with specific distributions, or by performing operations on existing tensors. Tensors are central to the implementation of neural networks and other machine learning algorithms in TensorFlow.

Variable

TensorFlow has a class called Variable, accessible by using tf.Variable. The tf. Variable class represents a tensor whose values are manipulated by operations such as read and modify. As you will see later in this chapter, tf.keras employs tf.Variable to store model parameters. Listing 5-1 shows a Python example to illustrate how to use a variable.

Constant

TensorFlow also provides support for constants, which are tensors whose values remain fixed once initialized and cannot be altered. The function tf.constant() returns a constant tensor. To create a constant in TensorFlow, use the following function:

tf.constant(value, dtype=None, shape=None, name='Const')

where value is the actual value or a list that is set as the constant; dtype is the data type of the resulting tensor represented by the constant; shape is an optional parameter and represents the dimensions of the resulting tensor; and name is the name of the tensor. If you do not specify the data type, tf.constant() will infer it from the value of the constant.

Listing 5-1 shows a simple code example that creates a tensor variable.

Listing 5-1. Creating a Tensor Variable

```
Filename: Listing_5_1.py
1    import tensorflow as tf
2
3    # create a tensor variable with zero filled with default
     datatype float32
4    a_tensor = tf.Variable(tf.zeros([2,2,2]))
5
6    # Create a 0-D array or scalar variable with data type tf.int32
7    a_scalar = tf.Variable(200, tf.int32)
8
9    # Create a 1-D array or vector with data type tf.int32
10   an_initialized_vector = tf.Variable([1, 3, 5, 7, 9, 11], tf.int32)
11
```

```
12   # Create a 2-D array or matrix with default data type which is
     tf.float32
13   an_initialized_matrix = tf.Variable([ [2, 4], [5, 25] ])
14
15   # Get the tensor's rank and shape
16   rank = tf.rank(a_tensor)
17   shape = tf.shape(a_tensor)
18
19   # Create a constant initialized with a fixed value.
20   a_constant_tensor = tf.constant(123.100)
21   print(a_constant_tensor)
22   tf.print(a_constant_tensor)
```

Line 1 of Listing 5-1 imports the TensorFlow package. Line 4 creates a tensor with shape [2,2,2] filled with zeros. By default, it creates a tensor of data type tf.float32 (if no data type is specified while creating the tensor, it will default to float32). However, the data type is inferred from the initial value.

Line 7 creates a scalar data with type int32, line 10 creates a vector with data type int32, and line 13 creates a 2×2 matrix with the default data type float32.

Line 16 shows how to get the tensor's rank (see Table 5-1), and line 17 shows how to obtain the shape.

Line 20 creates a constant tensor with a value initialized as 123.100. Its data type is interpreted by the value it is initialized with.

Lines 20 and 21 show two different ways of printing the tensor. Execute the code and notice the difference between the two print statements.

To evaluate a tensor, use the Tensor.eval() method, which creates an equivalent NumPy array with the same shape as the tensor. Note that the tensor is evaluated only when the default tf.Session is active.

This book does not focus extensively on TensorFlow. Instead, it concentrates on the specific TensorFlow features that are necessary for developing computer vision and deep learning models. To learn more about working with TensorFlow's Python functions, visit the official TensorFlow website. You can find the API specification at https://www.tensorflow.org/api_docs/python/tf. We will revisit TensorFlow throughout the rest of the book.

Our First Computer Vision Model with Deep Learning: Classification of Handwritten Digits

We have reached the stage where we can begin constructing and training our initial computer vision model. Our initial focus will be on creating a simple multilayer perceptron classifier, often compared to the "Hello World" program in deep learning. Through this process, you will gain hands-on experience in building an operational computer vision model. As in earlier sections, I will provide a comprehensive explanation of the TensorFlow code line by line. But first, let's discuss the computer vision model we aim to create and get an overview of the steps necessary to create it.

Model Overview

Our objective is to train a model to classify images of handwritten digits (0 to 9) using an artificial neural network.

We will construct a neural network that performs supervised learning. As discussed in Chapter 4, for supervised learning, we need a dataset that contains labeled data. In other words, we need images that are already labeled with the digits they represent. For example, if an image contains the handwritten digit 5, it will be labeled with 5. Similarly, all images we want to use in the training must be labeled with corresponding numbers.

Our dataset has ten classes, one class for each digit. The class index starts with 0. Therefore, our classes are in the range (0,9).

The labeled image dataset is divided into two parts, typically in a 70:30 ratio:

- *Training set*: The 70 percent labeled images are used for actual training.

  To achieve desirable outcomes, it is important to have balanced training data, ensuring nearly equal representation of all classes. If the training set lacks balanced classes, the majority class will have a stronger influence on the model, potentially leading to infrequent or nonexistent predictions for the minority class. To address this issue, various techniques can be employed to balance the classes, such as oversampling and undersampling. In *oversampling*, additional instances of the minority classes are introduced, thereby increasing their representation to approximate that of the majority classes. Conversely, in *undersampling*, images from the majority class are removed to align its quantity with that of the minority class.

207

There are other synthetic methods to balance the classes. The synthetic minority oversampling technique (SMOTE) is one such method but is not recommended for computer vision. However, the research paper at `https://arxiv.org/pdf/1710.05381.pdf` concludes that the undersampling performs on par with oversampling and therefore should be preferred for computational efficiency.

- *Test set*: 30 percent of the labeled data is typically used as a test set. Images from the test set are passed through the trained model, and the predicted results are compared to the labels to assess the model accuracy.

  Maintaining a distinct set of images between the training and test sets is crucial. Additionally, it is important to ensure that the test set comprises all classes in equal proportions.

We will perform the following tasks to build the model:

1. Download the image dataset containing handwritten digits with their labels from `https://storage.googleapis.com/tensorflow/tf-keras-datasets/mnist.npz`.

2. Configure a multilayer perceptron (MLP) classifier with four layers: the input layer, two hidden layers, and the output layer.

3. Fit the MLP model with the training set. (*Fitting* the model means training the model.)

4. Evaluate the trained model using the test set.

5. Predict using the model on a different dataset (not used in the training or test sets) and display the result.

Model Implementation

Finally, we have arrived at a point where we look at the TensorFlow code line by line to learn how to train a deep learning–based model for computer vision for classifying handwritten digits.

Let's explore Listing 5-2, which demonstrates how to train a deep learning–based computer vision model.

Listing 5-2. Four-Layer MLP for Classification of Images with Handwritten Digits

```
Filename: Listing_5_2.py
1     import tensorflow as tf, numpy as np, pandas as pd
2     import matplotlib.pyplot as plt
3     # Load MNIST data using built-in datasets download function
4     mnist = tf.keras.datasets.mnist
5     (x_train, y_train), (x_test, y_test) = mnist.load_data(path="/content/
      data/mnist.npz")
6
7     # Normalize the pixel values by dividing each pixel by 255
8     x_train, x_test = x_train / 255.0, x_test / 255.0
9
10    # Build the 4-layer neural network (MLP)
11    model = tf.keras.models.Sequential([
12      tf.keras.layers.Flatten(input_shape=(28, 28)),
13      tf.keras.layers.Dense(128, activation='relu'),
14      tf.keras.layers.Dense(60, activation='relu'),
15      tf.keras.layers.Dense(10, activation='softmax')
16    ])
17
18    # Compile the model and set optimizer, loss function, and metrics
19    model.compile(optimizer='adam',
20                  loss='sparse_categorical_crossentropy',
21                  metrics=['accuracy'])
22
23    # Finally, train or fit the model
24    trained_model = model.fit(x_train, y_train, validation_split=0.3,
      epochs=100)
25
26    # Visualize loss and accuracy history
27    plt.plot(trained_model.history['loss'], 'r--')
28    plt.plot(trained_model.history['accuracy'], 'b-')
29    plt.legend(['Training Loss', 'Training Accuracy'])
30    plt.xlabel('Epoch')
```

```
31   plt.ylabel('Percent')
32   plt.show();
33
34   # Evaluate the result using the test set.\
35   evalResult = model.evaluate(x_test,  y_test, verbose=1)
36   print("Evaluation", evalResult)
37   predicted = model.predict(x_test)
38   predicted_df = pd.DataFrame(predicted)
39   print("Probabilities of all classes predicted from the model:")
40   print(predicted_df)
41   print("Predicted class based on the highest probability:")
42   predicted_df.idxmax(axis=1)
43
44   print("===Confusion Matrix==")
45   confusion = tf.math.confusion_matrix(y_test, np.argmax(predicted,
     axis=1), num_classes=10)
46   tf.print(confusion)
47   tensor_values = confusion.numpy()
48   confusion_df = pd.DataFrame(tensor_values)
49   print(confusion_df)
```

To make it simpler for reference, the following explanation of Listing 5-2 has bold text markers to assist in navigating the code sections.

Imports: Line 1 imports the TensorFlow package, which gives access to the Keras deep learning library and several other deep learning–related functions. Line 2 imports matplotlib.

Dataset: Line 4 initializes the keras.datasets.mnist module, which provides a builtin function to download the Modified National Institute of Standards and Technology (MNIST) handwritten digit image data. The MNIST database is a large collection of handwritten digits that is widely used for training various computer vision systems. More information about the dataset is and tools to download it is available at the official Keras website, https://keras.io/api/datasets/mnist/. However the official website of MNIST dataset is https://yann.lecun.com/exdb/mnist/.

Line 5 downloads the MNIST dataset. The load_data() function in the mnist module downloads the digits database and returns a tuple of NumPy arrays. By default, it will download the database in your home directory location, ~/.keras/datasets, with a default file name of mnist.npz. You can download it to any other location by providing an absolute file path, for example, in the function load_data(path='/absolute/path/ mnist.npz'). Make sure that the directory already exists.

The load_data() function returns a tuple of NumPy arrays, as described here:

> x_train: Contains pixel values of images that we will use for training

> y_train: Contains the labels for each image in x_train

> x_test and y_test: Contain the pixel values of images and corresponding labels for the test dataset

Data normalization: We need to normalize the pixel values so that they are between 0 and 1. Dividing each pixel by 255 will normalize it, as shown in line 8. The x_train and x_test NumPy arrays are divided by a scalar 255 to normalize these arrays.

In this example, we are downloading a publicly available dataset using built-in functions in TensorFlow. If you have data in your local disk or any distributed file system, TensorFlow provides functions to load the data. You will see how to load the file from the local file system later in this chapter.

Neural network: Lines 11 to 16 encompass the definition of our neural network, which, for the sake of clarity, is expressed as a single statement but broken into multiple lines. Let's examine the various components of this definition:

- tf.keras.models.Sequential: This is a TensorFlow class that provides a function to create layers of the neural network. In this example, we are creating four layers and passing it as an array to the constructor of the Sequential class.

- tf.keras.layers: This module provides APIs to create different types of neural network layers. In this example:

  - tf.keras.layers.Flatten(input_shape=(28, 28)) defines the input layer by initializing the Flatten() function. The input images are 28×28 pixels with a single channel. The argument to this function is the input shape. This flatten function will create 28×28 = 784 neurons in the input layer. Remember, the number of neurons in the input layer is the same as the number of features (plus 1 if a bias is used). Our digit images are 28×28 pixels, and each pixel value is taken as an input feature; hence, the number of nodes in this layer is 784. We will see more examples with complex features later in this chapter. Let's keep things simple for now.

— `tf.keras.layers.Dense` creates a dense layer in the neural network. The dense layer takes two important parameters: the number of neurons and the activation function. Notice that we have three types of dense layers in the neural network in Listing 5-2:

Hidden layer 1: The number of neurons is 128, and the activation function is `relu`.

Hidden layer 2: The number of neurons is 60, and the activation function is `relu`.

Output layer (the last layer): The number of neurons is 10, and the activation function is `softmax`.

Activation function: Why is the activation function in the hidden layers `relu`? Recall from the "Activation Functions" section and Figure 5-10 that ReLU always generates output in the range from 0 to infinity and does not generate any negative number. The pixel values, after normalization, are in the range (0,1). Therefore, RELU makes perfect sense for this layer.

Why is `softmax` in the output layer? Remember, softmax generates probability distributions of the neuron outputs. The output layer generates probabilities of each class. In this example, for each input image, it will generate ten probabilities, one for each class. The sum of these probabilities will be equal to 1. The class with the highest probability is generally taken as the predicted class for the input image.

Number of neurons: Why do we have only ten neurons in the output layer? It's because we have only ten digits to be predicted, and the output layer for classification problems should have the same number of neurons as the number of classes to be predicted.

Optimizer: Lines 19 through 21 call the `compile()` function to build the neural network with the configuration we provided earlier. The function `compile()` takes the following:

- `optimizer = 'adam'`: The name of the optimization function that will try to find the minimum of the loss function.

- `loss = 'sparse_categorical_crossentropy'`: The loss function that will be optimized. This is a multiclass classification, and the `sparse_categorical_crossentropy` loss function is our choice.

- `metrics= ['accuracy']`: A list of metrics to be evaluated by the model during training and testing. Since we have a single output model and it's a classification problem, we pass only one metric, the "accuracy," in this list.

Model fitting: Line 24 actually fits the model. When this line executes, the model starts learning. The `model.fit()` takes these arguments:

- `x_train`: NumPy representation of normalized values of the pixels

- `y_train`: NumPy of the labels

- `validation_split = 0.3`: Tells the algorithm to hold 30 percent off the training data to use for validation

- `epochs = 100`: Number of training iterations

If you want to use your test dataset, or any other dataset that you have access to, for validation, instead of `validation_split`, you could use `validation_data=(x_val, y_val)`.

Hyperparameters: The question is, how many iterations or epochs should we use to train our model? Generally, it takes more than one iteration for the neural network to learn. This is one of those parameters that you will need to tune. When your model starts learning, you will see the output printed in the console (e.g., the PyCharm console, if you execute the code in PyCharm). It shows the loss and accuracy for each epoch. With each epoch, the loss should decrease and the accuracy should increase. If you start noticing that the loss no longer decreases or the accuracy no longer increases, you should set your epoch value at that level.

The hyperparameter of epoch needs to be predetermined before commencing the training process. As a result, it is challenging to determine the optimal number of epochs without actually executing the training. To overcome this challenge, TensorFlow offers a technique called *early stopping*, which is used during model training to prevent overfitting and determine the optimal number of training iterations. The concept behind early stopping is to monitor the performance of the model on a validation set during training and stop the training process if the performance starts to deteriorate.

The validation set is a separate portion of the labeled data that is not used for training but is used to evaluate the model's performance during training. The performance metric (such as accuracy or loss) on the validation set is tracked at regular intervals.

Early stopping works by comparing the current performance metric with the best observed performance so far. If the performance metric on the validation set does not improve or starts to worsen over a predefined number of consecutive epochs, the training is stopped early, assuming that the model has reached its optimal state and further training would only lead to overfitting.

By stopping the training early, early stopping helps in preventing the model from memorizing the training data too closely, improving its ability to generalize and perform well on unseen data.

In TensorFlow, early stopping can be implemented using callback functions, such as the `EarlyStopping` callback provided by the `tf.keras` module. This callback can be configured with parameters like the monitored metric, the number of epochs to wait for improvement, and the mode (minimization or maximization) of the monitored metric.

Figure 5-17 shows a sample training output with 100 epochs.

```
Train on 42000 samples, validate on 18000 samples

Epoch 1/100

42000/42000 [==============] - 5s 126us/sample - loss: 0.2858 - accuracy: 0.9165 - val_loss: 0.1709 - val_accuracy:
0.9484

Epoch 2/100

42000/42000 [==============] - 4s 90us/sample - loss: 0.1196 - accuracy: 0.9644 - val_loss: 0.1424 - val_accuracy:
0.9588




.....

Epoch 99/100

42000/42000 [==============] - 4s 91us/sample - loss: 0.0064 - accuracy: 0.9987 - val_loss: 0.3400 - val_accuracy:
0.9752

Epoch 100/100

42000/42000 [==============] - 4s 106us/sample - loss: 0.0027 - accuracy: 0.9991 - val_loss: 0.3492 - val_accuracy:
0.9742
```

Figure 5-17. *Sample console output with loss and accuracy per epoch*

Learning curve: Lines 27 to 32 plot graphs of loss versus epoch and graphs of accuracy versus epoch to understand how well the training occurred. The trained model maintains a history of losses and accuracy per epoch that is accessible by using `history['loss']` and `history['accuracy']`.

The example plot in Figure 5-18 shows that the loss, indicated by the red line, exhibits a consistent decline with each epoch until around the tenth epoch, at which point the loss begins to plateau, indicating that further iterations are unlikely to yield significant reductions. Consequently, to optimize computational resources, it is advisable to set the epoch to approximately 10 (in this example) and refrain from additional computations.

Similarly, the accuracy level in Figure 5-18 (blue line) increases and becomes flat after a few epochs. Both of these—loss and accuracy—help us determine the number of iterations for training the neural network.

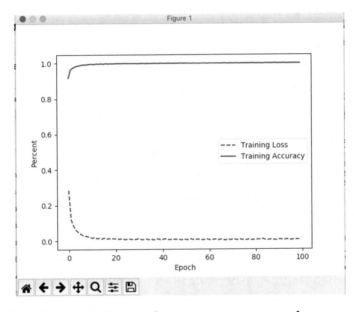

Figure 5-18. *Plot of training loss and accuracy versus epoch*

We can print all the keys within the `History` object by calling `history.keys()`. We may also want to plot the `val_acc` and `val_loss` graphs to see how the model evaluates against the 30 percent validation data.

Evaluation: Line 35 evaluates the model against the test dataset. We use the `evaluate()` function that takes these parameters:

- `x_test`: NumPy containing normalized pixel values of all test images

- `y_test`: NumPy containing labels for the test dataset

- `verbose =1`: Optional parameter to print the output

Line 36 prints the sample output, shown in Figure 5-19, which indicates that our model achieves an accuracy of 0.9787 or 97.87 percent (and a loss value of 0.2757) when evaluated on the test dataset. This level of accuracy is deemed satisfactory, indicating that our model performs well.

Evaluation [0.2757401464153796, 0.9787]

Figure 5-19. *Evaluation output*

If you have a test dataset, like the one we have in this example, we do not need to hold 30 percent off the training set, as in line 24. The parameter `validation_split=0.3` is optional if you want to perform the evaluation using the test data like we did in line 35.

Prediction: In Line 37, we use the trained model to predict the classes of input images that were not used in model training. Any new image (with a normalized NumPy of pixel values) can be fed to the model to predict its class.

To predict a class, we use the function `model.predict()`, which takes the image NumPy as a parameter. The output from the `predict()` function is a NumPy of arrays. The elements of this array are probabilities of each class. The index of the max probability is the predicted class of that image.

For example, the input image with a handwritten digit gets the prediction probabilities shown in Figure 5-20. Starting from zero, the sixth index (highlighted in yellow) has the highest probability of 0.99844. Therefore, the predicted class for the input image is 7, which matches with the handwritten digits, as shown on the left in the figure.

[1.8943774e-06, 4.848908e-06, 0.00090997526, 0.00060881954, 5.6300826e-07, 1.5920621e-07, *0.998444*, 3.4792773e-09, 1.1292449e-05, 1.8514449e-05]

Figure 5-20. *Input image and prediction probabilities*

In Line 38, we are converting the NumPy array of the predicted class probabilities into a pandas DataFrame. In Line 42, we are using the pandas function `idxmax(axis=1)` to get the column index with the highest probability. The argument `axix=1` indicates we are interested in the column index and not the row index.

Lines 44 through 49 are described in the "Evaluation Metrics" section later in this chapter in the context of the confusion matrix.

Congratulations! You have successfully constructed and trained your first neural network for computer vision. The next section explains techniques for evaluating the effectiveness of our model, distinguishing between good and subpar performance. After that, we will explore methods to fine-tune parameters, aiming to enhance our model by achieving lower loss and higher accuracy levels.

Model Evaluation

Once we have trained a model, it is crucial to evaluate its performance by analyzing the loss and accuracy metrics. However, it is important to note that high accuracy and low loss on the training data do not guarantee the same level of accuracy when encountering new data. To ensure reliable predictions, it is necessary to assess the model's performance using a separate set of test data, distinct from the training set. Here are some commonly employed evaluation methods that are prevalent in practice.

Overfitting

An overfit model is characterized by its ability to predict with a higher accuracy level when presented with training data compared to prediction accuracy when validation and test data are used as inputs. For instance, if a model achieves a high accuracy of 97 percent on the training set but exhibits a lower accuracy of 70 percent on the test or validation set, it is indicative of overfitting. Figure 5-21 provides a visual representation of overfitting, where the test accuracy quickly falls below the training accuracy.

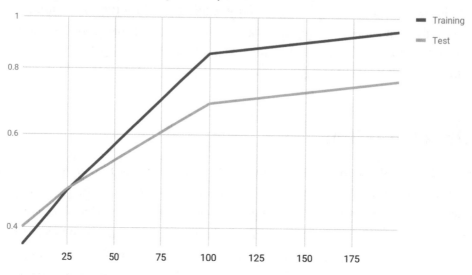

Figure 5-21. *Example of overfitting*

How do you avoid overfitting? There are a few ways to control or avoid overfitting:

- *Regularization*: See the section "Regularization" earlier in this chapter.

- *Dropout*: As mentioned earlier, dropout serves as an effective method of regularization. With dropout, neurons are randomly dropped out, which means the output of the dropped-out neurons is not fed as input in the next layer. The dropout is temporary and applies to a particular pass only. That means weight updates are not applied to the temporarily removed neurons during that pass.

  In TensorFlow, dropout is implemented by adding a layer, called the *dropout layer*, and specifying dropout rate or probability (e.g., 20 percent). The dropout layer can be added either in the input layer or in the hidden layer. For most practical purposes, we keep this dropout probability small to avoid losing important features.

  The example code snippet presented in Listing 5-3 illustrates the layout of a TensorFlow neural network with a dropout layer.

Listing 5-3. Code Fragment to Show the Dropout Layer

```
....
model = tf.keras.models.Sequential([
 tf.keras.layers.Flatten(input_shape=(28, 28)),
 tf.keras.layers.Dense(128, activation='relu'),
 tf.keras.layers.Dropout(0.2),
 tf.keras.layers.Dense(60, activation='relu'),
 tf.keras.layers.Dense(10, activation='softmax')])
.....
```

Underfitting

A model is said to be *underfitting* when it cannot capture the underlying trend from the training data. An underfit model simply means that the model does not fit the data well enough. It usually happens either when we have a small dataset or when the dataset is not a true representation of the actual scenario we are trying to model. The accuracy of an underfit model is not good for both training and test sets. This kind of model should be avoided.

A good way to avoid underfitting is to add more data to the training set or have enough data that has all the variations and trends that we are trying to model. Also, feature engineering to select the right features helps to reduce underfitting.

Evaluation Metrics

To evaluate the quality of the model, it's crucial to consider additional significant metrics. The following descriptions outline these metrics, which are derived by comparing the predicted outcome with the corresponding label values from the test dataset.

- *True positive rate (TPR) or sensitivity*: In case of binary classification, if the label value is positive and the model also predicts a positive value, it is called a *true positive (TP)*. The TPR is defined as follows:

  TPR = total number of all TPs / total number of all positive cases

- *True negative rate (TNR) or specificity*: In case of binary classification, if the label value is negative and the model also predicts a negative value, it is called a true negative or TN. The TNR is defined as follows:

  TNR = total number of all TNs / total number of negative cases

- *False positive rate (FPR) of fallout*: The FPR is defined as follows:

  FPR = total number of false positive cases / total number of negative cases

- *False negative rate (FNR) or miss rate*: The FNR is defined as follows:

  FNR = total number of false negative cases / total number of positive cases

- *Confusion matrix*: A confusion matrix is also called an *error matrix*. It shows the number of positives and negatives of each class in a grid form. For example, if we have two classes, dog and cat, the confusion matrix may look like this:

	cat (predicted)	dog (predicted)
cat (actual)	80	10
dog (actual)	8	92

In this example, the cat class has 80 true positives, 10 false positives, and 8 false negatives. Similarly, for the dog class, there are 92 true positives, 8 false positives, and 10 false negatives.

Listing 5-4 shows lines 45 to 49 from Listing 5-2. These code segments calculate the confusion matrix, convert it into a NumPy array, transform it into a Pandas DataFrame, and ultimately print the resulting confusion matrix.

Listing 5-4. Confusion Matrix Calculation

```
.....
44    (print("===Confusion Matrix==")
45    confusion = tf.math.confusion_matrix(y_test, np.argmax(predicted,
      axis=1), num_classes=10)
46    tf.print(confusion)
47    tensor_values = confusion.numpy()
48    confusion_df = pd.DataFrame(tensor_values)
49    print(confusion_df)
.....
```

Line 37 of Listing 5-2, predicted = model.predict(x_test), uses the test dataset to predict from the model. The output is a NumPy array of probabilities for each input. np.argmax(predicted, axis=1) gets the index of the max probabilities in the array. The index represents the predicted class.

In Listing 5-4, the `tf.math.confusion_matrix()` function in line 45 calculates the confusion matrix. It takes these arguments:

- `y_test`: The actual class labels of the test set.

- `np.argmax(predicted, axis=1)`: The predicted class

- `num_classes = 10`: Optional argument that represents the number of classes we want our model to predict

The `confusion_matrix()` function returns a tensor. If you print this tensor directly by using `print(confusion)`, it will not show you the values of the tensor. You will need to execute the tensor so that it calculates all the values before displaying to the console. That is the reason the tensor was first converted into a NumPy array and then transformed into Pandas DataFrame (Line 48) for the purpose of printing (Line 49).

Figure 5-22 shows a sample confusion matrix from the test set we used in this example.

	0	1	2	3	4	5	6	7	8	9
0	965	1	2	1	1	6	2	0	1	1
1	0	1116	3	1	0	0	3	1	11	0
2	4	2	1007	2	1	0	2	5	9	0
3	0	0	5	987	0	5	0	5	7	1
4	1	1	1	1	947	1	9	3	5	13
5	2	0	0	10	1	867	3	1	6	2
6	3	3	1	1	1	4	943	0	2	0
7	1	6	7	1	0	1	0	1003	3	6
8	0	0	3	4	1	2	0	1	961	2
9	3	3	2	2	10	3	0	4	5	977

Figure 5-22. *Confusion matrix output sample*

In Figure 5-22, the diagonal values represent the true positive values for the classes aligned with the diagonal. On the left and right of the diagonal, you can find the false positive values, while above and below the diagonal lie the false negative values. As an example, consider the number 965, which corresponds to the zeroth class. This value represents the true positive for that class. All the numbers in the same row are false positives, and all the numbers vertically below it are false negatives for the zeroth class.

- *Precision*: Precision is the ratio of total number of true positives and total number of predicted positives. The formula to calculate precision is presented below:

  Precision = total number of true positives / total number of predicted positives

  = true positives / (true positives + false positives)

  = TP / (TP + FP)

  Ideally, a model should have zero false positives (FP = 0). In this case, the precision equals 1, indicating a perfect score or 100 percent precision. Essentially, the higher the precision, the better the model's performance.

- *Recall*: Recall is the ratio of total number of true positives and total number of actual positives. Recall is the same as the true positive rate. The formula to calculate recall is as follows:

  Recall = total number of true positives / total number of positives

  = total number of true positives / (total number of true positives + total number of false negatives)

  = TP / (TP + FN)

  Ideally, a model should have zero false negatives (FN = 0). When this condition is met, the recall value is 1, indicating perfect performance or 100 percent recall. In other words, the higher the recall, the better the model's performance.

F1 score: When evaluating a model, it is important to consider both precision and recall. As just described, both of these metrics should be close to 100 percent for an ideal model. However, if one of these metrics is smaller than the other, judging the overall performance of the model can be challenging. The F1 score addresses this situation by combining precision and recall into a single composite metric, providing a balanced assessment of the model's quality. By considering both precision and recall, the F1 score helps you to make an informed judgment about how good or bad the model is. The F1 score is the harmonic mean of precision and recall and is calculated by using the following formula:

F1 score = 2 × precision × recall / (precision + recall)

- *Accuracy*: Accuracy is defined as the ratio of the total number correct predictions of both positive and negative classes and the total sample size. The formula to calculate the accuracy is shown below:

Accuracy = (TP + TN) / total sample count

= (TP + TN) / (P + N)

= (TP + TN) / (TP + TN + FP + FN)

The metrics covered in this section play a crucial role in the decision-making process regarding whether to deploy the model in a production environment or to fine-tune parameters and retrain the model. By analyzing these metrics, we can assess the model's performance and determine the necessary steps to optimize its effectiveness.

Hyperparameters

Hyperparameters are those parameters to the neural network model that we set before the learning process starts. These are considered external parameters as opposed to the parameters that the algorithm computes from the training data. Hyperparameters are not inferred by the algorithm while the model is being trained. These hyperparameters affect the overall performance of the model, including the accuracy and training execution time.

In the previous sections, we have already discussed the meaning and significance of various hyperparameters used in training neural networks for computer vision. Here are some of the commonly encountered hyperparameters that require tuning:

- Number of hidden layers in the network

- Number of neurons in the hidden layers

- Dropout and learning rates

- Optimization algorithms

- Activation functions

- Loss functions

- Epochs or number of iterations

- Split for validation set

- Batch size

- Momentum

TensorBoard

TensorBoard is a web-based visualization tool provided by TensorFlow that can be used to visualize and understand the training process and results of machine learning models. Here are several reasons why TensorBoard is popular and commonly used:

- *Model visualization*: TensorBoard allows you to visualize your model's architecture, such as the graph structure, layer shapes, and parameter statistics. This helps you understand and debug your model more effectively.

- *Training monitoring*: During the training process, TensorBoard can display real-time visualizations of various metrics like loss, accuracy, and other custom metrics. It enables you to track the progress of your model's training and identify potential issues or improvements.

- *Hyperparameter tuning*: TensorBoard can help you optimize your model's hyperparameters by visualizing and comparing different runs. You can examine how changing hyperparameters affects the performance and select the best configuration.

- *Embedding visualization*: If you're working with high-dimensional data or embeddings, TensorBoard provides embedding visualizations that allow you to explore and understand the relationships between data points in lower-dimensional spaces.

- *Debugging and profiling*: TensorBoard provides tools for debugging and profiling your model's performance. You can analyze the execution time of different operations, identify bottlenecks, and optimize your code for better performance.

Overall, TensorBoard enhances the workflow of machine learning practitioners by providing an intuitive and interactive interface for visualizing and analyzing models, training progress, and various metrics. It promotes better understanding, faster debugging, and more informed decision-making throughout the machine learning development cycle.

TensorBoard provides an HParams dashboard that helps us identify the best experiment or most promising sets of hyperparameters. Using the neural network example from the previous section, we can leverage the HParams dashboard to visualize different hyperparameters and gain insights into how they should be tuned for optimal performance.

Before we work through the following example, check whether you have TensorBoard installed. To verify if TensorBoard is installed in your virtual environment, execute the following command (ensure that you are in the same virtual environment, cv, that we used throughout this book):

```
(cv) username $: tensorboard --logdir my logdir
```

If everything goes well, you should see output similar to this:

```
TensorBoard 2.12.2 at http://localhost:6006/ (Press CTRL+C to quit)
```

Point your browser to http://localhost:6066, and you should see the TensorBoard web UI.

To visualize TensorBoard embedded within the Google Colab environment, load the TensorBoard extension and then run the tensorboard command as shown here:

```
%load_ext tensorboard
%tensorboard --logdir mydir
```

Experiments for Hyperparameter Tuning

The code example in Listing 5-5 demonstrates a simple experiment with only three hyperparameters for a simple neural network. The example is kept simple for learning purposes.

Our goal is to conduct experiments with the following parameters:

- Number of neurons in the first hidden layer

- Optimization functions

- Dropout rates

After the experiments are complete, we want to visualize the result in the TensorBoard web UI and use an HParams dashboard to analyze the result.

Listing 5-5 shows the code flow.

Listing 5-5. Hyperparameter Tuning and Visualization on HParams of TensorBoard

```
Filename: Listing_5_5.py
1    import tensorflow as tf
2    from tensorboard.plugins.hparams import api as hp
3
4    # Load MNIST data using built-in datasets download function
5    mnist = tf.keras.datasets.mnist
6    (x_train, y_train), (x_test, y_test) = mnist.load_data()
7
8    x_train, x_test = x_train / 255.0, x_test / 255.0
9
10   HP_NUM_UNITS = hp.HParam('num_units', hp.Discrete([16, 32]))
11   HP_DROPOUT = hp.HParam('dropout', hp.RealInterval(0.1, 0.2))
12   HP_OPTIMIZER = hp.HParam('optimizer', hp.Discrete(['adam', 'sgd']))
13
14   METRIC_ACCURACY = 'accuracy'
15
16   with tf.summary.create_file_writer('logs/hparam_tuning').as_default():
17     hp.hparams_config(
18       hparams=[HP_NUM_UNITS, HP_DROPOUT, HP_OPTIMIZER],
19       metrics=[hp.Metric(METRIC_ACCURACY, display_name='Accuracy')],
20     )
21
22
23   def train_test_model(hparams):
24       model = tf.keras.models.Sequential([
25           tf.keras.layers.Flatten(),
26           tf.keras.layers.Dense(hparams[HP_NUM_UNITS], activation=tf.
             nn.relu),
27           tf.keras.layers.Dropout(hparams[HP_DROPOUT]),
```

```
28        tf.keras.layers.Dense(10, activation=tf.nn.softmax),
29     ])
30     model.compile(
31         optimizer=hparams[HP_OPTIMIZER],
32         loss='sparse_categorical_crossentropy',
33         metrics=['accuracy'],
34     )
35
36     model.fit(x_train, y_train, epochs=5)
37     _, accuracy = model.evaluate(x_test, y_test)
38     return accuracy
39  def run(run_dir, hparams):
40    with tf.summary.create_file_writer(run_dir).as_default():
41      hp.hparams(hparams)  # record the values used in this trial
42      accuracy = train_test_model(hparams)
43      tf.summary.scalar(METRIC_ACCURACY, accuracy, step=1)
44
45  session_num = 0
46
47  for num_units in HP_NUM_UNITS.domain.values:
48    for dropout_rate in (HP_DROPOUT.domain.min_value, HP_DROPOUT.domain.
      max_value):
49      for optimizer in HP_OPTIMIZER.domain.values:
50        hparams = {
51            HP_NUM_UNITS: num_units,
52            HP_DROPOUT: dropout_rate,
53            HP_OPTIMIZER: optimizer,
54        }
55        run_name = "run-%d" % session_num
56        print('--- Starting trial: %s' % run_name)
57        print({h.name: hparams[h] for h in hparams})
58        run('logs/hparam_tuning/' + run_name, hparams)
59        session_num += 1
```

Lines 5 through 8 load the same MNIST digits data that we worked with before.
Line 10 sets the values for the number of neurons or units: 16 and 32.

Line 11 sets the dropout rates: 0.1 and 0.2.

Line 12 sets the optimization functions: adam and sgd.

Notice that the model.fit() function is called within a nested for loop (lines 47 through 59) for each combination of the three hyperparameters. The metrics output is written in a log file logs/hparam_tuning. The rest of the code structure is straightforward and does not need any explanation.

After the experiments are executed successfully, launch TensorBoard by using the following command (again, ensure you are in the virtual environment called cv):

```
(cv) username $: tensorboard -logdirlogs/hparam_tuning
```

You may have to pass the absolute path to the logs/hparam_tuning directory.

Launch the browser and point to http://localhost:6006. You should see the TensorBoard web UI. From the top-right drop-down, select HPARAMS. You should see the dashboard similar to the one in Figure 5-23.

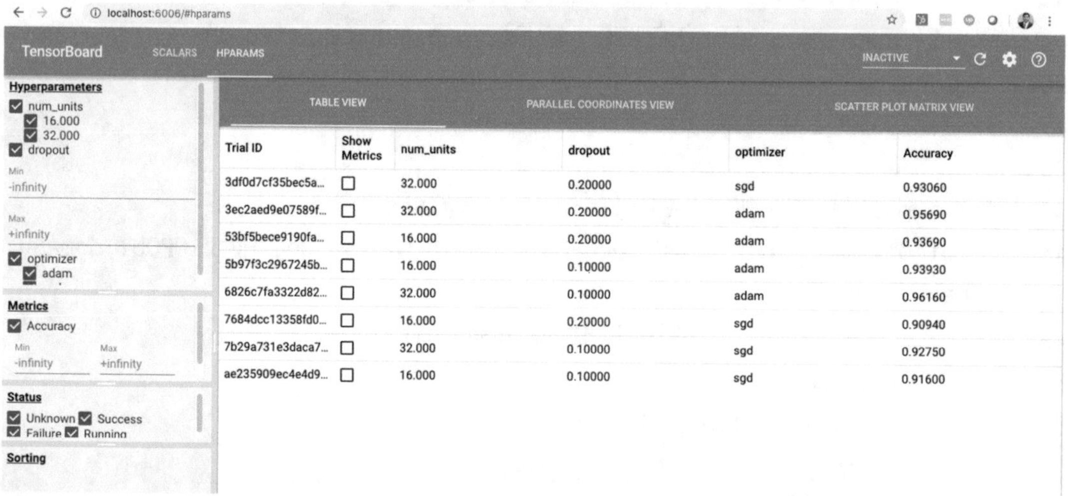

Figure 5-23. *TensorBoard showing the HPARAMS view containing accuracies corresponding to each combination of hyperparameters*

From this dashboard, you can see the combination of hyperparameters that gives the highest accuracy: 96.160 percent accuracy for 32 neurons, 0.1 dropout, and the use of Adam optimizer.

Alternatively, click the Parallel Coordinates View tab to change to the screen shown in Figure 5-24.

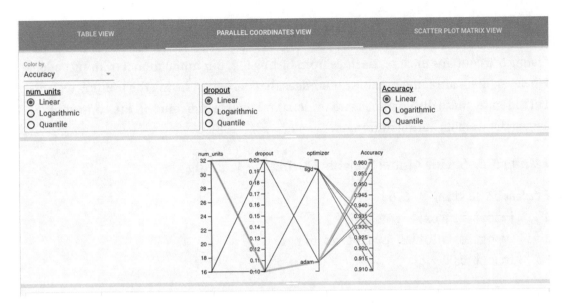

Figure 5-24. *Parallel Coordinates View tab of HPARAMS*

As shown in Figure 5-24, clicking the link to the highest accuracy (or any accuracy that you want to examine), you will see the green highlighted path that represents the combination of hyperparameters that generated the accuracy.

Saving and Restoring Model

Saving and loading trained models is crucial to avoiding the need for retraining every time you want to use them. Model training can be time-consuming, taking hours or even days depending on factors like data size, hardware capacity, and network configuration. Therefore, it is essential to save models during and after training. By doing so, you can resume training from the point it was interrupted, saving valuable time that would otherwise be lost.

In the following sections, we will delve into the process of training a neural network, saving model checkpoints during training, saving the final model for future use, loading the saved model, and effectively utilizing it in various applications. This enables us to harness the benefits of the trained model without undergoing the entire training process repeatedly.

Save Model Checkpoints During Training

Listing 5-6 contains most of the lines from Listing 5-2, our initial model training code. However, there are some notable differences that pertain to saving the training weights. Let's identify these distinct lines and understand their significance in the context of saving the weights during training.

Listing 5-6. Saving Model Weights During the Training

```
Filename: Listing_5_6.py
1    import tensorflow as tf
2    import matplotlib.pyplot as plt
3    import os
4
5    # The file path where checkpoint will be saved.
6    checkpoint_path = "cv_checkpoint_dir/mnist_model.ckpt"
7    checkpoint_dir = os.path.dirname(checkpoint_path)
8
9    # Create a callback that saves the model's weights.
10   cp_callback = tf.keras.callbacks.ModelCheckpoint(filepath=check
     point_path,
11   save_weights_only=True,
12   verbose=1)
13
14   # Load MNIST data using built-in datasets download function.
15   mnist = tf.keras.datasets.mnist
16   (x_train, y_train), (x_test, y_test) = mnist.load_data()
17
18   # Normalize the pixel values by dividing each pixel by 255.
19   x_train, x_test = x_train / 255.0, x_test / 255.0
20
21   # Build the ANN with 4-layers.
22   model = tf.keras.models.Sequential([
23     tf.keras.layers.Flatten(input_shape=(28, 28)),
24     tf.keras.layers.Dense(128, activation='relu'),
```

```
25    tf.keras.layers.Dense(60, activation='relu'),
26    tf.keras.layers.Dense(10, activation='softmax')
27    ])
28
29    # Compile the model and set optimizer, loss function, and metrics
30    model.compile(optimizer='adam',
31                    loss='sparse_categorical_crossentropy',
32                    metrics=['accuracy'])
33
34    # Finally, train or fit the model, pass callbacks to save the model
      weights.
35    trained_model = model.fit(x_train, y_train, validation_split=0.3,
      epochs=10, callbacks=[cp_callback])
36
37    # Visualize loss  and accuracy history
38    plt.plot(trained_model.history['loss'], 'r--')
39    plt.plot(trained_model.history['accuracy'], 'b-')
40    plt.legend(['Training Loss', 'Training Accuracy'])
41    plt.xlabel('Epoch')
42    plt.ylabel('Percent')
43    plt.show();
44
45    # Evaluate the result using the test set.
46    evalResult = model.evaluate(x_test,  y_test, verbose=1)
47    print("Evaluation Result: ", evalResult)
```

By comparing Listing 5-6 to Listing 5-2, we can identify the following different lines.

Line 3 imports the os package that provides file system–related functions that are used in saving the model to a file path.

Line 6 is the file name that will store our model weights.

Line 7 creates the operating system–specific file path object.

Line 10 initializes a TensorFlow callback class called ModelCheckpoint by passing the following arguments:

- filepath: This is the file path object that we created in line 7.

- save_weights_only: Instead of saving the entire model during the training, we should save the weights only. By default, this is set to False, which means save the entire model. By setting this to True, we let the neural network know that we want to save the weights only.

- verbose = 1: Prints the logs and runs the status on the console. Otherwise, the default 0 means silent.

The following are other arguments that we may want to pass based on what the intent is:

- save_best_only: This is False by default. If set to True, the algorithm will evaluate and save the best weights as determined by the metrics we pass.

- save_frequency: The default value is epoch, which means we want to save checkpoints at the end of every epoch. You can also pass an integer to indicate how frequently you want to save the checkpoints. For example, if you set save_frequency = 5, the checkpoints will be saved every fifth epoch.

You will notice that in Listing 5-6, almost all lines are the same as in Listing 5-2 except line 35, which fits the model. Line 35 has an additional argument to the fit() function, callbacks = [cp_callback], which saves the checkpoints during the model training.

Notice that we set epochs=10 as an argument to the fit() function, in Line 35 in Listing 5-6. Figure 5-25 and Figure 5-26 show sample output of loss and accuracy of this model. The model accuracy with test data is 0.9775, and the loss is 0.084755.

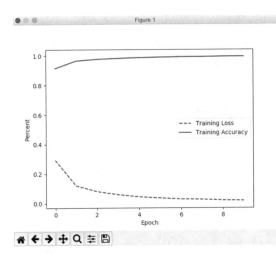

Figure 5-25. *Training loss and accuracy*

Evaluation Result:

[0.08475547107886305,
0.9775]

Figure 5-26. *Model evaluation with epoch=10*

Manually Save Weights

If you want to manually save the weights, instead of saving the checkpoint every epoch or periodically, you can simply add this function:

```
# Save the model weights
checkpoint_path = "cv_checkpoint_dir/mnist_model.ckpt"
model.save_weights(checkpoint_path)
```

Load the Saved Weights and Retrain the Model

If you want to load the saved weights either because you want to resume training after interruption or because you have more data (or for any other reason), simply add the following line after you have created/configured the neural network:

```
# Load saved weights
model.load_weights(checkpoint_path)
```

233

Make sure you have initialized your neural network like you did in lines 22 and 30 of Listing 5-6. It is important to note that the network architecture must be the same as the network that stored the checkpoints.

Saving the Entire Model

Call the `model.save()` function to save the entire model, including the model architecture, weights, and training configuration. Make sure to call the function `model.save()` after you call the `fit()` method. That is, call the `save()` function after line 35 of Listing 5-6. Here is the code snippet to save the entire model:

```
# Save the entire model to a directory path "path/to/save/model".
# You can also give the absolute pass to save the model.
tf.saved_model.save(model, "path/to/save/model")
```

The `tf.saved_model.save()` function saves the model in the `SavedModel` format, which is a serialized representation of the model architecture, weights, and training configuration.

After running this code, you will find the saved model in the specified directory. It will contain multiple files and directories that represent the saved model's assets, variables, and TensorFlow metadata.

Note that you can also save specific parts of the model, such as only the model's architecture or only the model's weights, using different functions like `model.save()` or `model.save_weights()`. However, using `tf.saved_model.save()` is recommended for saving the entire model in the `SavedModel` format, as it allows for easy deployment and compatibility with different TensorFlow platforms and versions.

Retraining the Existing Model

If you want to retrain an existing model with additional data or with the same data but different hyperparameters, you can employ the following code snippet:

```
# Load and create the existing model, including its weights and the
optimizer
model = tf.saved_model.load(' path/to/save/model')
```

```
# Show the model architecture
model.summary()
#Retrain the model
retrained_model = model.fit(x_train, y_train, validation_split=0.3,
epochs=10)
```

After retraining the model, make sure to save the new model using the `tf.saved_model.save()` function, as described in the previous section.

Using a Trained Model in Applications

If you have a trained model saved in the file system, you can easily load it and utilize the `predict()` function to make predictions. Here's an example:

```
# Load the saved model, including its weights and the optimizer
model = tf.saved_models.load(' path/to/save/model')
# Predict the class of the input image from the loaded model
predicted = model.predict(x_pixel_data)
print("Predicted", predicted)
```

Throughout our exploration, we have covered several key aspects of training neural networks for computer vision tasks. We began by demonstrating an example of training a neural network to predict digits using the MNIST image dataset. We then delved into saving the trained model, retraining it with additional data or different hyperparameters, and making predictions based on the retrained model.

Furthermore, we discussed the significance of monitoring model performance and introduced TensorBoard as a powerful tool for visualizing various model metrics and hyperparameters. By leveraging TensorBoard, we gained insights into the training process, identified areas for improvement, and made informed decisions to optimize our models.

By applying these techniques, we can build robust and effective computer vision models for a wide range of applications.

In the next section, we will explore convolutional neural networks (CNNs), a powerful architecture that combines traditional neural networks with an additional network responsible for automatic feature extraction. We will analyze the technique employed by CNNs to extract and select features from input images. Throughout this process, we will familiarize ourselves with commonly used terminologies associated with CNNs. Additionally, we will walk through a TensorFlow code implementation to train our own CNN model for image classification. As we proceed, I will provide a detailed explanation of each line of code to ensure a comprehensive understanding.

To demonstrate the practical application of CNNs, we will work on a specific example: classifying chest X-ray images to detect cases of pneumonia. By leveraging the power of CNNs, we aim to train a model that can automatically analyze and classify these medical images accurately.

Convolutional Neural Network

A *convolutional neural network (CNN)* is a special type of artificial neural network that differs from conventional artificial neural networks (ANNs) primarily through its *automatic feature engineering* capabilities. Whereas ANNs often require manual feature engineering, CNNs excel at automatically extracting and selecting relevant features from input data. This characteristic makes CNNs particularly well-suited for handling grid-like structured data, such as images.

By incorporating specialized layers known as *convolutional layers*, CNNs apply local operations on small regions of the input data. Through this process, the network learns to identify and extract essential features, such as edges, textures, or shapes, from the input data. The hierarchical architecture of CNNs enables them to progressively build higher-level representations by combining these learned features.

This automatic feature engineering aspect of CNNs significantly reduces the need for manual feature extraction and selection, allowing the network to learn and adapt to various patterns and complexities within the data. Consequently, CNNs have become a fundamental tool in computer vision tasks, such as image classification, object detection, and image segmentation.

Architecture of CNN

A conventional ANN or MLP consists of an input layer, one or more hidden layers, and an output layer. A CNN consists of a regular MLP -- consisting of an input layer, hidden layers and an output layer -- stacked on top of another network called convolutional network, that is responsible for automated feature extraction from input images. The input images are passed to the first layer of the convolutional network, as depicted in Figure 5-27.

The convolutional layer plays a crucial role in performing feature engineering on the input images. It applies a set of learnable filters or kernels to small regions of the input data, convolving them with the image to produce feature maps (as described further in the following sections). Each filter captures specific patterns or features present in the input images, such as edges or textures.

The output from the convolutional layer, which consists of these feature maps, is then fed into the "input" layer of the fully connected MLP. The MLP implements traditional deep learning algorithms, such as feedforward propagation, to classify the images based on the extracted features.

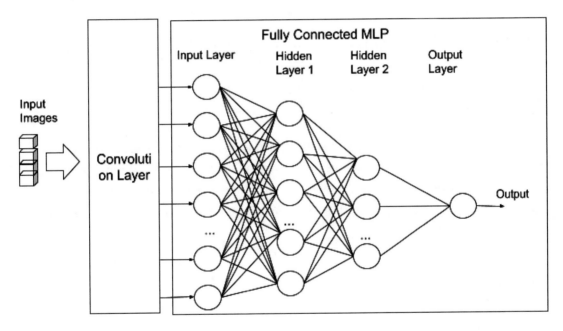

Figure 5-27. *CNN architecture*

By combining the automatic feature engineering capabilities of the multiple layers with the classification power of the fully connected MLP layers, CNNs can effectively learn and classify complex patterns and structures in images.

The convolutional network has two types of layers:

- *Convolutional*: This layer extracts features from the images (feature extraction).

- *Subsampling*: This layer selects from extracted features (feature selection).

Figure 5-28 depicts a complete CNN.

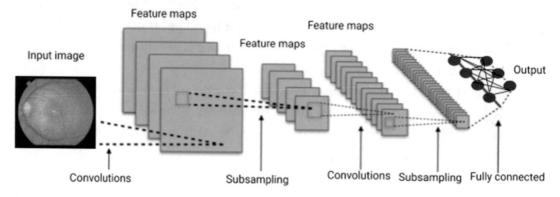

Figure 5-28. *CNN with convolution, subsampling, and fully connected MLP layers*

How a CNN Works

We saw in Chapter 2 that a computer sees a black-and-white image with a single channel as a 2D matrix of pixel values (as shown in Figure 5-29). A color image with RGB channels (three channels) is shown as a stack of these 2D matrices. These stacks of matrices form a 3D tensor (remember tensors?). Figure 5-30 shows a visual presentation of a 3D image tensor.

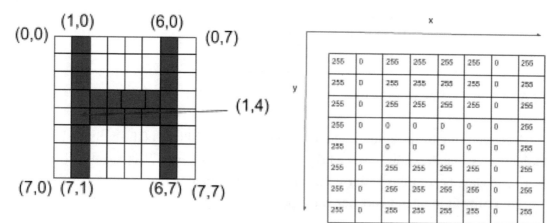

Figure 5-29. *A black-and-white image (left) is seen as a 2D matrix by a computer (right)*

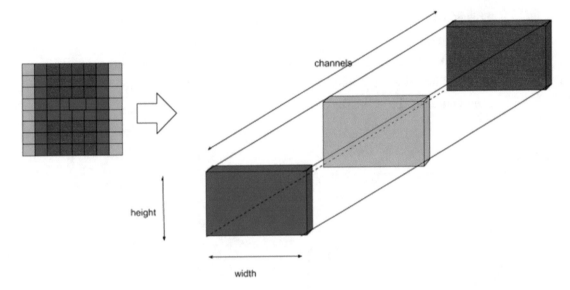

Figure 5-30. *Tensor representation of a three-channel color image as a stack of 2D matrices*

With this background of how images are represented as a tensor, let's examine the convolutional process.

Convolution

Imagine that we have an image that we glance over with a magnifying glass, keeping a note of important patterns we observe. This is a good analogy of how convolution works.

Here are the steps to extract important features from an image using convolution:

1. Divide the image into grids of size $k \times k$ pixels. This is called a *kernel*, which is represented as a $k \times k$ matrix.

2. Define one or more filters that are of the same dimensions as the kernel.

3. Take the first kernel (starting from the top-left corner of the 2D matrix) of one of the channels, do element-wise multiplication with the first filter, and add the multiplication results. Do the same with the other channels and sum the results of all three channels to get the pixel value of a newly created feature.

This is demonstrated in Figure 5-31. For this example, we take a 7×7×3 image with the kernel size 3×3. We have two sets of filters: W0 and W1 (shown in red). The filter W0 has a bias of 1, and filter W1 does not have any bias. The output feature is shown in the green color grid (on the far right).

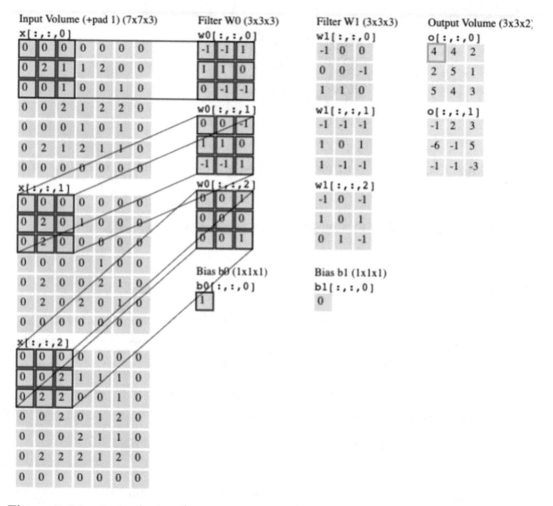

Figure 5-31. *Convolution (image courtesy of Andrej Karpathy)*

The output is calculated as shown here:

```
Channel 1 output = 0x(-1) + 0x(-1) + 0x1 + 0x1 + 2x1 +
1x0 + 0x0 + 0x(-1) + 1x(-1) = 2
Channel 2 output = 0x0 + 0x0 + 0x(-1) + 0x1 + 2x1 + 0x0
+ 0x(-1) + 2x(-1) + 0x1 = 0
Channel 3 output = 0x0 + 0x0 + 0x1 + 0x0 + 0x0 + 2x0 +
0x0 + 2x0 + 2x1 = 1
Feature value = channel 1 output + channel 2 output +
channel 3 output + bias
Feature value = 2 + 0 + 1 + 1 = 4
```

Again, the value 4 is shown highlighted in the top green grid's top-left corner in Figure 5-31 (far right).

4. The kernel is now moved to the right, and the feature value is calculated as explained earlier. When the kernel is moved all the way to the right, it is moved down to the next row starting from the leftmost pixels of that row. The number of steps to the horizontal and vertical directions the kernel is moved to scan the entire image is called the *stride*. The stride is expressed as *s* (for example, 2 or 3, etc.). A stride of 2 means the kernel will move two steps to the right, and when it reaches the right edge of the image, it moves down by 2 pixels.

5. When the entire image is scanned, a feature matrix is created. The dimensions of the feature matrix in our example are 3×3 (for a 7×7×3-pixel image, 3×3 kernel, and 2×2 stride). This feature matrix, also known as a *feature map*, is shown in Figure 5-31 in the top 3×3 green grid (to the right).

6. The same convolution process is repeated with the next set of filters, and a feature map is created. The bottom green grid in Figure 5-31 shows the feature map from the second filter.

7. This process is repeated for all the filters, and feature maps are generated from each filter.

Pooling/Subsampling/Downsampling

Convolution extracts features from the images. These features are represented as $n \times n$ matrices and are fed to another layer, called the *pooling layer*, which performs "downsampling," much like feature selection. Max pooling and average pooling are two popular methods to downsample the features.

Max Pooling

In the pooling layer, much like the convolution stage, the feature matrix is divided into grids of $k \times k$ (e.g., 2×2 pixels in Figure 5-32) kernels with stride s (e.g., stride 1 in the example). In the max pooling layer, the maximum pixel value from each kernel area is taken, and a downsampled matrix is generated. This process is repeated for each filter output from the previous layer.

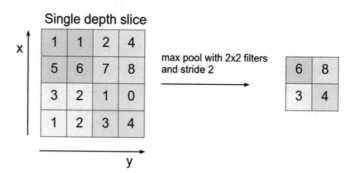

Figure 5-32. *Max pooling to downsample features (image courtesy of Andrej Karpathy)*

Average Pooling

Average pooling works the same way as max pooling except that in average pooling the average (not the max) of kernel pixels is taken to create the downsampled matrix.

A CNN typically consists of alternating convolutional and pooling layers along with a multilayer perceptron (as shown in Figure 5-33).

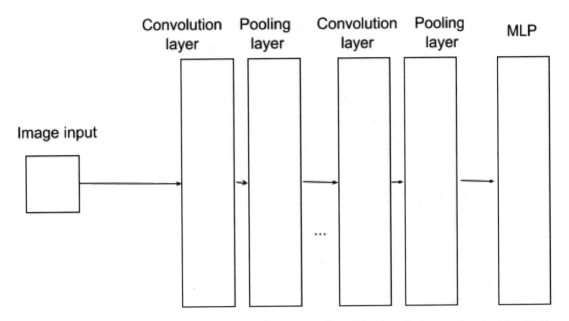

Figure 5-33. *CNN layers, alternating convolutional and pooling layers with MLP*

Summary of CNN Concepts

Here are the CNN concepts we have explored thus far:

- A CNN consists of alternating convolutional and pooling layers with MLP at the end. Every convolutional layer does not necessarily have a pooling layer.

- Convolution is a feature extraction process in the convolutional layer.

- A kernel of dimension $k \times k$ is defined to divide input images into grids.

- Filters, of the same dimension as the kernel, are multiplied with the pixels in the kernel, and the results are summed over each pixel and each image channel. An optional bias is added to the result to generate feature matrices.

- The pooling layer implements downsampling algorithms (max pooling or average pooling) to downsample the features.

- The process is repeated for each pair of convolutional-pooling layers where output from one pooling layer is fed as input to the next convolutional layer.

- The last convolutional/pooling layer feeds feature matrices to the input layer of the MLP.

- The MLP part of the network learns as a conventional MLP network.

Training a CNN Model: Pneumonia Detection from Chest X-rays

TensorFlow with Keras makes it easy to train a CNN model. With only a few lines of code, you can implement and train a CNN model.

In this section, we will delve into a code example that demonstrates how to train a model for pneumonia detection using chest X-rays.

Caution   The model presented here is a basic CNN intended for academic and learning purposes only. It should not be used for diagnosing any medical conditions.

Chest X-ray Dataset

We obtained chest X-ray images from a publicly available Kaggle-hosted website at `https://www.kaggle.com/paultimothymooney/chest-xray-pneumonia`. It's important to note that this dataset of images is released under the Creative Commons License described at `https://creativecommons.org/licenses/by/4.0/` and is subject to its terms.

The dataset consists of two sets of images, one set that includes chest X-rays of disease-free lungs (normal) and another set that includes X-rays of pneumonia-infected lungs. These two sets of images are stored in separate directories; all normal images are stored in the NORMAL directory, and all pneumonia images are stored in the PNEUMONIA directory. Furthermore, the dataset is divided into training, test, and validation sets.

After downloading the images from the Kaggle website, save them in the local disk. Figure 5-34 shows a sample directory structure.

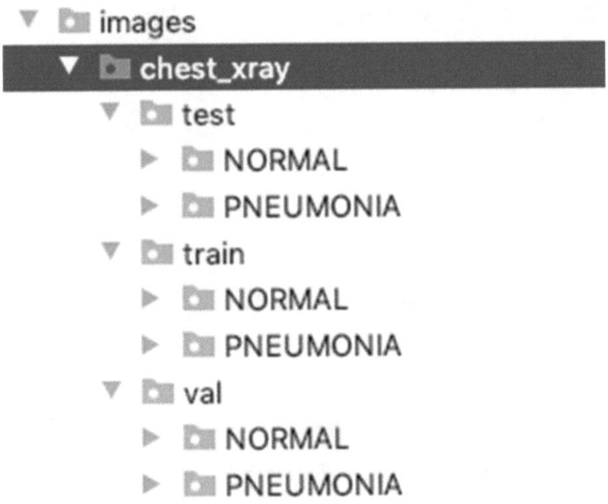

Figure 5-34. *Directory structure of chest X-ray images*

Figure 5-34 illustrates the directory structure of the parent directory chest_xray, which includes subdirectories named test, train, and val. Each of these subdirectories further contains two subdirectories: NORMAL and PNEUMONIA. The NORMAL directory stores images representing normal chest X-rays, while the PNEUMONIA directory stores images depicting X-rays with pneumonia. This directory organization serves as a method of labeling images with their respective classes.

Code Structure

While organizing code for simplicity and ease of understanding, it is important to note that there are more effective ways to structure it, making it more reusable and following object-oriented principles. In production-quality work, it is highly recommended to parameterize the code for flexibility and maintainability, avoiding any hard-coded values. However, for the purpose of learning, I have used some hard-coded values to make the code easy to understand.

CNN Model Training

Listing 5-7 shows the code sample for training a CNN model for predicting pneumonia from chest X-rays.

Listing 5-7. Code to Train CNN Model to Predict Pneumonia from Chest X-rays

```
FileName: Listing_5_7.py
1    import numpy as np
2    import pathlib
3    import cv2
4
5    import tensorflow as tfimport matplotlib.pyplot as plt
6
7
8    # Section 1: Loading images from directories for training and test
9    trainig_img_dir ="/content/chest_xray/train"
10    test_img_dir ="/content/chest_xray/test"
11
12   # ImageDataGenerator class provides mechanism to load both small and
     large dataset.
13   # Instruct ImageDataGenerator to scale to normalize pixel values to
     range (0, 1)
14   datagen = tf.keras.preprocessing.image.ImageDataGenerator(resca
     le=1./255.)
15   #Create training image iterator that will be loaded in small batch
     size. Resize all images to a standard size.
16   train_it = datagen.flow_from_directory(trainig_img_dir, batch_size=8,
     target_size=(1024,1024))
17   #Create test image iterator that will be loaded in small batch size.
     Resize all images to a standard size.
18   test_it = datagen.flow_from_directory(test_img_dir, batch_size=8,
     target_size=(1024, 1024))
19
20   # Lines 22 through 24 are optional to explore your images.
21   # Notice, next() function call returns both pixel and labels values as
     numpy arrays.
22   train_images, train_labels = train_it.next()
23   test_images, test_labels = test_it.next()
24   print('Batch shape=%s, min=%.3f, max=%.3f' % (train_images.shape,
     train_images.min(), train_images.max()))
```

```
25
26   # Section 2: Build CNN and train with training dataset.
27   # You could pass argument parameters to build_cnn() function to set
     some of the values
28   # such as number of filters, strides, activation function, number of
     layers, etc.
29   def build_cnn():
30       model =  tf.keras.models.Sequential()
31       model.add(tf.keras.layers.Conv2D(64, (3, 3), activation='relu',
         strides=(2,2), input_shape=(1024, 1024, 3)))
32       model.add(tf.keras.layers.MaxPooling2D((2, 2)))
33       model.add(tf.keras.layers.Conv2D(32, (3, 3), strides=(2,2),activat
         ion='relu'))
34       model.add(tf.keras.layers.MaxPooling2D((2, 2)))
35       model.add(tf.keras.layers.Conv2D(16, (3, 3), strides=(2,2),activat
         ion='relu'))
36       model.add(tf.keras.layers.Flatten())
37       model.add(tf.keras.layers.Dense(16, activation='relu'))
38       model.add(tf.keras.layers.Dense(2, activation='softmax'))
39       return model
40
41   # Build CNN model
42   model = build_cnn()
43   # Compile the model with optimizer and loss function
44   model.compile(optimizer='adam',
45                   loss='categorical_crossentropy',
46                   metrics=['accuracy'])
47
48   # Fit the model using the fit() function and passing the data iterator
     to it so that it iteratively loads large number of images in batches
49   history = model.fit(train_it, epochs=100,
50                       validation_data=test_it, validation_steps=32)
51
52   # Section 3: Save the CNN model to disk for later use.
53   model_path = "models/pneumoniacnn"
54   tf.saved_model.save(model, model_path)
```

```
55
56   # Section 4: Display evaluation metrics
57   print(history.history.keys())
58   plt.plot(history.history['accuracy'], label='accuracy')
59   plt.plot(history.history['val_accuracy'], label = 'val_accuracy')
60   plt.plot(history.history['loss'], label='loss')
61   plt.plot(history.history['val_loss'], label = 'val_loss')
62
63   plt.xlabel('Epoch')
64   plt.ylabel('Metrics')
65   plt.ylim([0.5, 1])
66   plt.legend(loc='lower right')
67   plt.show()
68   test_loss, test_acc = model.evaluate(test_images,  test_labels,
     verbose=2)
69   print(test_acc)
```

The code in Listing 5-7 for CNN model training is logically divided into the following four sections:

- *Loading images (lines 9 through 24)*: We have our training and test images stored in directories as described earlier. To load these images for the purpose of training and validation, we used a powerful class, ImageDataGenerator, provided by Keras. Here is the line-by-line explanation of how we used this class:

  Line 9 and 10 are the directories that have training and test images in their subdirectories.

  Line 14 initializes the ImageDataGenerator class. We passed the argument rescale = 1/255 because we want to normalize the pixel values to be in the range between 0 and 1. This normalization is done by multiplying each pixel of the images by 1/255. We call this line datagen as indicated by the variable name.

  Line 16 is calling the flow_from_directory() function of the datagen object. This function loads images from the directory training_img_directory, in a batch mode (e.g., batch_size = 32), and resizes the images to a size indicated by target_size (e.g., 1024×1024px). This is a highly scalable function and will be able to load millions of images without loading all of them in memory. It will load at a time as many images as indicated by the batch_size argument. Resizing all images to a standard size is important for most machine learning exercises. Note that the default resize value of this function is 256. If you omit the resize argument, all your input images will be resized to 256×256.

Line 17 does the same as line 16 except that it is loading the images from the test directory. Although we have validation data in our directory (the dataset downloaded from the Kaggle website contains validation images), the number is small, and therefore we are using the test dataset for validation.

The function flow_from_directory() returns an iterator. If you iterate over this iterator, you will get a tuple of two NumPy arrays: arrays of image pixel values and arrays of labels.

Note that labels are interpreted from the subdirectories the images are read from. For example, all images from the NORMAL directories will get the label NORMAL, and similarly images belonging to the PNEUMONIA subdirectory will get the PNEUMONIA label. But wait! Aren't these labels supposed to be numeric? These directory names are sorted by their names and indexed, starting from 0. In our case, NORMAL will be indexed as 0 and PNEUMONIA as 1. But, it does not stop here. The function flow_from_directory() takes an additional argument called class_mode. By default the value of class_mode is categorical. You could also pass a value to it as binary or sparse. The differences between these three are as follows:

– categorical will return 2D one-hot encoded labels.

– binary will return 1D binary labels.

– sparse will return 1D integer labels.

Lines 22 through 24 are optional and not needed for training the model. I provided them to show you how you could explore the values from the iterator returned from the `flow_from_directory()` function.

- *CNN configuration and training (lines 29 through 50)*: Lines 29 through 39 implement a function to build a CNN. These lines are our main focus in this section, so let's break down what is going on here.

Line 30 creates a sequential neural network to which we stack up layers. Recall that we used the same `tf.keras.model.Sequential` class to create the sequential model. The `add()` function of the `model object` is used to add layers in sequential order—the layer added first is executed first and so on.

Line 31 adds our first layer to the network. If you recall from our previous discussion on CNN, our first layer of the CNN must be a convolutional layer that takes the input (image pixel values). Here we are using the `Conv2D` class to define our convolutional layer. We are passing five important parameters to `Conv2D()`:

- filters, which in this example is 64.

- The kernel dimension, which in this example is 3×3 pixels and passed as a tuple (3,3).

- The activation function, which in our case is `relu` (as the pixel values range from 0 to 1 and are never negative).

- The next parameter is to set the strides, which is by default (1,1) if not set. In our case, we set it to (2,2).

- The final parameter is to set the input size. Since our images are resized to 1024×1024 pixels, colored (with three channels), the `input_shape` is (1024,1024,3).

Line 32 adds the pooling layer, `MaxPooling2D`. Recall that the convolution and pooling layers may be alternated, except for the layer before the MLP layers. We are passing the argument to set the size of the grid or kernel. In our example, it is set to be (2,2).

Lines 33, 34, and 35 are again our convolutional and pooling layers. You can have as many convolution and pooling layers as are required to achieve the desired accuracy levels.

The outputs from the convolutional layer, line 35, are fed to the first layer of the MLP. Recall that the first layer of the MLP is called the input layer, followed by hidden layers, and finally the output layer.

Line 36 flattens the output from line 35.

Line 37 is the hidden layer of the MLP and has been explained in the ANN section at the beginning of this chapter.

Line 38 is the final layer, the output layer. As explained previously, we are using the activation function softmax as we are solving a classification problem involving two classes: normal and pneumonia.

Line 42 simply calls the `build_cnn()` function and creates a `model object`.

Line 44 compiles the model, as we saw earlier with ANNs. Notice the difference between line 44 and line 31 of Listing 5-6 in the loss function. Here we are using the loss function `categorical_crossentropy` as opposed to `sparse_categorical_crossentropy` that we used in Listing 5-6.

Finally, we are starting the training in line 49. Notice that we are calling the function `fit()` as we did in Listing 5-6. The `fit()` function automatically detects that the input is a generator from the class `ImageDataGenerator` that loads images in a small batches. The function `fit()` takes an important parameter called `steps_per_epoch`, which is the number of batches it will complete in each epoch. Here is the official definition:

`steps_per_epoch`: The total number of steps (batches of samples) to yield from `generator` (the data loader) before declaring one epoch finished and starting the next epoch. It should typically be equal to the number of samples of your dataset divided by the batch size. For example, if you have 1,000 files in your training set and your `batch_size` is 8, you should set `steps_per_epoch` equal to 1000/8 = 125.

Another important parameter to this function is `validation_steps`, which is defined as follows:

`validation_steps`: This is relevant only if `validation_data` is a generator. It is the total number of steps (batches of samples) to yield from the generator (the data loader) before stopping.

- *Saving the CNN model to disk (lines 53 and 54)*: Line 54 saves the trained model to the directory specified in line 53. You could save the training checkpoints as well.

- *Evaluation and visualization (lines 57 through 69)*: We plot a graph of training loss, validation loss, training accuracy, and test accuracy against epochs. Line 68 evaluates the model and simply prints the accuracy in line 69.

Figure 5-35 shows a sample output while the model runs. Figure 5-36 shows a sample plot of training and validation metrics. As the graph shows, the losses of both training and validation decrease as the number of epochs increases. Also, the accuracy improves over epochs.

```
Epoch 1/10
16/16 [==============================] - 126s 8s/step - loss: 0.6689 - accuracy: 0.6953 - val_loss: 0.6374 - val_accuracy: 0.6719
Epoch 2/10
16/16 [==============================] - 113s 7s/step - loss: 0.4902 - accuracy: 0.7500 - val_loss: 0.5442 - val_accuracy: 0.7344
Epoch 3/10
16/16 [==============================] - 100s 6s/step - loss: 0.3313 - accuracy: 0.8281 - val_loss: 0.2979 - val_accuracy: 0.8438
Epoch 4/10
16/16 [==============================] - 136s 8s/step - loss: 0.3130 - accuracy: 0.8516 - val_loss: 0.2127 - val_accuracy: 0.9219
Epoch 5/10
16/16 [==============================] - 107s 7s/step - loss: 0.2858 - accuracy: 0.8672 - val_loss: 0.3694 - val_accuracy: 0.7656
Epoch 6/10
16/16 [==============================] - 102s 6s/step - loss: 0.2343 - accuracy: 0.9219 - val_loss: 0.2187 - val_accuracy: 0.8906
Epoch 7/10
16/16 [==============================] - 130s 8s/step - loss: 0.3260 - accuracy: 0.8828 - val_loss: 0.1669 - val_accuracy: 0.9531
Epoch 8/10
16/16 [==============================] - 94s 6s/step - loss: 0.1941 - accuracy: 0.9297 - val_loss: 0.4719 - val_accuracy: 0.7812
Epoch 9/10
16/16 [==============================] - 101s 6s/step - loss: 0.3174 - accuracy: 0.8828 - val_loss: 0.1896 - val_accuracy: 0.9375
Epoch 10/10
16/16 [==============================] - 102s 6s/step - loss: 0.2728 - accuracy: 0.8594 - val_loss: 0.3509 - val_accuracy: 0.7969
```

Figure 5-35. *Sample output from the CNN model training*

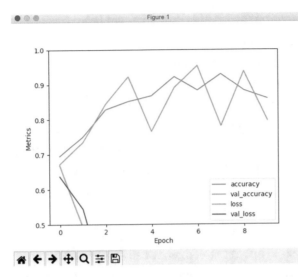

Figure 5-36. *Sample plot of metrics (loss and accuracy over epoch) for training and valuation*

Pneumonia Prediction

Listing 5-8 shows how to use the previously trained CNN model to predict pneumonia from a new set of images.

Listing 5-8. Code for Predicting Pneumonia by Using the Trained CNN Model

```
FileName: Listing_5_8.py
1    import numpy as np, pandas as pd
2    import tensorflow as tf
3    import os
4
5    model_path = "/content/models/pneumoniacnn"
6    input_dir = "/content/chest_xray/val"
7    classes = ["NORMAL", "PNEUMONIA"]
8    image_dimensions = [1024, 1024]
9    # Load the trained model, including its weights and the optimizer
10   model = tf.saved_model.load(model_path)
11
12   # Predict the class of the input image from the loaded model
13   for file in os.listdir(input_dir):
```

```
14      image_dir = os.path.join(input_dir, file)
15      if os.path.isdir(image_dir):
16        for img in os.listdir(image_dir):
17          image_path = os.path.join(image_dir, img)
18          image = tf.io.read_file(image_path)
19          image = tf.io.decode_image(image, channels=3)
20          image = tf.image.resize(image, image_dimensions)
21          image = image / 255.0  # Normalize pixel values between 0 and 1
22          image = tf.expand_dims(image, axis=0)  # Add batch dimension
23          predicted = model(image)
24          max_probability = np.max(predicted)
25          max_index = np.argmax(predicted)
26          print("Image:", image_path, " Predicted", classes[max_index], "
            Actual:", file, " Probability: ", max_probability)
```

The code for classifying or predicting images for the presence of pneumonia is divided into three parts.

- *Loading saved models (line 10)*: Recall from Listing 5-7 that we saved the trained model in the directory models/pneumoniacnn. Line 10 loads the saved model from the disk using the TensorFlow function saved_model.load(). This function loads the trained model, metadata, weights, and the optimizer.

- *Loading images (lines 13 through 22)*: In this example, we are working with image predictions sourced from the val directory. The val directory has two subdirectories, NORMAL and PNEUMONIA, which serve as labels for the contained images. Lines 13 to 16 establish two for loops to iterate through these directories and obtain the paths of the images within them (line 17). Using the TensorFlow io.read_file() function, line 18 loads the image from the respective directory. Line 19 utilizes the io.decode_image() function, which automatically detects the image format (BMP, GIF, JPEG, or PNG) and converts the input bytes string into a tensor accordingly. During model training, the images were resized to 1024×1024. Therefore, line 20 resizes the input image to match this dimension. Line 21 performs pixel value normalization, while line 22 converts the image tensor into a batch of size 1.

- *Predicting pneumonia (Line 23)*: The predicted class probabilities are generated by the model(image) function and stored as a NumPy array. In line 24, the maximum probability from the array is obtained, while in line 25 the index corresponding to the maximum probability is retrieved. Subsequently, in line 26, the obtained result is printed.

Figure 5-37 shows a sample prediction output.

```
Image: person1951_bacteria_4882.jpeg  Predicted PNEUMONIA  Actual: PNEUMONIA  Probability:  0.99987674
Image: person1949_bacteria_4880.jpeg  Predicted PNEUMONIA  Actual: PNEUMONIA  Probability:  0.9980684
Image: person1950_bacteria_4881.jpeg  Predicted PNEUMONIA  Actual: PNEUMONIA  Probability:  0.99479514
Image: person1954_bacteria_4886.jpeg  Predicted PNEUMONIA  Actual: PNEUMONIA  Probability:  0.99556977
Image: person1946_bacteria_4875.jpeg  Predicted PNEUMONIA  Actual: PNEUMONIA  Probability:  0.99975246
Image: person1952_bacteria_4883.jpeg  Predicted PNEUMONIA  Actual: PNEUMONIA  Probability:  0.9927549
Image: person1947_bacteria_4876.jpeg  Predicted PNEUMONIA  Actual: PNEUMONIA  Probability:  0.99962413
Image: person1946_bacteria_4874.jpeg  Predicted PNEUMONIA  Actual: PNEUMONIA  Probability:  0.9998703
Image: NORMAL2-IM-1442-0001.jpeg  Predicted NORMAL  Actual: NORMAL  Probability:  0.9954625
Image: NORMAL2-IM-1431-0001.jpeg  Predicted PNEUMONIA  Actual: NORMAL  Probability:  0.9109347
Image: NORMAL2-IM-1430-0001.jpeg  Predicted PNEUMONIA  Actual: NORMAL  Probability:  0.864686
Image: NORMAL2-IM-1427-0001.jpeg  Predicted PNEUMONIA  Actual: NORMAL  Probability:  0.8725711
Image: NORMAL2-IM-1437-0001.jpeg  Predicted PNEUMONIA  Actual: NORMAL  Probability:  0.92962617
Image: NORMAL2-IM-1440-0001.jpeg  Predicted NORMAL  Actual: NORMAL  Probability:  0.5443065
Image: NORMAL2-IM-1436-0001.jpeg  Predicted PNEUMONIA  Actual: NORMAL  Probability:  0.94846416
Image: NORMAL2-IM-1438-0001.jpeg  Predicted NORMAL  Actual: NORMAL  Probability:  0.51886207
```

Figure 5-37. *Sample prediction output*

A CNN is an important algorithm extensively employed in the field of computer vision. In this section, you have gained knowledge about the fundamental concepts underlying CNNs and their operational mechanisms. Additionally, we have explored code examples that illustrate the training of customized CNN models for the purpose of pneumonia prediction.

Examples of Popular CNNs

The CNN we built in Listing 5-7 is not a production-quality network. We built a simple network to learn the basics. Let's look at some of the popular networks that were proven successful globally.

LeNet-5

The LeNet-5 CNN architecture, introduced in 1998 by LeCun et al. in their paper "Gradient-Based Learning Applied to Document Recognition," was mainly used for recognizing handwritten and machine-generated characters (optical character recognition [OCR]) from documents. The architecture is simple and straightforward and hence used widely in teachings. Here are the salient features of the LeNet-5 architecture:

- It is a CNN consisting of seven layers.

- Out of these seven layers, there are three convolution layers (C1, C3, and C5).

- There are two subsampling layers (S2 and S4).

- There is one fully connected layer (F6) and one output layer.

- The convolutional layers use 5×5 convolution kernels with stride 1.

- The subsampling layers are 2×2 average pooling layers.

- The entire network uses the TanH activation function except for the output layer, which uses softmax.

Figure 5-38 shows the LeNet-5 network.

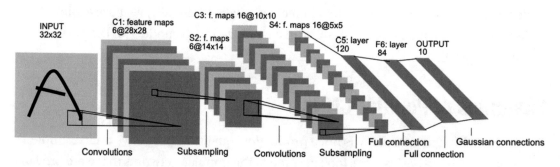

Figure 5-38. *LeNet-5 (image courtesy of https://yann.lecun.com/exdb/publis/pdf/lecun-01a.pdf)*

Here's an exercise for you: modify the TensorFlow code from Listing 5-7 and implement LeNet-5.

AlexNet

AlexNet is a convolutional neural network architecture designed by Alex Krizhevsky et al. It became popular when AlexNet competed in the ImageNet Large Scale Visual Recognition Challenge (ILSVRC) in 2012 and achieved a top-five error of 15.3 percent, more than 10.8 percentage points lower than that of the runner-up. AlexNet is a deep network, and despite being computationally expensive, it became feasible because of the use of GPUs.

The features of AlexNet are as follows:

- It is a deep CNN containing eight layers.

- The input size is 224×224×3 color images.

- The first five layers are a combination of convolutional and max pooling layers with the following configurations:

  - *Convolution layer 1*: Kernel 11×11, filters 96, strides 4×4, activation ReLU

  - *Pooling layer 1*: MaxPooling with kernel size 3×3, strides 2×2

  - *Convolution layer 2*: Kernel 5×5, filters 256, strides 1×1, activation ReLU

  - *Pooling layer 2*: MaxPooling with kernel size 3×3, strides 2×2

  - *Convolution layer 3*: Kernel 3×3, filters 384, strides 1×1, activation ReLU

  - *Convolution layer 4*: Kernel 3×3, filters 384, strides 1×1, activation ReLU

  - *Convolution layer 5*: Kernel 3×3, filters 384, strides 1×1, activation ReLU

  - Pooling layer 5: MaxPooling with kernel size 3×3, strides 2×2

- The last three layers are a fully connected MLP.

- All convolution layers use ReLU activation functions.

- The output layer uses softmax activation.

- There are 1,000 classes in the output layer.

- The network has 60 million parameters and 650,000 neurons, and it takes about 3 days to train on a GPU.

Figure 5-39 shows an illustration of AlexNet.

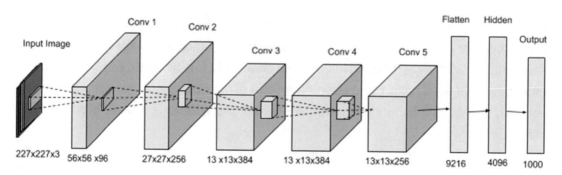

Figure 5-39. *AlexNet with five convolution layers and three fully connected MLPs*

VGG-16

The popular VGG-16 deep neural network won the ILSVRC competition in 2014. VGG was designed by researchers at Oxford Visual Geometry Group (VGG). Their publication is available at `https://arxiv.org/abs/1409.1556`.

Here is a list of the salient features of VGG-16:

- It is a CNN that consists of 16 layers.

- It has 13 convolutional layers and 3 fully connected dense layers.

- The 16 layers have the following features:

  - *Convolution layer 1*: Input size 224×224x3, kernel 3×3, filters 64, activation ReLU

  - *Convolution layer 2*: Kernel 3×3, filters 64, activation ReLU

  - *Pooling layer*: MaxPooling, kernel size 2×2 and strides 2×2

  - *Convolution layer 3*: Kernel 3×3, filters 128, activation ReLU

  - *Convolution layer 4*: Kernel 3×3, filters 128, activation ReLU

  - *Pooling layer*: MaxPooling, kernel size 2×2 and strides 2×2

  - *Convolution layer 5*: Kernel 3×3, filters 256, activation ReLU

  - *Convolution layer 6*: Kernel 3×3, filters 256, activation ReLU

  - *Convolution layer 7*: Kernel 3×3, filters 256, activation ReLU

  - *Pooling layer*: MaxPooling, kernel size 2×2 and strides 2×2

- – *Convolution layer 8*: Kernel 3×3, filters 512, activation ReLU

- – *Convolution layer 9*: Kernel 3×3, filters 512, activation ReLU

- – *Convolution layer 10*: Kernel 3×3, filters 512, activation ReLU

- – *Pooling layer*: MaxPooling, kernel size 2×2 and strides 2×2

- – *Convolution layer 11*: Kernel 3×3, filters 512, activation ReLU

- – *Convolution layer 12*: Kernel 3×3, filters 512, activation ReLU

- – *Convolution layer 13*: Kernel 3×3, filters 512, activation ReLU

- – *Pooling layer*: MaxPooling, kernel size 2×2 and strides 2×2

- – *Fully connected layer 14 (MLP input layer)*: Flatten dense layer with input size 25088

- – *Fully connected hidden layer 15*: Dense layer with input size 4096

- – *Fully connected output layer*: Dense layer for 1,000 classes

- • This network has 138 million parameters.

Figure 5-40 shows the VGG-16 network.

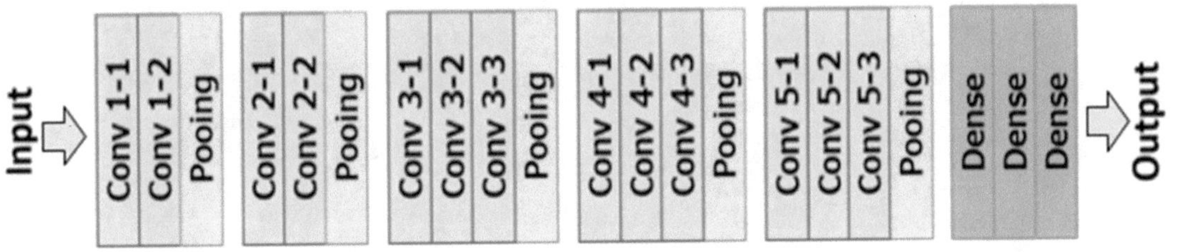

Figure 5-40. *VGG-16 architecture with 16 layers (13 convolutional layers and 3 dense layers)*

Here's another exercise for you: modify Listing 5-7 and implement the VGG-16 network using TensorFlow.

Summary

This chapter provided a fundamental understanding of artificial neural networks and convolutional neural networks. We implemented TensorFlow-based code to train our ANN and CNN models, evaluated model performance, and applied the saved models to image classification tasks. Additionally, we explored the process of hyperparameter tuning and leveraged the HParams dashboard in TensorBoard to visualize and analyze models' performance. We wrapped up with an overview of three popular CNN architectures, LeNet-5, AlexNet, and VGG-16, to understand their unique characteristics and applications.

Throughout the chapter, the primary focus was solving classification problems, where our models were trained to classify input images into different classes.

Building on this foundation, Chapter 6 introduces and engages the techniques and methodologies key to detecting objects within images. We will explore the fundamental principles, algorithms, and architectures used in object detection tasks, providing a foundational framework for understanding this important aspect of computer vision.

Deep Learning in Object Detection

Chapter 5 covered how to classify images using a standard multilayer perceptron (MLP) and a convolutional neural network (CNN). In classification tasks, our focus is predicting the class of the entire image, without considering the specific objects present within it. In this chapter, we will explore how to detect objects and determine their locations within the image.

Our learning objectives in this chapter are as follows:

- To explore popular deep learning algorithms utilized in object detection
- To train custom object detection models using TensorFlow on GPU
- To use trained models to predict objects within images

The concepts introduced in this chapter will be applied in the subsequent three chapters to develop practical computer vision applications in real-world scenarios.

Object Detection

Object detection involves two distinct sets of activities: locating objects and classifying objects. Locating objects within the image is called *object localization*, which is typically performed by drawing bounding boxes around the objects. Before the deep learning algorithms became popular, object localization was performed by marking each pixel in the image that contained the object. For example, object detection was performed using techniques such as edge detection and drawing contours (both of which are covered in Chapter 3) and HOGs (covered in Chapter 4). Using these techniques for object localization was compute-intensive, slow, and often lacked accuracy.

© Shamshad Ansari 2023
S. Ansari, *Building Computer Vision Applications Using Artificial Neural Networks*,
https://doi.org/10.1007/978-1-4842-9866-4_6

As discussed in Chapter 5, *object classification* refers to assigning a specific label or category to an object within an image.

Deep learning–based object detection techniques have demonstrated superior speed and accuracy compared to algorithms that do not use deep learning. Although the learning process is computationally demanding in deep learning–based object detection, the actual detection is fast and well-suited for real-time object detection. Examples of applications utilizing deep learning–based object detection include

- Autonomous vehicles

- Airport security systems

- Video surveillance systems

- Defect detection in industrial production processes

- Industrial quality assurance systems

- Facial recognition technology

Deep learning algorithms for object detection have evolved over time. In this chapter, we will explore two different variations of convolutional neural networks used in object detection: two-step convolutions and single-step convolutions. A *region-based convolutional neural network (R-CNN)* is a two-step algorithm. *You only look once (YOLO)* and *single-shot multibox detection (SSD)* are examples of single-step algorithms for object detection.

Before we delve deeper into object detection algorithms, it is important to establish the definition of a fundamental metric widely employed in object detectors.

Intersection over Union

Intersection over union (IoU), also known as *Jaccard index*, is one of the most commonly used evaluation metrics in object detection algorithms. It is used to measure the identity of two arbitrary shapes.

In object detection, we create training sets by drawing bounding boxes around objects for labeling. These bounding boxes in the training set are also known as the *ground truth*. During the model learning, the object detection algorithm predicts bounding boxes and compares them against the ground truth. IoU is used to evaluate how closely the predicted bounding box overlaps with the ground truth.

The IoU between a predicted bounding box A and a ground truth box B is calculated by using the formula shown in Figure 6-1.

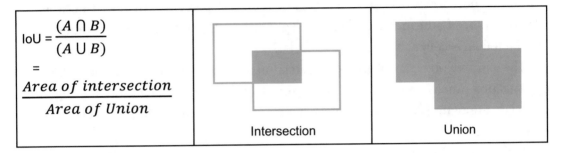

Figure 6-1. *IoU*

When we label an image, we typically draw rectangular boxes around the objects within the image. This rectangular region surrounding the object is the ground truth. In Figure 6-2, the ground truth is shown by the green rectangular box.

When the algorithm learns, it predicts the bounding boxes surrounding the object. In Figure 6-2, the red rectangular region is the predicted bounding box.

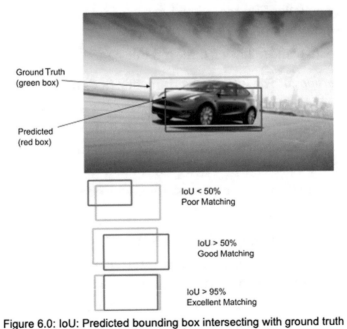

Figure 6.0: IoU: Predicted bounding box intersecting with ground truth

Figure 6-2. *IoU, predicted bounding box intersecting with ground truth*

The learning algorithm computes the IoU between the ground truth and the predicted bounding boxes. The match between the predicted and ground truth is considered poor if the IoU between them is less than 50 percent. If the IoU is between 50 and 95 percent, the match is considered good. An IoU greater than 95 percent is considered an excellent match.

The learning objective of an object detection algorithm is to optimize the IoU. This optimization process involves fine-tuning the algorithm's parameters and training the model on annotated datasets to achieve higher IoU values, indicating improved object detection performance.

Let's now explore various deep learning algorithms used in object detection. We will also review their strengths and weaknesses and how they compare with each other.

Region-Based Convolutional Neural Network

R-CNN was the first successful model that used a large convolutional neural network to detect objects in images. The detection method is described by Ross Girshick et al. in their 2014 paper titled "Rich Feature Hierarchies for Accurate Object Detection and Semantic Segmentation" (https://arxiv.org/pdf/1311.2524.pdf). Figure 6-3 demonstrates the R-CNN method.

R-CNN comprises of the following three modules:

- *Region proposal*: The R-CNN algorithm first finds regions in the image that might contain objects. These regions are called *region proposals*. They are called *proposals* because these regions may or may not contain objects, and the objective of the learning function is to eliminate those regions that do not contain objects. These region proposals are the bounding boxes around the objects (as shown in Figure 6-3, diagram 2).

  The R-CNN system proposed by Girshick et al. is agnostic to the algorithm that finds the region proposal. That means you could use any algorithm, such as HOG, to find the regions. They used an algorithm known as *selective search*, which looks at the image through grids of different sizes. For each grid size, the algorithm attempts to group together adjacent pixels by comparing the texture, color, or pixel values to identify objects. The algorithm uses this method to create region proposals. In summary, the algorithm creates a set of bounding boxes of potential target objects.

- *Feature extraction*: The region proposals are cropped out of the image and resized. These cropped images are then fed to a standard CNN to extract features (see Figure 6-3, diagram 3). According to the original paper, the AlexNet deep learning CNN was used for feature extraction. From each region, 4,096-dimensional feature vectors were extracted.

- *Classifier*: The extracted features are classified by using the standard classification algorithms, such as the linear SVM model (diagram 4 of Figure 6-3).

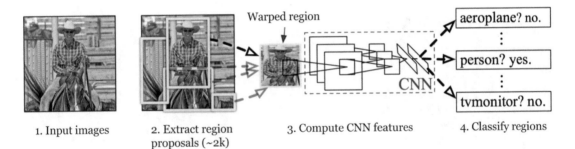

1. Input images

2. Extract region proposals (~2k)

3. Compute CNN features

4. Classify regions

Figure 6-3. *R-CNN model (Source: Girshick et al.,* `https://arxiv.org/pdf/1311.2524.pdf`*)*

Although R-CNN was the first successful deep learning–based object detection system, it suffers a serious issue with respect to performance. Its time performance problem is because of the following:

- Each region proposal is passed to the CNN for feature extraction. This may amount to approximately 2,000 passes per image.

- Three different models need to be trained: the CNN for feature extraction, the classifier model to predict the image class, and the regression model to tighten the bounding boxes. The training is compute-intensive and adds to the computation time.

- Each of the region proposals needs to be predicted. Because of the number of regions, the predictions from the CNN are slow.

Fast R-CNN

To overcome the limitations of R-CNN, Ross Girshick from Microsoft published a paper in 2015 titled "Fast R-CNN" that proposed a single model to learn and output regions and classifications directly (https://arxiv.org/pdf/1504.08083.pdf).

A Fast R-CNN also uses an algorithm, for example edge boxes, to generate region proposals. Unlike an R-CNN, which crops and resizes region proposals, the Fast R-CNN processes the entire image. Instead of classifying each region, the Fast R-CNN pools the CNN features corresponding to each region proposal.

Figure 6-4 shows the Fast R-CNN architecture. It takes the entire image as input and generates a set of region proposals. The last layer of the deep CNN has a special layer called the *region of interest (ROI)* pooling layer. The ROI pooling layer extracts a fixedlength feature vector from the feature map specific for a given input candidate region.

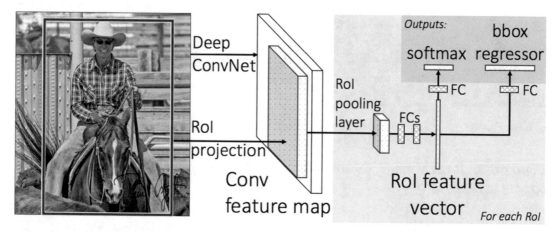

Figure 6-4. *Fast R-CNN architecture (Source: Ross Girshick, https://arxiv.org/ pdf/1504.08083.pdf)*

Each ROI feature vector from the ROI pool is fed to a fully connected MLP that generates two sets of outputs—one for the object class and the other for the bounding boxes. The softmax activation function predicts the object class, and a linear regressor generates the bounding boxes corresponding to the predicted class. The process is repeated for each region of interest from the ROI pool.

As the original paper describes, Ross Girshick applied the Fast R-CNN with VGG16 to the Microsoft COCO dataset to establish a preliminary baseline. The COCO dataset (https://cocodataset.org/) is a large-scale object detection, segmentation, and captioning dataset available in the public domain for free. The Fast R-CNN training set consists of 80,000 images, and the training was iterated for 240,000 epochs. The model quality was assessed as follows:

- The mean average precision (mAP) with the PASCAL object dataset: 35.9 percent

- The average precision (AP) with the COCO dataset: 19.7 percent

Compared to an R-CNN, the Fast R-CNN is much faster to train and make predictions. However, it still needs a set of candidate region proposals with each input image, and a separate model predicts the regions.

Faster R-CNN

Shaoqing Ren et al. at Microsoft Research published a paper in 2016 titled "Faster R-CNN: Towards Real-Time Object Detection with Region Proposal Networks" (https://arxiv.org/pdf/1506.01497.pdf). This paper describes an improved version of the Fast R-CNN from the training speed and detection accuracy perspectives. Except for the region proposal method, a Faster R-CNN is architecturally similar to a Fast R-CNN.

A Faster R-CNN architecture consists of a *region proposal network (RPN)* that shares the full-image convolutional features with the detection network, thus enabling nearly cost-free region proposals. An RPN is a fully convolutional network. It simultaneously predicts object bounds and objectness scores at each position of the image. The RPN is trained end to end to generate high-quality region proposals. These region proposals are used by the Fast R-CNN for detection. This is illustrated in Figure 6-5.

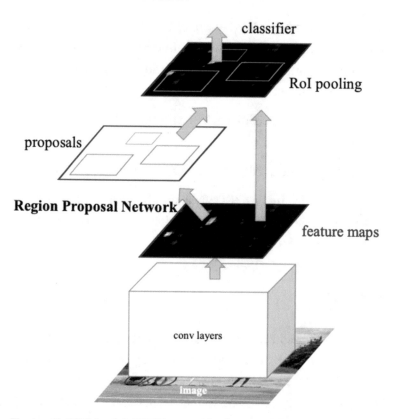

***Figure 6-5.** Faster R-CNN with RPN, a unified network for faster object detection (Source: Shaoqing Ren et al., https://arxiv.org/pdf/1506.01497.pdf)*

A Faster R-CNN consists of two parts: the RPN and the Fast R-CNN.

Region Proposal Network

An RPN is a deep CNN that takes an image input and generates output as a set of rectangular object proposals. Each rectangular proposal has an objectness score.

Figure 6-6 shows how an RPN generates region proposals. We take the convolutional feature map generated by the last shared convolutional layer and slide a small network. This small network takes as input an $n \times n$ spatial window of the input convolutional feature map. Each sliding window is mapped to a lower-dimensional feature, such as a 256-dimensional feature for AlexNet or a 5,126-dimensional feature for VGG-16.

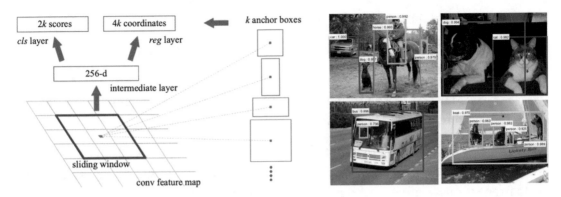

Figure 6-6. *Region detection using sliding window and anchor (Source: Shaoqing Ren, et al, https://arxiv.org/pdf/1506.01497.pdf)*

This feature is fed into two sibling fully connected layers—a box-regression layer for predicting bounding boxes and a box-classification layer for predicting object classes.

Multiple region proposals are predicted at each sliding window location. Assuming the maximum number of proposals at each window location is k, the total number of bounding boxes coordinates will be 4k, and the number of object classes will be 2k (one for the probability of being an object and the other for the probability of not being an object). These region boxes at each window are called *anchors*.

Fast R-CNN

The second part of the Faster R-CNN is the detection network. This part is exactly the same as the Fast R-CNN (as described earlier). The Fast R-CNN takes input from the RPN to detect objects in images.

Mask R-CNN

The Mask R-CNN extends the Faster R-CNN. The Faster R-CNN is widely used for object detection tasks because of its speed of detection. We have already seen that, for a given image, Faster R-CNN predicts the class label and bounding box coordinates for each object in the image. The Mask R-CNN adds an extra branch for predicting an object mask along with the object class and bounding box coordinates (review the concept of masking in Chapter 3).

Here is how the Mask R-CNN differs from its predecessor, the Faster R-CNN:

- The Faster R-CNN has two outputs: a class label and bounding box coordinates.

- The Mask R-CNN has three outputs: a class label, bounding box coordinates, and object mask.

Ross Girshick et al. explained Mask R-CNN in their 2017 paper titled "Mask R-CNN" (https://arxiv.org/pdf/1703.06870.pdf). In the Mask R-CNN, each pixel is classified into a fixed set of categories without differentiating object instances. It introduces a concept called *pixel-to-pixel alignment* between the output and input layers of the neural network. The class of each pixel determines the masks in the ROI.

Figure 6-7 illustrates the Mask R-CNN network architecture.

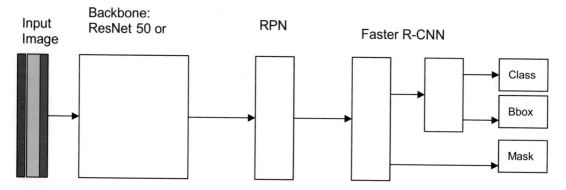

Figure 6-7. *Mask R-CNN with additional mask prediction branch in a Faster R-CNN*

As shown in Figure 6-7, the network consists of three modules—backbone, RPN, and output head.

Backbone

The backbone is the standard deep neural network. The original Mask R-CNN paper describes using ResNet-50 and ResNet-101. The backbone's main role is feature extraction.

In addition to ResNet, a *feature pyramid network (FPN)* is used to extract the finer feature details of the image. The FPN consists of decreasing size layers of a CNN in which case each forward layer has fewer number of neurons. As shown in Figure 6-8, each higher layer passes the features to the lower layers, and predictions are done at each layer. The size of the higher layer is smaller, which means the feature size will be smaller than the previous layers. This approach captures features of the image at different scales, thus allowing it to detect smaller objects in the image.

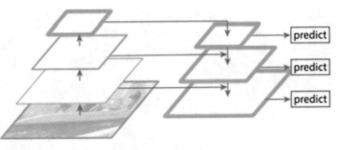

Figure 6-8. *FPN (Source: Tsung-Yi Lin, et al.)*

FPN is an add-on to the backbone network and is typically performes independently of the ResNet or other backbone network. FPN can be added not only to the Mask R-CNN but also to the Fast R-CNN to be able to detect objects of different sizes.

RPN

As described earlier, the RPN module is used for generating region proposals. The RPN architecture in the case of Mask R-CNNs is the same as in the case of Faster R-CNNs.

Output Head

As shown in Figure 6-8, the last module consists of the Faster R-CNN with an additional output branch. Therefore, a total of three outputs are generated by this module. The outputs—an object class and bounding box coordinates—are the same as in the case of Faster R-CNNs. The third output is the object mask, which is a list of pixels defining the object contours.

What Is the Significance of the Masks?

The Mask R-CNN (like the Faster R-CNN) generates object classes and the bounding boxes. The combination of these two helps us locate the objects within the image. The mask output from the network is used in object segmentation. This object segmentation is popularly used in optical character recognition (OCR) to extract text from documents.

Another example of usage of a Mask R-CNN is in airport security where travelers' bags are scanned and visualized with masking. Figure 6-9 shows a typical display of a Mask R-CNN.

Figure 6-9. *Display of images with bounding boxes and masks (Source: Ross Girshick et al.)*

Mask R-CNN in Human Pose Estimation

An interesting use of a Mask R-CNN is in estimating the human pose. The network can be extended to model locations of keypoints as a one-hot mask. Keypoints are defined as the points of interest on the image. For humans, these keypoints represent major joints such as an elbow, shoulder, or knee. The keypoints are selected such that they do not change with the rotation, movement, shrinkage, translation, and distortion. The Mask R-CNN is trained to predict K masks, one for each of K keypoint types (e.g., left shoulder, right elbow). See Figure 6-10.

Figure 6-10. *Display of human pose estimation using keypoint prediction (Source: Ross Girshick, et al.)*

To train a network to estimate human pose, the training images are marked with K keypoints of an instance object. For each keypoint, the training target is a one-hot $m \times m$ binary mask where only a single pixel is labeled as the foreground.

According to the original paper, the authors used a variant of ResNet-FPN architecture as the feature extraction backbone. The head architecture (or output module) was similar to the regular Mask R-CNN. The keypoint head consisted of a stack of eight 3×3 512-D convolution layers, followed by a deconvolutional layer and 2× bilinear upscaling. This produced an output resolution of 56×56. It was estimated that a relatively high-resolution output (compared to masks) is required for keypoint-level localization accuracy.

Single-Shot Multibox Detection

An R-CNN and its variants are two-stage detectors. They have two dedicated networks: one network generates the region proposals to predict bounding boxes, and the other network predicts object classes. These two-stage detectors are fairly accurate, but they come with a high computational cost. This means these detectors are not suitable for detecting objects in streaming videos in real time.

A single-shot object detector predicts both the bounding boxes and the object classes in a single forward pass of the network.

Single-shot multibox detection (SSD) was explained by Wei Liu et al. in a 2016 paper titled "SSD: Single Shot MultiBox Detector" (`https://arxiv.org/pdf/1512.02325.pdf`). First we will review how SSD works, and later in this chapter, we will train a custom SSD model using TensorFlow.

SSD Network Architecture

An SSD neural network consists of two components: base network and prediction/detection network.

- *Base network*: The base network is a deep convolutional network that is truncated before any classification layer. For example, remove the fully connected layer of ResNet or VGG to create the base network for SSD. The base network is used for feature extraction from the input images.

- *Detection network*: To the base network, attach some extra convolutional layers that will actually do the prediction of bounding boxes and object classes. The detection network has the following characteristics -- multiscale feature maps and default boxes and aspect ratios.

Multiscale Feature Maps for Detection

The convolutional layers attached to the end of the base network are designed in such a way that these layers decrease in size progressively. This allows it to predict objects at multiple scales. This can be visualized as shown in Figure 6-11.

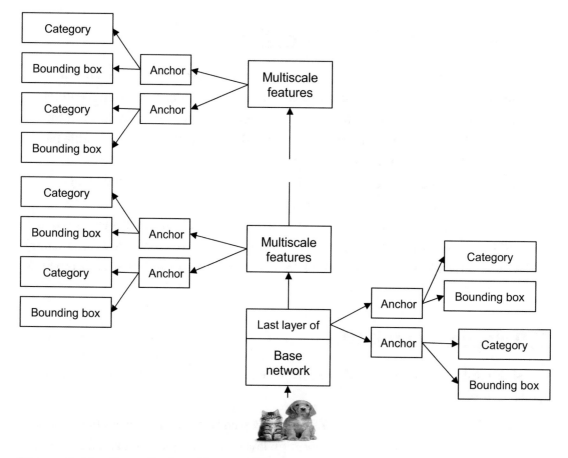

Figure 6-11. *Convolutional layers of decreasing size to predict object class categories and bounding boxes at scales*

As shown in Figure 6-11, every detection layer and, optionally, the last layer of the base network predicts offsets of the four coordinates of the bounding boxes and object class categories. Bounding boxes and objects are predicted through *anchor boxes*, described next.

Anchor Boxes and Convolutional Predictors for Detection

Anchors are one or more rectangular shapes set at each convolution point of the feature map. In Figure 6-12, there are five rectangular anchors (shown in red outlines) set at a point (shown in blue).

In SSD, typically five anchor boxes are selected at each point. Each of these anchors acts as a detector. That means there are typically five detectors at each location of the feature map, and each one of them detects five different objects (or no object). The varying size of these detectors allows them to detect objects of different sizes. Smaller detectors will detect smaller objects, and larger detectors are capable of detecting larger objects.

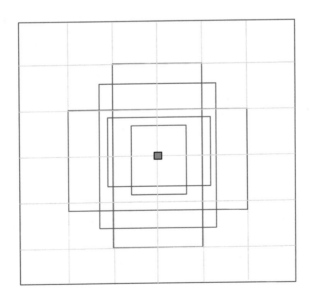

Figure 6-12. *Anchor boxes*

At each convolution point on the feature map (shown in blue in Figure 6-12), the algorithm predicts offsets of bounding boxes relative to anchor boxes. It also predicts the class scores that indicate the presence of a class instance in each of these boxes.

Default Boxes and Aspect Ratios

It is important to note that these anchors are chosen beforehand as constants. In SSD, a set of fixed "default anchors" is mapped at each convolution point.

Assume that there are K number of boxes at each location; we compute C class scores and the four offset coordinates relative to the default box. This will result in a total of $(C+4)\times K$ filters around each convolution point. Assuming the feature size of $m\times n$, the output tensor size will be $(C+4)\times K\times m\times n$.

These default anchors are applied at each of the detection convolutional layers (as shown in Figure 6-11). The size of these convolutional layers decreases progressively, allowing these layers to generate several feature maps of different resolutions.

Figure 6-13 shows the overall network architecture.

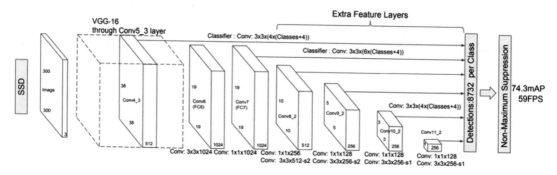

Figure 6-13. *Truncated VGG backbone with additional convolutional layers for detection (Source: Liu et al., https://arxiv.org/pdf/1512.02325.pdf)*

Training

The training process of an SSD model for object detection involves simultaneously predicting object class scores and bounding box coordinates at multiple scales across the image. The training process for optimizing loss functions for bounding box detection and the strategy of object matching is described below:

Matching Strategy

During network training, the algorithm determines which default boxes correspond to the ground truth and then trains the network accordingly. To match the default boxes with the ground truth, the algorithm uses IoU to determine the overlap. This IoU-based overlap is also called the Jaccard overlap. An IoU threshold of 0.5 is considered to determine whether the default box overlaps any ground truth. This overlapping using

276

IoU is performed at each layer, allowing the network to learn at scale. The SSD learning process starts with the default boxes as predictions and attempts to regress closer to the ground truth bounding boxes. Figure 6-14 illustrates the concept of overlapping and selection of default boxes.

Figure 6-14. *Matching of default box with ground truth box*

Training Objective

SSD's learning objective is to optimize a loss function, which is the weighted sum of the localization loss (loc) and the confidence loss (conf) over all matched default boxes.

Choosing Scales and Aspect Ratios for Default Boxes

The decreasing size of the detection layers of the SSD network allows it to learn different object scales. As the training moves forward, the size of the feature map decreases. How does the algorithm determine the size of default boxes for each layer?

For each layer, the algorithm calculates the scale using the following formula:

$$S_k = S_{min} + \{(S_{max} - S_{min})/(m - 1)\} (k - 1), \text{ where } k \in [1, m]$$

where m is the size of the feature map, $S_{min} = 0.2$ for the lowest layer, and $S_{max} = 0.9$ for the highest layer. All other layers in between are evenly spaced. Recall that five default boxes are used in SSD. These default boxes are set for different aspect ratios: $a_r \in \{1, 2, 3, \frac{1}{2}, \frac{1}{3}\}$. The width and height of each default boxes are calculated using the following formula:

$$\text{Width} = S_k \sqrt{a_r}$$

$$\text{Height} = S_k / \sqrt{a_r}$$

For an aspect ratio of 1, another box of scale $S'_k = \sqrt{(S_k S_{k+1})}$ is calculated. That means six default boxes per feature map are determined. The center of the default boxes are set using this formula: ($(i+0.5)/ |fk|$, $(j+0.5)/ |fk|$), where $|fk|$ is the size of the k-th square feature map, i, j $\in$ [0, |fk|).

By combining predictions for all default boxes with different scales and aspect ratios from all locations of many feature maps, a diverse set of predictions is generated. This covers various input object sizes and shapes.

We will see in the upcoming "YOLO" section that YOLO uses K-means clustering to dynamically select anchor boxes. Also, in YOLO, these anchors are called *priors* or *bounding box priors*.

Hard Negative Mining

At each layer and for each feature map, many default boxes are created. After matching with the ground truth (where IoU $\geq$ 0.5), the majority of these default boxes will not overlap with the ground truth. These nonoverlapping default boxes (IoU < 0.5) are called *negative boxes*, and those matching with ground truth are positive boxes. In most cases, the number of negatives is way higher than the number of positives. This causes class imbalance, which will skew the predictions. To balance the classes, negative boxes are sorted, the topmost probable negative boxes are taken, and the rest are discarded to make the negative:positive ratio at most 3:1. It has been found that this ratio leads to faster optimization.

Data Augmentation

SSD is robust to various input object sizes and shapes. To make it robust, each training image is sampled by one of the following options:

- Use the entire original image.

- Sample a patch so that the minimum IoU is 0.1, 0.3, 0.5, 0.7, or 0.9.

- Randomly sample a patch.

The characteristics of each sample are the following:

- The size of each sampled patch is [0.1, 1] of the original image size.

- The aspect ratio is between ½ and 2.

- Keep the overlapped part of the ground truth box if the center of it is in the sampled patch.

After these sampling steps, each sampled patch is resized to a fixed size and is horizontally flipped with a probability of 0.5, in addition to applying some photometric distortions.

Nonmaximum Suppression

At the time of inference, a large number of boxes are generated during the forward pass of the SSD learning process. Processing all of these bounding boxes will be compute-intensive and time-consuming. Therefore, it is important to get rid of those bounding boxes, which have low confidence of containing objects and have low IoU. Only the top N bounding boxes having the maximum IoU and confidence are selected, and nonmaximum boxes are dropped or suppressed. This eliminates duplicates and ensures that only the most likely predictions are retained by the network.

SSD Results

SSD is a fast, robust, and accurate model. With the VGG-16 base architecture, SSD compares favorably to its state-of-the-art object detector counterparts in terms of both accuracy and speed. The SSD-512 model (the highest-resolution network using 512×512 input images) is at least three times faster and more accurate compared to the state-ofthe-art Faster R-CNN on the PASCAL VOC and COCO datasets. The SSD-300 model performs real-time object detection more accurately in streaming video at 59 frames per second (FPS) speed, which is faster than the first version of YOLO. In Chapter 7, we will explore how to detect objects in videos using SSD.

YOLO

YOLO is a fast, real-time, and multi-object detection algorithm. YOLO consists of a single convolutional neural network that predicts simultaneously the bounding boxes and class probabilities of objects within them. YOLO trains on the full image, and the network is set up to solve regression problems to detect objects. Therefore, YOLO does not need a complex processing pipeline, which makes it extremely fast.

A base network runs 45 frames per second on an Nvidia Titan X GPU. The speed is higher with faster versions of GPU, and it could go up to 150 FPS. This makes YOLO suitable for detecting objects in streaming videos in real time with less than 25 milliseconds of latency. Furthermore, YOLO achieves more than twice the mAP of other real-time systems.

YOLO was introduced by Joseph Redmon, Santosh Divvala, Ross Girshick, and Ali Farhadi in 2016 in their paper titled "You Only Look Once: Unified, Real-Time Object Detection" (https://arxiv.org/pdf/1506.02640.pdf).

The detection process, as illustrated in Figure 6-15 and described in the original paper, is as follows:

1. The input image is divided into S×S grids.

2. If the center of the object falls within a grid, that grid is responsible for detecting that object.

3. Each grid cell predicts B number of bounding boxes and a confidence score for these bounding boxes.

4. The confidence score is calculated using the following formula:

   Confidence score = probability of objectness × IOU between the predicted box and the ground truth

   If the bounding box does not contain any object, the confidence score is zero.

5. For each bounding box, the network makes five predictions: x, y, w, h, and confidence, where

   - The (x, y) coordinates represent the center of the box relative to the bounds of the grid cell.

   - w and h are the width and height relative to the whole image.

   - The confidence prediction represents the IOU between the predicted box and any ground truth box.

6. At the same time, the network predicts, for each grid cell, a class conditional probability C conditioned on the grid cell containing an object. Only one conditional probability per grid cell is predicted, regardless of how many bounding boxes B are predicted.

7. To obtain the class-specific confidence score for each box, the following formula is applied:

   Class confidence score = Pr(Classi|Object) × Pr(Object) × IOU between prediction and ground truth

   where Pr(Classi|Object) represents the probability of a class given the object within the grid cell.

8. These predictions are encoded as an S × S × (B × 5 + C) tensor.

   The inventors of YOLO used the following settings for evaluation:

   - Dataset: PASCAL Visual Object Classes, `http://host.robots.ox.ac.uk/pascal/VOC/` (PASCAL VOC)

   - S = 7

   - B = 2

   - C = 20, as PASCAL VOC had 20 object classes

The final prediction yielded a 7 × 7 × (2 × 5 + 20) = 7 × 7 × 30 tensor.

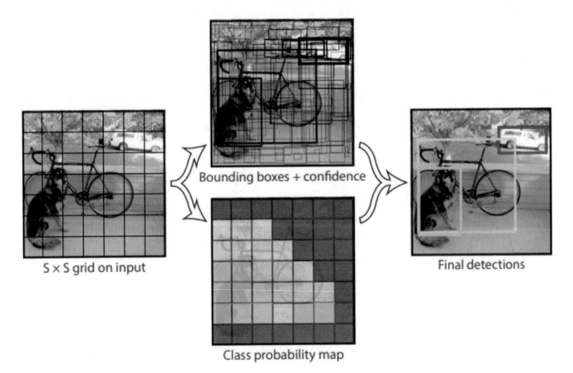

Figure 6-15. *Illustration of YOLO object detection (Source: Joseph Redmon et al.)*

YOLO Network Design

The YOLO network architecture was inspired by GoogLeNet for image classification. A slightly modified GoogLeNet for YOLO consists of 24 convolutional layers with max pooling followed by two fully connected layers. Notice the output tensor or dimension 7×7×30 generated from the last layer shown in a full network in Figure 6-16.

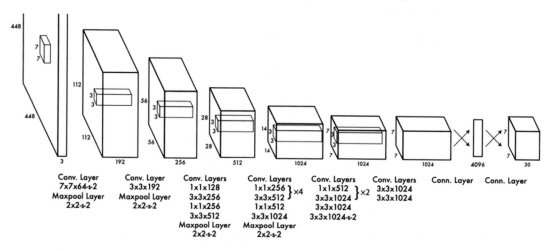

Figure 6-16. *YOLO neural network architecture (Source: Joseph Redmon et al.)*

Limitations of YOLO

Although YOLO is one of the fastest object detection algorithms, it has a few limitations:

- It struggles with small objects that come in groups such as flocks of birds.

- It can predict only one class of objects within a cell grid.

- It does not predict well if the object has an unusual aspect ratio that was not seen in the training set.

- Its accuracy is less than some of the state-of-the-art algorithms, such as the Faster R-CNN.

YOLO9000 or YOLOv2

YOLOv2 is an improved version of YOLO. It improves detection accuracy and speed compared to YOLO. It was trained to detect more than 9,000 object classes; therefore, the name YOLO9000 was given to it. This improvement and the detection algorithm are described in the paper titled "YOLO9000: Better, Faster, Stronger" published in December 2016 by Joseph Redmon and Ali Farhadi (`https://arxiv.org/pdf/1612.08242.pdf`).

YOLOv2 was designed to overcome some of the limitations, especially the precision and recall levels, of YOLO. Furthermore, it is able to detect objects with unseen aspect ratios.

Here is a list of improvements made in YOLOv2 to achieve a better, faster, and stronger result:

- *Batch normalization*: YOLOv2 adds batch normalization on all the convolutional layers in YOLO. Recall that batch normalization helps regularize the model. By using batch normalization, YOLOv2 shows a mAP improvement of more than 2 percent.

- *High-resolution classifier*: YOLOv2 is fine-tuned to learn from higher-resolution input images. At 448×448 resolution, the network output is improved by 4 percent mAP.

- *Convolution with anchor boxes*: YOLOv2 removed the fully connected layers and uses fully convolutional layers. It also introduced anchor boxes to predict bounding boxes. Although there is a slight decrease in accuracy, by using anchor boxes, YOLOv2 is able to detect more than 1,000 objects per image compared to 98 in YOLO.

- *Dimension cluster*: The size of the anchor boxes is determined by using K-means clustering of the VOC 2017 training set. A value of k=5 provides the best trade-off between average IOU/model complexity. The average IOU is 61.0 percent.

- *Fine-grained features*: YOLOv2 uses a pass-through layer that concatenates the higher-resolution features by stacking adjacent features into different channels instead of spatial locations. This approach gives a modest 1 percent performance increase.

- *Multiscale training*: YOLOv2 is able to detect objects in images of different sizes. Instead of fixing the input image size, YOLOv2 changes the network on the fly every few iterations. For example, every ten batches the network randomly chooses a new image dimension. This means the same network can predict detections at different resolutions. At low resolutions, YOLOv2 operates as a cheap and fairly accurate detector.

  The 288×288 YOLOv2 network runs at more than 90 FPS with mAP almost as good as Fast R-CNN. This makes it ideal for smaller GPUs, high-frame-rate video, or multiple video streams. At high resolution, YOLOv2 is a state-of-the-art detector with 78.6 mAP on VOC 2007 while still operating at faster than real-time speeds.

- *DarkNet instead of GoogLeNet*: YOLOv2 uses a CNN called DarkNet-19. This network has 19 convolutional layers and 5 max pooling layers. Darknet-19 only requires 5.58 billion operations, as opposed to 30.67 billion in VGG or 8.52 billion in YOLO, to process an image. Yet it achieves 72.9 percent top-one accuracy and 91.2 percent top-five accuracy on ImageNet. Figure 6-17 shows the Darknet-19 network architecture.

Type	Filters	Size/Stride	Output
Convolutional	32	3 × 3	224 × 224
Maxpool		2 × 2/2	112 × 112
Convolutional	64	3 × 3	112 × 112
Maxpool		2 × 2/2	56 × 56
Convolutional	128	3 × 3	56 × 56
Convolutional	64	1 × 1	56 × 56
Convolutional	128	3 × 3	56 × 56
Maxpool		2 × 2/2	28 × 28
Convolutional	256	3 × 3	28 × 28
Convolutional	128	1 × 1	28 × 28
Convolutional	256	3 × 3	28 × 28
Maxpool		2 × 2/2	14 × 14
Convolutional	512	3 × 3	14 × 14
Convolutional	256	1 × 1	14 × 14
Convolutional	512	3 × 3	14 × 14
Convolutional	256	1 × 1	14 × 14
Convolutional	512	3 × 3	14 × 14
Maxpool		2 × 2/2	7 × 7
Convolutional	1024	3 × 3	7 × 7
Convolutional	512	1 × 1	7 × 7
Convolutional	1024	3 × 3	7 × 7
Convolutional	512	1 × 1	7 × 7
Convolutional	1024	3 × 3	7 × 7
Convolutional	1000	1 × 1	7 × 7
Avgpool		Global	1000
Softmax			

Figure 6-17. *Darknet-19 (Source: Joseph Redmon et al.,* `https://arxiv.org/`
`pdf/1612.08242.pdf`*)*

- *Joint classification and detection*: YOLOv2 can learn from a dataset
  containing labels for both classification and detection. During the
  training, when the network sees images labeled for detection, it
  performs the full YOLOv2 loss function optimization. And, when
  it sees images for classification, it backpropagates losses using
  the classification part of the network. The dataset for YOLOv2 was
  created by combining datasets from COCO and ImageNet. The
  network, capable of learning from both classification and detection
  dataset, makes a stronger model compared to plain YOLO.

The following table summarizes the YOLOv2 improvements and their effect on accuracy and speed (compared to plain YOLO):

	Modifications	Effects
Better	Batch normalization	2% mAP improvement
	High-resolution classifier	4% mAP improvement
	Convolution with anchor boxes	Capable of detecting more than 1,000 objects per image
	Dimension cluster	4.8% mAP improvement
	Fine-grained features	1% mAP improvement
	Multiscale training	1.1% mAP improvement
Faster	Darknet-19	33% computation decrease, 0.4 percent mAP improvement
	Convolutional prediction layer	0.3% mAP improvement
Stronger	Joint classification and detection	Able to detect more than 9,000 objects

YOLOv3

Another improved version of YOLO is YOLOv3, which provides some improvements to YOLOv2. YOLOv3 is described in the paper titled "YOLOv3: An Incremental Improvement" published in April 2018 by Joseph Redmon and Ali Farhadi (https://arxiv.org/pdf/1804.02767.pdf).

The features and improvements of YOLOv3 are described here:

- *Bounding box prediction*: There is no change in YOLOv3 compared to YOLOv2 when it comes to detecting bounding boxes. YOLOv3 uses sum of squared error loss during the training. It also predicts an objectness score for each bounding box using logistic regression. The objectness score is taken as 1 if the bounding box prior overlaps a ground truth object by more than any other bounding box prior. Only one bounding box prior is assigned for each ground truth object.

If the bounding box prior is not the best but does overlap a ground truth object by more than some threshold, the prediction is ignored. The inventors of YOLOv3 used a threshold of 0.5. The system assigns only one bounding box prior for each ground truth object.

- *Object class prediction*: The network predicts multiple classes of an object within a bounding box. The softmax activation function is not suitable for predicting multilabel classes. Therefore, YOLOv3 uses a regression classifier instead of softmax.

- *Predictions across scales*: YOLOv3 predicts bounding boxes at three different scales. It still uses the K-means cluster to determine bounding box priors. It has nine clusters and three scales arbitrarily selected, and then it divides the clusters evenly across scales.

  For example, on the COCO dataset, the nine clusters were as follows: (10×13), (16×30), (33×23), (30×61), (62×45), (59×119), (116×90), (156×198), (373×326).

- *Feature extractor*: As a feature extraction backbone, YOLOv3 uses an improved version of Darknet-19. This network was given the name Darknet-53. It has 53 convolutional layers. Figure 6-18 shows the Darknet-53 network architecture.

	Type	Filters	Size	Output
	Convolutional	32	3 × 3	256 × 256
	Convolutional	64	3 × 3 / 2	128 × 128
	Convolutional	32	1 × 1	
1×	Convolutional	64	3 × 3	
	Residual			128 × 128
	Convolutional	128	3 × 3 / 2	64 × 64
	Convolutional	64	1 × 1	
2×	Convolutional	128	3 × 3	
	Residual			64 × 64
	Convolutional	256	3 × 3 / 2	32 × 32
	Convolutional	128	1 × 1	
8×	Convolutional	256	3 × 3	
	Residual			32 × 32
	Convolutional	512	3 × 3 / 2	16 × 16
	Convolutional	256	1 × 1	
8×	Convolutional	512	3 × 3	
	Residual			16 × 16
	Convolutional	1024	3 × 3 / 2	8 × 8
	Convolutional	512	1 × 1	
4×	Convolutional	1024	3 × 3	
	Residual			8 × 8
	Avgpool		Global	
	Connected		1000	
	Softmax			

Figure 6-18. *Darknet-53 used in YOLOv3 (Source: https://arxiv.org/pdf/1804.02767.pdf)*

- *Training*: There was no change in the training approach in YOLOv3 compared to YOLOv2. The training was performed on the full image, with multiscaled data, batch normalization, and mixed classification and detection labels.

Here are the YOLOv3 results:

- For the overall mAP, YOLOv3 performance drops significantly due to a much wider network (53 layers compared to 19 in YOLOv2).

- YOLOv3 with 608×608-resolution images got 33.0 percent mAP in 51ms inference time, while RetinaNet-101–50–500 only got 32.5 percent mAP in 73ms inference time.

- YOLOv3's accuracy level is on par with SSD variants with a 3× faster detection time.

YOLOv4

YOLOv4 is an improved version of YOLO, built upon the success and lessons learned from YOLOv3. The original paper of YOLOv4, by Alex et al, is located at `https://arxiv.org/pdf/2004.10934.pdf`.

YOLOv4 introduces several significant improvements over YOLOv3:

- *Backbone architecture*: YOLOv4 utilizes a more advanced backbone network called CSPDarknet53, which is based on the Darknet architecture. It incorporates Cross-Stage Partial Network (CSPNet) modules, enabling more efficient and accurate feature extraction.

- *Neck architecture*: YOLOv4 introduces a neck architecture known as Spatial Pyramid Pooling (SPP) and Path Aggregation Network (PANet). SPP allows the model to capture contextual information at different scales, while PANet aggregates features from multiple levels to enhance the detection accuracy.

- *Improved anchor generation*: YOLOv4 employs an anchor-free method called Anchor-Free YOLO (YOLOv4-csp), which eliminates the need for anchor boxes. This approach simplifies the bounding box prediction process and improves the accuracy for small objects.

- *Use of advanced techniques*: YOLOv4 incorporates various advanced techniques such as Mish activation function, Complete Intersection over Union (CIOU) loss function, and self-adversarial training. These techniques contribute to better gradient flow, enhanced localization precision, and improved overall performance.

- *Training strategies*: YOLOv4 employs different training strategies such as random shapes training, mosaic data augmentation, and modified data augmentation policies. These strategies enhance the model's ability to generalize and perform well under various scenarios.

YOLOv7

The YOLO object detection algorithm has undergone multiple iterations, each building upon the strengths of its predecessors. The most recent version represents a significant advancement over the earlier renditions. YOLOv4 stands as the ultimate achievement within the official Darknet project's series of iterations.

However, the landscape of YOLO-based models expanded beyond the official lineage. YOLOv5, along with several other iterations such as YOLOR and YOLOX, emerged as notable examples. These models, though unofficial in the context of the original Darknet project, contributed significantly to the advancement of the YOLO approach.

YOLOv7 has set a new benchmark by outperforming all existing object detectors in terms of both speed and accuracy, operating across a wide spectrum ranging from 5 FPS to 160 FPS. Notably, it has an unparalleled accuracy with a 56.8 percent average precision (AP), making it the frontrunner among real-time object detectors operating at 30 FPS or higher on the GPU V100 platform.

The official YOLOv7 paper, "YOLOv7: Trainable bag-of-freebies sets new state-of-the-art for real-time object detectors" (`https://arxiv.org/pdf/2207.02696.pdf`) was released in July 2022 by Chien-Yao Wang, Alexey Bochkovskiy, and Hong-Yuan Mark Liao. The source code was released as open source under the GPL-3.0 license, a free copyleft license, and can be found in the official YOLOv7 GitHub repository, `https://github.com/WongKinYiu/yolov7`.

YOLOv7 introduces a significant advancement in real-time object detection accuracy while maintaining efficient inference. In benchmark comparisons, YOLOv7 outperforms other established object detectors by reducing approximately 40 percent of parameters and 50 percent of computation in cutting-edge real-time object detection systems. This leads to quicker inference speeds and heightened detection precision.

In essence, YOLOv7 presents a faster and more robust network architecture that incorporates a more efficient feature integration approach. This, in turn, results in heightened accuracy in object detection, a stronger loss function, and improved efficiency in label assignment and model training. Consequently, YOLOv7 demands significantly less-expensive computational hardware compared to alternative deep learning models. Moreover, it can be trained rapidly on smaller datasets, eliminating the need for pretrained weights.

YOLOv7 Architectural Features

The YOLOv7 paper introduces the following major changes:

- YOLOv7 architecture
  - Extended efficient layer aggregation networks (E-ELAN)
  - Model scaling for concatenation-based models
- Trainable bag-of-freebies
  - Planned re-parameterized convolution
  - Coarse for auxiliary and fine for lead loss

E-ELAN

The E-ELAN is the computational block in the YOLOv7 backbone. It has been designed by analyzing the following factors that impact speed and accuracy:

- Memory access cost
- I/O channel ratio
- Element-wise operation
- Activations
- Gradient path

E-ELAN architecture enables the framework to learn better. It is based on the ELAN computational block. E-ELAN uses expand, shuffle, merge cardinality to achieve the ability to continuously enhance the learning ability of the network without destroying the original gradient path. See Figure 6-19 for a reference architecture of ELAN and E-ELAN.

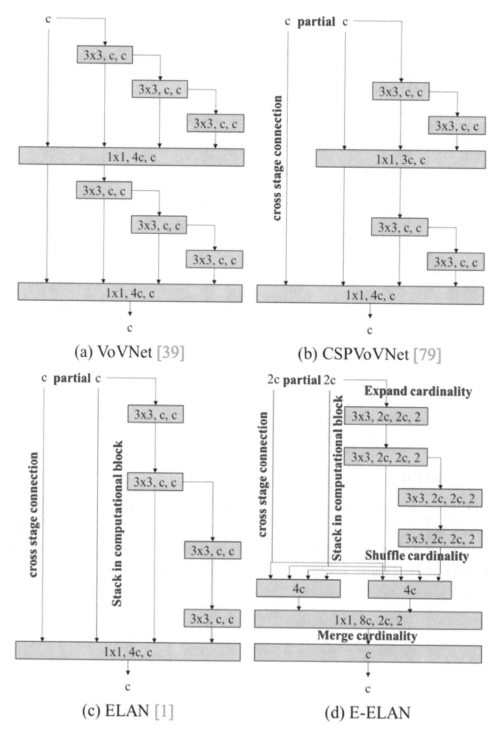

Figure 6-19. *VoVNet, CSPVoVNet, ELAN, and E-ELAN architecture (Source:*
https://arxiv.org/pdf/2207.02696.pdf)

Model Scaling for Concatenation-Based Models

Model scaling primarily aims to modify specific attributes of a model, resulting in the creation of models across various scales. This adaptation addresses the requirement for different inference speeds based on varying needs. For example, the scaling process for EfficientNet involves adjustments in width, depth, and resolution. Conversely, in the case of scaled-YOLOv4, the scaling model revolves around modifying the number of stages.

In conventional methods employing concatenation-based architectures, such as ResNet or PlainNet, it's essential to examine scaling factors collectively rather than in isolation. In other words, when enhancing model depth, this adjustment can trigger a proportional shift between input and output channel ratios within a transition layer. Consequently, this alteration might result in a reduction of the model's hardware utilization.

This is the rationale behind YOLOv7's introduction of compound model scaling within a concatenation-based model framework. The compound scaling approach preserves the inherent characteristics that the model possessed during its initial design, thereby ensuring the retention of the optimal structure.

This is the operational principle of compound model scaling: To illustrate, adjusting the depth factor of a computational block necessitates a corresponding modification in the output channel of that block. Subsequently, the width factor scaling is executed with an equivalent adjustment applied to the transition layers. Figures 6-20 and 6-21 depict this concept.

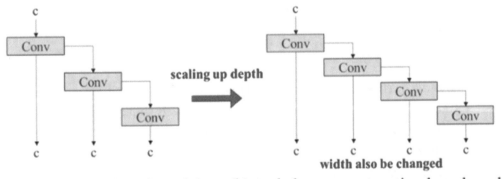

(a) concatenation-based model   (b) scaled-up concatenation-based model

Figure 6-20. *Illustration of concatenation-based and scaled-up concatenation-based models (Source: https://arxiv.org/pdf/2207.02696.pdf)*

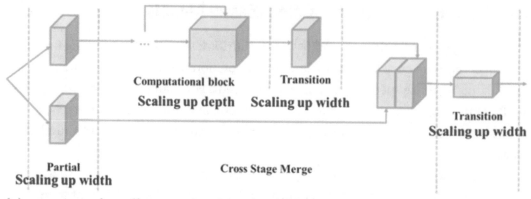

(c) compound scaling up depth and width for concatenation-based model

Figure 6-21. *Illustration of compound scaled-up concatenation-based model (Source: https://arxiv.org/pdf/2207.02696.pdf)*

Planned Re-parameterized Convolution

Re-parameterization is a post-training technique aimed at enhancing the model's performance. While it extends the training duration, it leads to enhanced inference outcomes. There are two categories of re-parameterization employed for model refinement:

- *Model-level re-parameterization*: This can be accomplished through the following two approaches:

  - Training multiple models using distinct training data while maintaining consistent settings. Subsequently, averaging their weights yields the ultimate model.

  - Calculating the average of model weights at various epochs.

- *Module level re-parameterization*: This method has gained a lot of traction in research. The model training process is split into multiple modules. The outputs are ensembled to obtain the final model. The authors in the YOLOv7 paper show the best possible ways to perform module-level ensemble. See Figure 6-22.

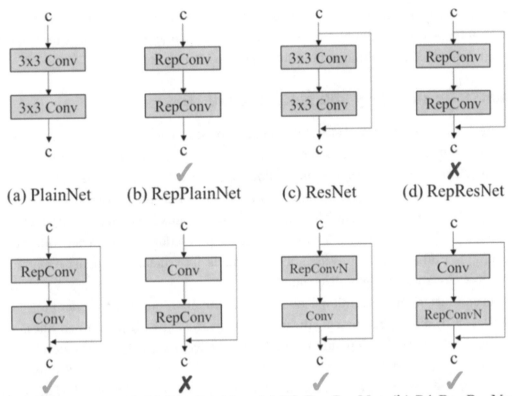

(a) PlainNet   (b) RepPlainNet   (c) ResNet   (d) RepResNet

(e) P1-RepResNet (f) P2-RepResNet (g) P3-RepResNet (h) P4-RepResNet

Figure 6-22. *Planned re-parameterized model (Source:* `https://arxiv.org/pdf/2207.02696.pdf`*)*

As shown in Figure 6-22, the 3×3 convolution layer of the E-ELAN computational block is replaced with the RepConv layer. Experiments were carried out by switching or replacing the positions of RepConv, 3×3 Conv, and Identity connection. The residual bypass arrow shown is an identity connection, which is a 1×1 convolutional layer. This experiment allows determination of what configuration works and what does not.

RepConv actually combines 3×3 convolution, 1×1 convolution, and identity connection in one convolutional layer. After analyzing the combination and corresponding performance of RepConv and different architectures, the YOLOv7 authors found that the identity connection in RepConv destroyed the residual in ResNet and the concatenation in DenseNet, which provided more diversity of gradients for different feature maps. For this reason, they use RepConv without identity connection

(RepConvN) to design the architecture of planned re-parameterized convolution. In their thinking, when a convolutional layer with residual or concatenation is replaced by re-parameterized convolution, there should be no identity connection.

Coarse for Auxiliary and Fine for Lead Loss

As previously described, YOLO architecture comprises a backbone, a neck, and a head. The head contains the predicted outputs. YOLOv7 uses multihead architecture.

In YOLOv7, the head responsible for final output is called the *lead head*, and the head used to assist training in the middle layers is called the *auxiliary head*.

With the help of an assistant loss, the weights of the auxiliary heads are updated.

Label assigner is a mechanism that merges the predictions from the network with the ground truth, subsequently assigning soft labels. It's crucial to emphasize that this process yields soft and generalized labels.

In contrast to the conventional label assignment technique, which generates rigid labels by directly referencing the ground truth and applying predefined rules, the approach of creating dependable soft labels involves employing calculation and optimization methods. These methods take into account both the quality and distribution of prediction outputs in conjunction with the ground truth. The following two approaches are used in creating soft labels as the training target:

- *Lead head guided label assigner*: Primarily relies on the prediction outcome of the lead head and the ground truth to compute its values. These values are then employed to generate soft labels using an optimization process. This collection of soft labels serves as the training target for both the auxiliary head and the lead head within the model.

- *Coarse-to-fine lead head guided label assigner*: Also leverages the lead head's predicted outcome and the ground truth to produce soft labels. However, this approach generates two distinct sets of soft labels: coarse labels and fine labels. The fine labels align with the soft labels created by the lead head guided label assigner. On the other hand, the coarse labels are generated by relaxing the constraints of the positive sample assignment process, effectively designating more grid cells as positive targets. Figures 6-23 depicts the lead guided assigner and coarse-to-fine lead guided assigner.

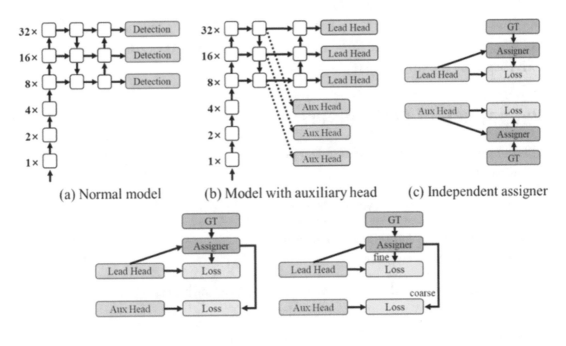

(d) Lead guided assigner (e) Coarse-to-fine lead guided assigner

Figure 6-23. *Example of normal model (a), model with auxiliary head (b), independent assigner (c), lead guided assigner (d) and coarse-to-fine lead guided assigner (e) (Source: https://arxiv.org/pdf/2207.02696.pdf)*

In this chapter, we will explore how to train a custom YOLOv7 model and perform object detection in images and videos.

Comparison of Object Detection Algorithms

Thus far in this chapter, we have explored three distinct algorithm classes for object detection: R-CNN and its variants, SSD and YOLO. These algorithms were trained on two popular datasets—VOC and COCO—and benchmarked for speed and accuracy. The comparison provided in this section can be used as a guide to decide the suitability and applicability of one algorithm versus the other in building systems for object detection. The performance metrics and benchmarking results have been mostly taken from the paper "Object Detection with Deep Learning: A Review" written by Zhong-Qiu Zhao, Peng Zheng, Shou-tao Xu, and Xindong Wu and published in April 2019 (https://arxiv.org/pdf/1807.05511.pdf).

Comparison of Architecture

Table 6-1 provides a comparison of object detection algorithms in terms of the neural network architecture they use.

Table 6-1. *Comparison of Neural Network Architecture of Object Detectors*

Object Detector	Region Proposal	Activation Function	Loss Function	Softmax Layer
R-CNN	Selective search	SGD	Hinge loss (classification), bounding box regression	Yes
Fast R-CNN	Selective search	SGD	Class log loss + bounding box regression	Yes
Faster R-CNN	RPN	SGD	Class log loss + bounding box regression	Yes
Mask R-CNN	RPN	SGD	Class log loss + bounding box regression + semantic sigmoid loss	Yes
SSD	None	SGD	Class sum-squared error loss + bounding box regression	No
YOLO	None	SGD	Class sum-squared error loss + bounding box regression + object confidence + background confidence	Yes
YOLOv2	None	SGD	Class sum-squared error loss + bounding box regression + object confidence + background confidence	Yes
YOLOv3	None	SGD	Class sum-squared error loss + bounding box regression + object confidence + background confidence	Logistic classifier
YOLOv4	None	Mish	CIOU (Complete Intersection over Union)	Logistic classifier

Comparison of Performance

Table 6-2 provides a performance comparison of the object detection algorithms trained on the Microsoft COCO dataset. The training was conducted on an Intel i7-6700K CPU with a single core and an Nvidia Titan X GPU.

Table 6-2. *Performance Comparison of Object Detection Models*

Object Detector	Trained On	mAP	Test Speed (Sec/Image)	Frames per Second (FPS)	Suitable for Real-Time Videos?
R-CNN	COCO 2007	66.0%	32.84	0.03	No
Fast R-CNN	COCO 2007 and 2012	66.9%	1.72	0.60	No
Faster R-CNN (VGG-16)	COCO 2007 and 2012	73.2%	0.11	9.1	No
Faster R-CNN (RestNet-101)	COCO 2007 and 2012	83.8%	2.24	0.4	No
SSD300	COCO 2007 and 2012	74.3%	0.02	46	Yes
SSD512	COCO 2007 and 2012	76.8%	0.05	19	Yes
YOLO	COCO 2007 and 2012	73.4%	0.02	46	Yes
YOLOv2	COCO 2007 and 2012	78.6%	0.03	40	Yes
YOLOv3 608×608	COCO 2007 and 2012	76.0%	0.029	34	Yes
YOLOv3 416×416	COCO 2007 and 2012	75.9%	0.051	19	Yes
YOLOv4 512×512	COCO 2017	61.71%	0.058	17	Yes

Training Object Detection Model Using TensorFlow

We are now prepared to write code to build and train our own object detection models. We will use the TensorFlow API and write code in Python. Object detection models are very compute-intensive and require a lot of memory and a powerful processor. Most general-purpose laptops or computers may not be able to handle the computations necessary to build and train an object detection model. For example, a MacBook Air with 32GB RAM and an eight-core CPU is not able to train a detection model involving about 7,000 images. Thankfully, Google provides a limited amount of GPU-based computing for free. It has been proven that these models run many folds faster on a GPU than on a CPU. Therefore, it is important to learn how to train a model on a GPU. For the purposes of demonstration and learning, we will use the free version of Google GPU. Let's first define what our learning objective is and how we want to achieve it.

- *Objective*: Learn how to train an object detection model using Keras and TensorFlow.

- *Dataset*: The Oxford-IIIT Pet dataset, which is freely available at `https://robots.ox.ac.uk/~vgg/data/pets/`. The dataset consists of 37 categories of pets with roughly 200 images for each class. The images have large variations in scale, pose, and lighting. They are already annotated with bounding boxes and labeled.

- *Execution environment*: We will use Google Colaboratory (`https://colab.research.google.com`), or Colab for short. We will utilize the GPU hardware accelerator that comes free with Colab. Google Colab is a free Jupyter notebook environment that requires no setup and runs entirely in the cloud. Jupyter notebook is an open source web-based application to write and execute Python programs. To learn more about how to use a Jupyter notebook, visit `https://jupyter.org`. The documentation is available at `https://jupyter-notebook.readthedocs.io/en/stable/`. We will learn the Colab notebook environment as we work through the code.

We will train the detection model with TensorFlow 2.12.0 on Google Colab, and after the model is trained, we will download and use it with a local environment running on our laptop.

TensorFlow on Google Colab with GPU

Google Colab offers a complimentary Jupyter notebook environment for machine learning education and training. It provides resources such as approximately 13GB of RAM, 80GB of disk space, and an Nvidia GPU, allowing for continuous usage up to 12 hours. If your session expires or you reach the 12-hour limit, you have the option to re-create the Colab runtime.

When you execute code in Colab, it runs on a virtual machine that is created specifically for your personal account. Once the session expires, the VM is terminated, and any data saved in the virtual disk is lost. However, Colab offers a convenient method to mount the Google Drive directory to the Colab virtual disk, enabling you to store data on the permanent Google Drive. This allows you to retrieve the data whenever you create a new Google Colab session.

Let's begin by utilizing Google Colab and configuring its settings.

Accessing Google Colab

To access Google Colab, you need a Google (or Gmail) account. If you don't already have one, you can sign up for an account at `https://accounts.google.com`.

To begin using Google Colab, open your web browser and navigate to `https://colab.research.google.com`. If you are already signed in to your Google account, you will have immediate access to Colab. If you're not signed in, do so using your Google account credentials to gain access to Google Colab.

Connecting to the Hosted Runtime

Click the Connect button located at the top right of the screen, below the user and setting icons, and then click "Connect to a hosted runtime" (see Figure 6-24). At this point, a Colab session is created.

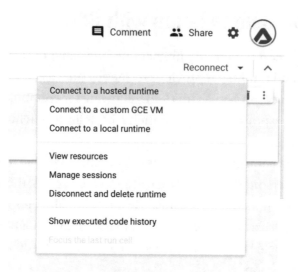

Figure 6-24. *Connecting to a hosted runtime*

Selecting a GPU Hardware Accelerator

Click Edit and then Notebook Settings (see Figure 6-25) to open a modal window. Select GPU as the hardware accelerator. Make sure you have Python 3 selected for the runtime type. Click the Save button (see Figure 6-26).

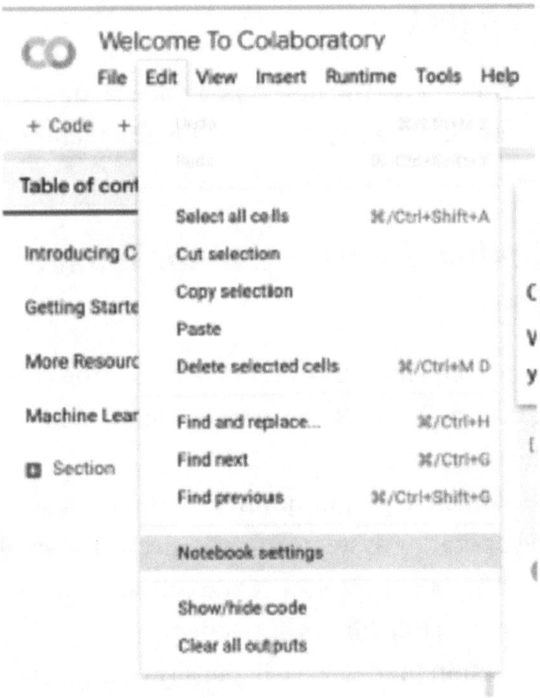

Figure 6-25. *Accessing notebook settings*

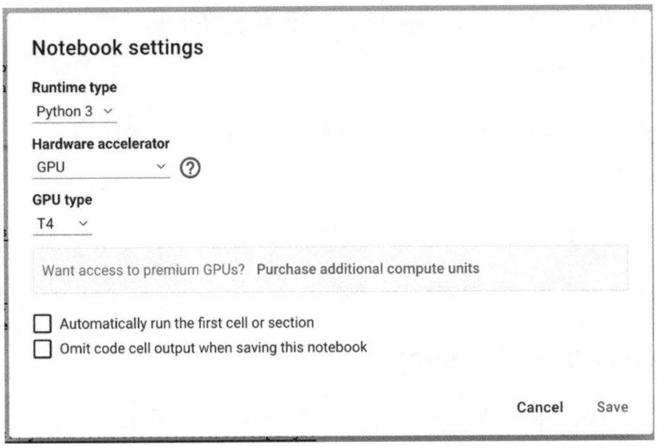

Figure 6-26. *Selecting GPU as the accelerator*

Creating a Colab Project

Click File and then New Notebook. The new notebook will open in a new browser tab. Give this notebook a meaningful name, such as Object Detection Model Training. By default, this notebook is saved in your Google Drive.

Setting Up the Runtime Environment for TensorFlow and Model Training

Click +Code to insert a code cell into the notebook. Notice the code block with an empty cell in the main area of the notebook. You can write Python code within this cell and execute it by clicking the execute icon.

Google Colab is an interactive programming environment and does not give direct access to the underlying operating system. You can invoke the shell using %%shell, which remains active within a single block of code cells it is invoked from. You can invoke the shell from as many code blocks as needed.

To train an object detection model, we will perform the following tasks:

1. Install and set up necessary libraries.

2. Download and install the TensorFlow `models` project.

3. Download the pet dataset and convert the files to TFRecord format.

4. Download a pretrained object detection model for transfer learning.

5. Configure the training pipeline.

6. Train and monitor the SSD model.

7. Finalize and export the model.

8. Visualize the model metrics and evaluation results using TensorBoard.

Installing and Setting Up Libraries

In Colab, you can execute a shell command by prefixing it with an exclamation mark (!). This feature enables you to run terminal commands directly within the Colab notebook environment. Listing 6-1 shows a list of libraries we will need for the object detection training.

Listing 6-1. Installing the Necessary Libraries and Packages

```
Filename: Listing_6_1
1    !pip install numpy==1.23
2    !pip install pillow==9.5
3    !apt install -y protobuf-compiler
```

As of this writing, TensorFlow version 2.12.0 requires NumPy version 1.23. However, Google Colab comes with a higher version of NumPy by default. To ensure compatibility with TensorFlow, it is essential to install NumPy version 1.23 explicitly, as shown in line 1.

Additionally, for seamless integration with the current version of TensorFlow, we are installing version 9.5 of the Pillow library (line 2), which is used for image processing tasks.

Please note that this NumPy and Pillow version requirements might change in the future, and you may need to update or remove lines 1 and 2 of Listing 6-1 accordingly.

In line 3, we are installing protobuf-compiler. The Protocol Buffers compiler, often referred to as "protoc" or "protobuf-compiler," is a command-line tool that comes with the Protocol Buffers library. It is used to compile .proto files into language-specific code, such as Python, C++, Java, Go, and more.

Installing TensorFlow's models Project

The TensorFlow models project is an official GitHub repository maintained by Google and the TensorFlow community. It contains a collection of pretrained machine learning models and tools related to the TensorFlow library. These models cover a wide range of tasks, including image classification, object detection, natural language processing, and generative models. The repository is a valuable resource for researchers and developers working in the field of machine learning and artificial intelligence. However, since the TensorFlow ecosystem evolves rapidly, it's best to refer to the official repository on GitHub for the latest information and content.

Listing 6-2 shows how to clone and install the models project.

Listing 6-2. Downloading and Installing TensorFlow models Project

```
Filename: Listing_6_2
1    %%shell
2    git clone --depth 1 https://github.com/tensorflow/models.git
```

```
3    cd models/research/
4    protoc object_detection/protos/*.proto --python_out=.
5    cd /content
6    git clone https://github.com/cocodataset/cocoapi.git
7    cd cocoapi/PythonAPI
8    make
9    cp -r pycocotools /content/models/research
10   cd
11   cd /content/models/research/
12   cp object_detection/packages/tf2/setup.py .
13   python -m pip install .
```

In line 1 of Listing 6-2, we are using %%shell, which is known as the magic command because using %%shell at the beginning of a code cell causes the entire cell's content to be interpreted as shell commands rather than Python code. This means we can execute any shell command supported by the underlying operating system directly in the cell.

In line 2, we clone the TensorFlow's Models repository from GitHub. The --depth 1 option for the clone command specifies that only the latest commit and its immediate parent commit should be retrieved. In other words, it performs a "shallow clone," fetching only the most recent commit history rather than the entire repository history. The cloned contents are stored in the directory models, which is created within the working directory, typically /content.

Line 3 changes from the current working directory to the /models/research directory.

In line 4, we execute the protoc command that converts all .proto files in the object_detection/protos directory into Python files.

Line 5 switches the working directory to the home directory, which is /content in Colab.

In line 6, we download cocoapi to the home directory, /content. Line 8 then performs a make operation, and line 9 copies the pycocotools subfolder to the /content/models/research directory. Next, we copy setup.py from object_detection/packages/tf2/ to the research directory (line 12).

Finally, in line 13, we run the pip install command to install the object detection API.

If everything goes well, the object detection API should be successfully installed. To test the installation, execute the code shown in Listing 6-3.

Listing 6-3. Testing Object Detection API Installation

```
Filename: Listing_6_3
1    %%shell
2    cd models/research/
3    python object_detection/builders/model_builder_tf2_test.py
```

Line 1 starts with the magic command, and line 2 changes the directory to the research directory. Line 3 executes the text by running the Python script `model_builder_tf2_test.py`.

After the test is completed, you should see output similar to the following:

```
...
[       OK ] ModelBuilderTF2Test.test_create_center_net_deepmac
[ RUN        ] ModelBuilderTF2Test.test_create_center_net_model0 (customize_
head_params=True)
INFO:tensorflow:time(__main__.ModelBuilderTF2Test.test_create_center_net_
model0 (customize_head_params=True)): 0.62s
I0726 13:49:07.605441 138969566220288 test_util.py:2462] time(__main__.
ModelBuilderTF2Test.test_create_center_net_model0 (customize_head_
params=True)): 0.62s
[       OK ] ModelBuilderTF2Test.test_create_center_net_model0 (customize_
head_params=True)
[ RUN        ] ModelBuilderTF2Test.test_create_center_net_model1 (customize_
head_params=False)
INFO:tensorflow:time(__main__.ModelBuilderTF2Test.test_create_center_net_
model1 (customize_head_params=False)): 0.33s
I0726 13:49:07.931119 138969566220288 test_util.py:2462] time(__main__.
ModelBuilderTF2Test.test_create_center_net_model1 (customize_head_
params=False)): 0.33s
[       OK ] ModelBuilderTF2Test.test_create_center_net_model1 (customize_
head_params=False)
[ RUN        ] ModelBuilderTF2Test.test_create_center_net_model_from_
keypoints
INFO:tensorflow:time(__main__.ModelBuilderTF2Test.test_create_center_net_
model_from_keypoints): 0.31s
```

```
I0726 13:49:08.246370 138969566220288 test_util.py:2462] time(__main__.
ModelBuilderTF2Test.test_create_center_net_model_from_keypoints): 0.31s
[       OK ] ModelBuilderTF2Test.test_create_center_net_model_from_
keypoints
[ RUN      ] ModelBuilderTF2Test.test_create_center_net_model_mobilenet
...
...
[       OK ] ModelBuilderTF2Test.test_unknown_faster_rcnn_feature_extractor
[ RUN      ] ModelBuilderTF2Test.test_unknown_meta_architecture
INFO:tensorflow:time(__main__.ModelBuilderTF2Test.test_unknown_meta_
architecture): 0.0s
I0726 13:49:42.312576 138969566220288 test_util.py:2462] time(__main__.
ModelBuilderTF2Test.test_unknown_meta_architecture): 0.0s
[       OK ] ModelBuilderTF2Test.test_unknown_meta_architecture
[ RUN      ] ModelBuilderTF2Test.test_unknown_ssd_feature_extractor
INFO:tensorflow:time(__main__.ModelBuilderTF2Test.test_unknown_ssd_feature_
extractor): 0.0s
I0726 13:49:42.314031 138969566220288 test_util.py:2462] time(__main__.
ModelBuilderTF2Test.test_unknown_ssd_feature_extractor): 0.0s
[       OK ] ModelBuilderTF2Test.test_unknown_ssd_feature_extractor
----------------------------------------------------------------------
Ran 24 tests in 37.431s

OK (skipped=1)
```

Downloading the Oxford-IIIT Pet Dataset

Let's insert another code cell into the notebook. We will download the annotated and labeled pet dataset from the official website to a directory in our Colab workspace. Listing 6-4 contains the code that downloads the pet dataset and annotations.

Listing 6-4. Downloading and Uncompressing the Images and Annotations of the Pet Dataset

```
FileName: Listing_6_4
1    %%shell
2    mkdir -p computer_vision/petdata
```

```
3    cd computer_vision/petdata
4    wget http://www.robots.ox.ac.uk/~vgg/data/pets/data/images.tar.gz
5    wget http://www.robots.ox.ac.uk/~vgg/data/pets/data/annotations.tar.gz
6    tar -xvf annotations.tar.gz
7    tar -xvf images.tar.gz
```

Line 1 invokes the shell. We need to do this in every cell block if we want to use any shell command.

Line 2 creates a new subdirectory, petdata, in the computer_vision directory.

Line 3 switches to the petdata subdirectory. We will download the pet dataset in the petdata directory.

Line 4 downloads the pet images, and line 5 downloads the annotations.

Lines 6 and 7 uncompress the downloaded images and annotations files.

When you execute this code block, you will see the images and annotations downloaded in the petdata directory. Images will be stored in the images subdirectory, and the annotations will be stored in the annotations subdirectory within the petdata directory.

Generating TensorFlow TFRecord Files

TFRecord is a simple format for storing a sequence of binary records. The data in TFRecord is serialized and stored in smaller chunks (e.g., 100MB to 200MB), which makes it more efficient to transfer across networks and read serially. You will learn more about TFRecord, its format, and how to convert images and associated annotations in the TFRecord file format in Chapter 9. For now, we will use a Python script provided in the research directory of the TensorFlow source code we downloaded from GitHub. The script is located at the path research/object_detection/dataset_tools/create_pet_tf_record.py.

Object detection algorithms take TFRecord files as input to the neural network. TensorFlow provides a Python script to convert the Oxford pet image annotation files to a set of TFRecord files. Listing 6-5 does the conversion of both training and test sets to TFRecords.

Listing 6-5. Converting Image Annotation Files to TFRecord Files

```
FileName: Listing_6_5
1    %%shell
2    mkdir -p /content/computer_vision/petdata/tfrecords
3    cd /content/models/research
4    python object_detection/dataset_tools/create_pet_tf_record.py \
5    --label_map_path=object_detection/data/pet_label_map.pbtxt \
6    --data_dir=/content/computer_vision/petdata \
7    --output_dir=/content/computer_vision/petdata/tfrecords
```

Line 1 invokes the magic shell.

Line 2 creates a new subdirectory, tfrecords, in the petdata directory.

Line 3 changes the working directory to the research directory.

Lines 4 through 7 run the Python script, create_pet_tf_record.py, which takes the following parameters:

- label_map_path: This file has the mapping of an ID (starting from 1) and corresponding class name. For the pet dataset, the mapping file is available in the object_detection/data/pet_label_map.pbtxt file. We will explore, in Chapter 9, how to generate this mapping file. But for now, let's just use what is already available. This is a JSON-formatted file with two keys, id and name. A few sample entries of the mapping file are shown here:

```
item {
  id: 1
  name: 'Abyssinian'
}
item {
  id: 2
  name: 'american_bulldog'
}
...
```

- data_dir: This is the parent directory of the images and annotations subdirectories.

- output_dir: This is the destination directory where the TFRecord files will be stored. You can supply any existing directory name. After conversion of images and annotations, the TFRecord files will be saved in this directory.

After this code block executes, it creates a set of *.record files in output_directory. The script, create_pet_tf_record.py, creates both the training and validation sets.

- *Training set*: The output directory, output_dir, should now contain ten training files and ten validation TFRecords. The number of *.record files may be different depending on the input size. The *.record files of training set are named as pet_faces_train.record-?????-of-00010. The regular expression ????? takes values sequentially from 00001 through 00010.

- *Validation set*: The evaluation dataset is named as pet_faces_val.record-?????-of-00010.

Downloading a Pretrained Model for Transfer Learning

Transfer learning in computer vision is a machine learning technique that uses a pretrained model on a large dataset a starting point for a new task or dataset. Instead of training a model from scratch, transfer learning allows us to transfer the knowledge learned by the pretrained model to the new problem domain.

Transfer learning in object detection using TensorFlow involves leveraging pretrained models, such as those available in the TensorFlow Model Zoo, to accelerate and improve the training process for object detection tasks. TensorFlow provides several pretrained object detection models, such as the Single Shot MultiBox Detector (SSD) and Faster R-CNN, which are trained on large-scale datasets like Common Objects in Context (COCO) or Open Images.

A collection of SSD-based models pretrained on the COCO 2017 dataset is available at https://github.com/tensorflow/models/blob/master/research/object_detection/g3doc/tf2_detection_zoo.md. Table 6-3 shows a summary of various pretrained object detection models available at the Model Zoo website.

***Table 6-3.** Pretrained Object Detection Models with Performance Metrics*

Model Name	Speed (ms)	COCO mAP	Outputs
SSD MobileNet v2 320x320	19	20.2	Boxes
SSD MobileNet V1 FPN 640x640	48	29.1	Boxes
SSD MobileNet V2 FPNLite 320x320	22	22.2	Boxes
SSD MobileNet V2 FPNLite 640x640	39	28.2	Boxes
SSD ResNet50 V1 FPN 640x640 (RetinaNet50)	46	34.3	Boxes
SSD ResNet50 V1 FPN 1024x1024 (RetinaNet50)	87	38.3	Boxes
SSD ResNet101 V1 FPN 640x640 (RetinaNet101)	57	35.6	Boxes
SSD ResNet101 V1 FPN 1024x1024 (RetinaNet101)	104	39.5	Boxes
SSD ResNet152 V1 FPN 640x640 (RetinaNet152)	80	35.4	Boxes
SSD ResNet152 V1 FPN 1024x1024 (RetinaNet152)	111	39.6	Boxes
Faster R-CNN ResNet50 V1 640x640	53	29.3	Boxes
Faster R-CNN ResNet50 V1 1024x1024	65	31.0	Boxes
Faster R-CNN ResNet50 V1 800x1333	65	31.6	Boxes
Faster R-CNN ResNet101 V1 640x640	55	31.8	Boxes
Faster R-CNN ResNet101 V1 1024x1024	72	37.1	Boxes
Faster R-CNN ResNet101 V1 800x1333	77	36.6	Boxes
Faster R-CNN ResNet152 V1 640x640	64	32.4	Boxes
Faster R-CNN ResNet152 V1 1024x1024	85	37.6	Boxes
Faster R-CNN ResNet152 V1 800x1333	101	37.4	Boxes
Faster R-CNN Inception ResNet V2 640x640	206	37.7	Boxes
Faster R-CNN Inception ResNet V2 1024x1024	236	38.7	Boxes
Mask R-CNN Inception ResNet V2 1024x1024	301	39.0/34.6	Boxes/Masks

To demonstrate our training approach, we'll utilize transfer learning from the SSD ResNet50 V1 FPN 640x640 (RetinaNet50) model from the following URL:

```
http://download.tensorflow.org/models/object_detection/tf2/20200711/ssd_
resnet50_v1_fpn_640x640_coco17_tpu-8.tar.gz
```

Feel free to download any of the models listed in Table 6-3 that best match your speed and mAP preferences.

The command set in Listing 6-6 downloads the SSD RetinaNet50 model.

Listing 6-6. Downloading a Pretrained SSD RetinaNet50 Object Detection Model

```
FileName: Listing_6_6
1    %%shell
2    cd computer_vision
3    mkdir pre-trained-model
4    cd pre-trained-model
5    wget http://download.tensorflow.org/models/object_detection/
     tf2/20200711/ssd_resnet50_v1_fpn_640x640_coco17_tpu-8.tar.gz
6    tar -xvf ssd_resnet50_v1_fpn_640x640_coco17_tpu-8.tar.gz
```

In Listing 6-6, line 1 activates the magic shell.

Line 2 changes the working directory to computer_vision.

Line 3 creates a subdirectory, pre-trained-model, in the parent computer_vision directory.

Line 5 uses the wget command to download the RetinaNet50 pretrained model from the Model Zoo website.

Line 6 uses the tar command to decompress the RetinaNet50 model file into a directory named /content/pre-trained-model/ssd_resnet50_v1_fpn_640x640_coco17_tpu-8.

In the Google Colab window, expand the left panel and check the Files tab. You should see something similar to the directory structure shown in Figure 6-27.

Figure 6-27. *Pretrained model directory structure*

Configuring the Object Detection Pipeline

The *training pipeline* is a configuration file that specifies various settings and hyperparameters for training an object detection model. It contains information about the model architecture, dataset paths, batch size, learning rate, data augmentation, evaluation settings, and more. It is usually written in the Protocol Buffers format (`.config` file).

The schema for the training pipeline is available in the location `object_detection/protos/pipeline.proto` in the `research` directory under the TensorFlow `models` project.

The JSON-formatted training pipeline is broadly divided into five sections, as shown here:

```
model: {
      (... Add model config here...)
}
train_config : {
      (... Add train_config here...)
}
```

```
train_input_reader: {
        (... Add train_input configuration here...)
}
eval_config: {
        (... Add eval_configuration here...)
}
eval_input_reader: {
        (... Add eval_input configuration here...)
}
```

The sections of the training pipeline is explained below:

- model: This section defines the architecture and configuration of the object detection model.

  - ssd or faster_rcnn: The type of model (Single Shot MultiBox Detector or Faster R-CNN)

- train_config: This section includes various settings related to the training process.

  - batch_size: The number of images used in each training batch

  - fine_tune_checkpoint: Path to the pretrained model checkpoint to initialize the model before training

  - num_steps: The number of training steps to be performed during the training process

  - data_augmentation_options: Options for data augmentation during training (e.g., random brightness, rotation, etc.)

- train_input_reader: This section includes the configuration for the training data input.

  - tf_record_input_reader: Path to the training TFRecord file (train.tfrecord)

  - label_map_path: Path to the label_map.pbtxt file

- eval_config: This section includes the settings for the evaluation process.

  - num_examples: The number of examples to be used for evaluation

- max_evals: Maximum number of evaluations to perform

- eval_input_reader: This section includes the configuration for the evaluation data input.

  - tf_record_input_reader: Path to the evaluation TFRecord file (eval. tfrecord)

  - label_map_path: Path to the label_map.pbtxt file

In Figure 6-27, notice the file pipeline.config in the pretrained model's directory, ssd_resnet50_v1_fpn_640x640_coco17_tpu-8. We need to edit the pipeline.config file to align the training settings accordingly. Download the pipeline.config file (right-click and Download) from the Colab, save it in the local computer drive, and edit it to configure the training pipeline for the model.

Let's examine the modifications that need to be made to the pipeline.config file, indicated by highlighting them in bold. The comments next to the lines that require editing offer further guidance on the specific values to be configured based on the given situation.

```
model {
  ssd {
    num_classes: 37 # Set this to the number of different label classes
    image_resizer {
      fixed_shape_resizer {
        height: 640
        width: 640
      }
    }
    feature_extractor {
      type: "ssd_resnet50_v1_fpn_keras"
      depth_multiplier: 1.0
      min_depth: 16
      conv_hyperparams {
        regularizer {
          l2_regularizer {
            weight: 0.00039999998989515007
          }
```

```
      }
      initializer {
        truncated_normal_initializer {
          mean: 0.0
          stddev: 0.029999999329447746
        }
      }
      activation: RELU_6
      batch_norm {
        decay: 0.996999979019165
        scale: true
        epsilon: 0.0010000000474974513
      }
    }
    override_base_feature_extractor_hyperparams: true
    fpn {
      min_level: 3
      max_level: 7
    }
  }
box_coder {
  faster_rcnn_box_coder {
    y_scale: 10.0
    x_scale: 10.0
    height_scale: 5.0
    width_scale: 5.0
  }
}
matcher {
  argmax_matcher {
    matched_threshold: 0.5 # Increase this for a better match.
    unmatched_threshold: 0.5 # Sum of matched and unmatched threshold
    must be equal to 1.
    ignore_thresholds: false
    negatives_lower_than_unmatched: true
```

```
      force_match_for_each_row: true
      use_matmul_gather: true
    }
  }
  similarity_calculator {
    iou_similarity {
    }
  }
  box_predictor {
    weight_shared_convolutional_box_predictor {
      conv_hyperparams {
        regularizer {
          l2_regularizer {
            weight: 0.00039999998989515007
          }
        }
        initializer {
          random_normal_initializer {
            mean: 0.0
            stddev: 0.009999999776482582
          }
        }
        activation: RELU_6
        batch_norm {
          decay: 0.996999979019165
          scale: true
          epsilon: 0.0010000000474974513
        }
      }
      depth: 256
      num_layers_before_predictor: 4
      kernel_size: 3
      class_prediction_bias_init: -4.599999904632568
    }
  }
```

```
anchor_generator {
  multiscale_anchor_generator {
    min_level: 3
    max_level: 7
    anchor_scale: 4.0
    aspect_ratios: 1.0
    aspect_ratios: 2.0
    aspect_ratios: 0.5
    scales_per_octave: 2
  }
}
post_processing {
  batch_non_max_suppression {
    score_threshold: 9.99999993922529e-09
    iou_threshold: 0.6000000238418579
    max_detections_per_class: 100
    max_total_detections: 100
    use_static_shapes: false
  }
  score_converter: SIGMOID
}
normalize_loss_by_num_matches: true
loss {
  localization_loss {
    weighted_smooth_l1 {
    }
  }
  classification_loss {
    weighted_sigmoid_focal {
      gamma: 2.0
      alpha: 0.25
    }
  }
  classification_weight: 1.0
  localization_weight: 1.0
}
```

```
    encode_background_as_zeros: true
    normalize_loc_loss_by_codesize: true
    inplace_batchnorm_update: true
    freeze_batchnorm: false
  }
}
train_config {
  batch_size: 64 # Increase/Decrease this value depending on the available
  memory (Higher values require more memory and vice-versa)
  data_augmentation_options {
    random_horizontal_flip {
    }
  }
  data_augmentation_options {
    random_crop_image {
      min_object_covered: 0.0
      min_aspect_ratio: 0.75
      max_aspect_ratio: 3.0
      min_area: 0.75
      max_area: 1.0
      overlap_thresh: 0.0
    }
  }
  sync_replicas: true
  optimizer {
    momentum_optimizer {
      learning_rate {
        cosine_decay_learning_rate {
          learning_rate_base: 0.03999999910593033
          total_steps: 25000
          warmup_learning_rate: 0.013333000242710114
          warmup_steps: 2000
        }
      }
```

```
      momentum_optimizer_value: 0.8999999761581421
    }
    use_moving_average: false
  }
  fine_tune_checkpoint: "/content/pre-trained-model/ssd_resnet50_v1_
  fpn_640x640_coco17_tpu-8/checkpoint/ckpt-0" # Path to checkpoint of
  pretrained model
  num_steps: 25000
  startup_delay_steps: 0.0
  replicas_to_aggregate: 8
  max_number_of_boxes: 100
  unpad_groundtruth_tensors: false
  fine_tune_checkpoint_type: "detection" # Set this to "detection" since we
  want to be training the full detection model
  use_bfloat16: false # Set this to false if you are not training on a TPU
  fine_tune_checkpoint_version: V2
}
train_input_reader {
  label_map_path: "PATH_TO_BE_CONFIGURED"  # Path to label map file
  tf_record_input_reader {
    input_path: "PATH_TO_BE_CONFIGURED"  # Path to training TFRecord file
  }
}
eval_config {
  metrics_set: "coco_detection_metrics"
  use_moving_averages: false
}
eval_input_reader {
  label_map_path: "PATH_TO_BE_CONFIGURED" # Path to label map file
  shuffle: false
  num_epochs: 1
  tf_record_input_reader {
    input_path: "PATH_TO_BE_CONFIGURED"  # Path to testing TFRecord
  }
}
```

As the `pipeline.config` file was preserved during the training of the pretrained model we downloaded for transfer learning, we will retain most of its contents intact, making alterations only to the specific highlighted parts. The parameters that require adjustment, based on the settings in your Colab environment, are as follows:

> `num_classes: 37`: Represents the 37 categories of pets in our dataset.

> `fine_tune_checkpoint: "/content/pre-trained-model/ssd_resnet50_v1_fpn_640x640_coco17_tpu-8/checkpoint/ckpt-0"`: This is the path where we stored the pretrained model checkpoint. Notice in Figure 6-27 that the file name of the model checkpoint is `model.ckpt.data-00000-of-00001`, but in the `fine_tune_checkpoint` configuration we provide only up to `model.ckpt` (you must not include the full name of the checkpoint file). To get the path of this checkpoint file, in the Colab file browser, right-click the file name and click Copy Path.

> `num_steps: 25000`: This is the number of steps the algorithm should execute. You may need to tune this number to get a desirable accuracy level.

> `fine_tune_checkpoint_type: "detection"`: By default, this is set to `"classification"` but needs to be changed to `"detection"` for object detection training.

> `use_bfloat16: false`: This is set to `true` by default for training the model on TPU-based hardware.

> `train_input_reader → label_map_path: /content/computer_vision/models/research/object_detection/data/pet_label_map.pbtxt`: This is the path of the file that contains the mapping of ID and class name. For the pet dataset, this is available in the `research` directory.

> `train_input_reader → input_path: /content/computer_vision/petdata/pet_faces_train.record-?????-of-00010`: This is the path of the TFRecord file for the training dataset. Notice that we used a regular expression (?????) in the training set path. This is important to include all training TFRecord files.

eval_input_reader → label_map_path: /content/computer_
vision/models/research/object_detection/data/pet_label_
map.pbtxt: This is the same as the training label map.

eval_input_reader → input_path: /content/computer_
vision/petdata/pet_faces_eval.record-?????-of-00010: This
is the path of the TFRecord file for the evaluation dataset. Notice
that we used a regular expression (?????) in the evaluation set
path. This is important to include all evaluation TFRecord files.

It is important to note that pipeline.config has the parameter override_base_
feature_extractor_hyperparams set to true; do not change this setting or else the
model will not run.

After editing the pipeline.config file, we need to upload it to Colab. We can upload
it to any directory location, but in this case, we are uploading it to its original location
from where we downloaded it. We will first remove the old pipeline.config file and
then upload the updated one.

To delete the old pipeline.config file from the Colab directory location, right-click
it and then click Delete. To upload the updated pipeline.config file from your local
computer, right-click the Colab directory (ssd_resnet50_v1_fpn_640x640_coco17_
tpu-8), click Upload, and browse and upload the edited version of the pipeline.config
file from the local computer.

Executing the Model Training

We are prepared to start the training process. To initiate the training, simply execute
Listing 6-7.

Listing 6-7. Executing the Model Training

```
1   %%shell
2   export PYTHONPATH=$PYTHONPATH:/content/models/research
3   export PYTHONPATH=$PYTHONPATH:/content/models/research/slim
4   cd /content/models/research
5   PIPELINE_CONFIG_PATH=/content/computer_vision/pre-trained-model/ssd_
    resnet50_v1_fpn_640x640_coco17_tpu-8/pipeline.config
6   MODEL_DIR=/content/computer_vision/pet_detection_model/
7   NUM_TRAIN_STEPS=100
```

```
8     SAMPLE_1_OF_N_EVAL_EXAMPLES=1
9     python object_detection/model_main_tf2.py \
--pipeline_config_path=${PIPELINE_CONFIG_PATH} \
--model_dir=${MODEL_DIR} \
--num_train_steps=${NUM_TRAIN_STEPS} \
--sample_1_of_n_eval_examples=${SAMPLE_1_OF_N_EVAL_EXAMPLES} \
--alsologtostderr
```

TensorFlow provides a Python script, model_main_tf2.py, to trigger the model training. This script is located in the directory models/research/object_detection. This script takes the following parameters:

- pipeline_config_path: This is the path of the pipeline.config file.

- model_dir: This is the directory where the trained model will be saved.

- num_train_steps: This is the number of steps we want our network to train. This will override the num_steps parameter in the pipeline. config file.

- sample_1_of_n_eval_examples: This determines one out of how many samples the model should use for evaluation.

Execute this code block in Colab and wait for the model to complete the learning from the training image set. While the model is learning, you will see the iteration losses printed in the Colab console. If everything goes well, you will have a trained object detection model saved in the model_dir directory (/content/computer_vision/pet_detection_model/ in our example).

Evaluating the Model

Recall that back in Listing 6-2 we installed pycocotools from cocoapi in lines 6 through 9. pycocotools is a Python library that provides tools and utilities for working with the Common Objects in Context (COCO) dataset. The COCO dataset is a widely used benchmark for object detection, segmentation, and captioning tasks. It contains a large collection of images with complex scenes and a rich set of annotations, including object instances, keypoints, and captions.

pycocotools allows researchers and developers to interact with the COCO dataset and evaluate the performance of object detection and segmentation models using standard evaluation metrics like AP and mAP.

Let's begin with a brief explanation of the evaluation process. While the training process is underway, it periodically creates checkpoint files inside the `model_dir` directory. These checkpoints represent snapshots of the model at specific training steps. When a new set of these checkpoint files is generated, the evaluation process comes into play. It utilizes these checkpoint files to assess the model's performance in detecting objects on the validation dataset. The evaluation results are then summarized through various metrics, providing insights into the model's progress over time.

The following are the steps to execute the evaluation process:

1. Ensure that `pycocotools` is installed correctly.

2. In the `pipeline.config` file, set `"coco_detection_metrics"` for the `metrics_set` key under the `eval_config` block:

```
eval_config {
  metrics_set: "coco_detection_metrics"
  use_moving_averages: false
}
```

3. Execute the evaluation command shown in Listing 6-8.

Listing 6-8. Executing the Model Evaluation

```
1   %%shell
2   python /content/models/research/object_detection/model_main_tf2.py \
3   --model_dir=/content/computer_vision/pet_detection_model \
4   --pipeline_config_path=/content/computer_vision/pre-trained-model/
    ssd_resnet50_v1_fpn_640x640_coco17_tpu-8/pipeline.config \
5   --checkpoint_dir=/content/computer_vision/pet_detection_model
```

Executing Listing 6-8 will take several minutes, the duration of which varies based on the number of validation images present in the dataset. During the evaluation process, it will perform periodic checks (defaulting to every 300 seconds) and utilize the most recent checkpoint files from `model_dir` to evaluate the model's performance.

The evaluation results are stored as TensorFlow event files (`events.out.tfevents.*`) within the `model_dir/eval_0` directory. These files are used to visualize the model performance, as explained in the following section.

Visualizing the Training Result in TensorBoard

As discussed in Chapter 5, TensorBoard is a web-based visualization tool provided by TensorFlow that is designed to help users understand, debug, and optimize their machine learning models. It allows you to monitor and visualize various aspects of your model's performance during training and evaluation, making it easier to gain insights and improve the model's effectiveness.

To see the training statistics and the model result, launch the TensorBoard dashboard using the code in Listing 6-9 in Colab. `--logdir` is the directory where we are saving the model checkpoints.

Listing 6-9. Launching the TensorBoard Dashboard to See the Training Results

```
1    %load_ext tensorboard
2    %tensorboard --logdir /content/computer_vision/pet_detection_model
```

Line 1 loads the TensorBoard notebook extension. This extension displays the TensorBoard dashboard embedded within the Colab screen.

Click the Images tab of the TensorBoard dashboard to display the evaluation results, as shown in Figure 6-28.

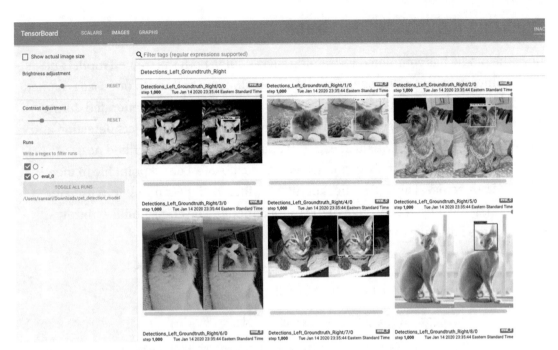

Figure 6-28. *Model training result in TensorBoard dashboard*

Alternatively, if you want to evaluate the model offline in your local computer and not on Colab, you can download the entire `pet_detection_model` directory where we saved the model checkpoints. The `final_model` directory, to which we exported our trained model, does not contain the full model statistics and training results. Therefore, you must download the entire `pet_detection_model` directory.

In your computer terminal (or command prompt), launch TensorBoard by passing the path to the `pet_detection_model` directory. Make sure you are in the virtual environment (as explained in Chapter 1). Here is the command:

```
(cv) username$ tensorboard --logdir ~/Downloads/pet_detection_model
```

After the previous command is successfully executed, open your web browser and go to `http://localhost:6006` to see the TensorBoard dashboard. Click the Images tab in the top menu to see the evaluation output with bounding boxes on the images, as shown in Figure 6-28.

Exporting the TensorFlow Graph

Once the model has completed its training successfully, both the trained model and its checkpoints are saved in the specified `model_dir` directory, which in our case is named `pet_detection_model`. This directory contains all the checkpoints generated during the training process. However, to utilize this model effectively for object detection and bounding box prediction, we must export it as a final model, as described next.

First, identify the latest checkpoint. Determine the latest checkpoint file in the `model_dir` directory. This is the checkpoint file with the highest number in its name, representing the most recent state of the trained model. The checkpoints typically consist of the following three files:

- `model.ckpt-${CHECKPOINT_NUMBER}.data-00000-of-00001`

- `model.ckpt-${CHECKPOINT_NUMBER}.index`

- `model.ckpt-${CHECKPOINT_NUMBER}.meta`

Take the checkpoint with the maximum `${CHECKPOINT_NUMBER}` value. Our model ran for 100 steps, so our max checkpoint files should look like the following:

- `model.ckpt-100.data-00000-of-00001`

- `model.ckpt-100.index`

- `model.ckpt-100.meta`

Listing 6-10 facilitates the export of our object detection–trained model to a directory that we specify.

Listing 6-10. Exporting the TensorFlow Graph

```
1    %%shell
2    export PYTHONPATH=$PYTHONPATH:/content/models/research
3    export PYTHONPATH=$PYTHONPATH:/content/models/research/slim
4    cd /content/models/research
5    python /content/models/research/object_detection/exporter_main_v2.py \
     --input_type image_tensor --pipeline_config_path /content/computer_vision \
       pre-trained-model/ssd_resnet50_v1_fpn_640x640_coco17_tpu-8/pipeline.config \
     --trained_checkpoint_dir /content/computer_vision/pet_detection_model \
      --output_directory /content/computer_vision/pet_detection_model/final_model
```

Line 5 exports the TensorFlow graph by calling the script exporter_main_v2.py, which is located in the directory models/research/object_detection. This script takes the following parameters:

- input_type: For our model, it will be image_tensor.

- pipeline_config_path: This is the same pipeline.config file path that we used before.

- trained_checkpoint_dir: This is the path where the trained model checkpoint is saved.

- output_directory: This is the directory where the exported graph will be saved. Figure 6-29 shows the output directory structure after the export script is executed.

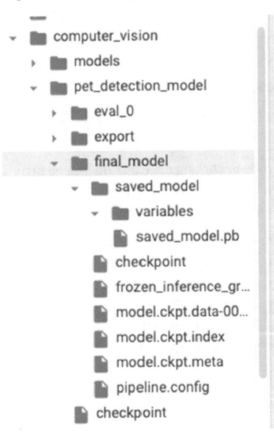

Figure 6-29. *Model exported in the directory* final_model

Downloading the Object Detection Model

Google Colab does not support downloading an entire directory. Downloading individual files is possible, but doing so for each file in a directory can be inefficient and time-consuming. Instead, you can save the fully trained model to your personal Google Drive, which provides a more convenient and efficient way to store and access your model for later use.

After a few hours of continuous usage or upon session expiration, Google Colab will terminate your virtual machine and delete all the data within that session. If you fail to download your model, you risk losing it. However, you have the option to directly save your models and data to Google Drive, which provides a secure and reliable storage solution. To ensure your model's safety, especially during extended training sessions, it is advisable to save all your necessary data and model files to Google Drive before commencing the training process. This precaution will safeguard your progress and results, allowing you to access them at any time from your Google Drive.

Here are the steps to mount your Google Drive:

1. In the left panel, click Files.

2. Select Mount Drive. This action will insert some code into the notebook area.

3. Execute the code by clicking the Play icon located in the code block. This will mount your Google Drive, enabling you to access and save files to it from your Colab notebook.

Click the authorization link to generate an authorization code. You may need to sign in to your Google account again. Copy the authorization code, paste it in the notebook, and press the Enter key. See Figure 6-31. After the drive is mounted, you will see a list of directories in the left panel on the Files tab (as shown in Figure 6-30). Notice that the example Google Drive in Figure 6-31 has a directory called computervision that was already created in the drive. Feel free to create any directory you want.

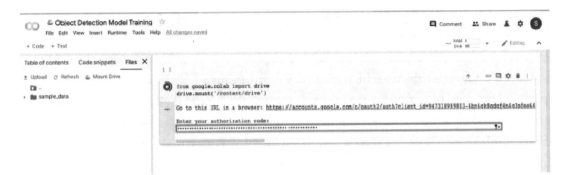

Figure 6-30. *Google Drive mounting*

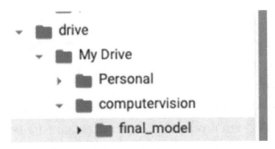

Figure 6-31. *Google Drive directory structure*

Move the `final_model` directory to the Google Drive directory.

To save the trained object detection model to the Google Drive directory, simply drag `final_directory` from the Colab directory to the Google Drive directory.

You must also copy to the Google Drive the following checkpoint files:

- `model.ckpt-100.data-00000-of-00001`

- `model.ckpt-100.index`

- `model.ckpt-100.meta`

To download the model from Google Drive, log in to your Google Drive and download the trained model to your local computer. You should download the entire `final_model` directory.

Detecting Objects Using Trained Models

As previously discussed, training a model is not a regular task. Once we've achieved a sufficiently strong model with high accuracy or mAP (mean average precision), there might be no need to undergo retraining, as long as the model maintains its precision in predictions. However, it's important to note that model training is resource-intensive, demanding several hours or days, even when utilizing GPUs, to develop a quality model. In certain cases, it's advantageous and cost-effective to conduct the training of your computer vision models in the cloud, harnessing the power of GPUs. Subsequently, once the model is prepared, it can be downloaded for local use on your computer or application server. This localized deployment of the model allows for the identification of objects within images in a cost-effective way.

This section explains the process of creating object detection predictors on your personal computer. This entails employing the model previously trained on Google Colab. As in previous chapters, we will use PyCharm as the IDE. While it's certainly feasible to use Colab for writing the object detection predictor, this approach isn't optimal from the standpoint of production deployment considerations.

The following are the general steps for crafting the object detection program, covered in depth in the next two sections:

1. Download and install the TensorFlow `models` project from the GitHub repository.

2. Write the Python code that will utilize the exported TensorFlow graph (exported model) to predict objects within new images that were not included in the training or test sets.

Installing TensorFlow's models Project

The installation process of the TensorFlow `models` project is the same as the process for Google Colab described earlier in this chapter. The only difference may be in the Protobuf installation, as it is platformdependent software. Before we start, make sure that the PyCharm IDE is configured to use the virtual environment we created in Chapter 1. We will execute commands in PyCharm's terminal window. If you choose to execute commands using the operating system's shell, make sure you have activated the virtual environment for the shell session. (See Chapter 1 to review virtualenv.) Here is the full set of steps to install and configure the `models` project:

1. Install the libraries that are needed to build and install the `models` project. Execute the commands shown in Listing 6-11 in the terminal or at the command prompt (from within the virtualenv).

Listing 6-11. Commands to Install Dependencies

```
1    pip install --user Cython
2    pip install --user contextlib2
3    pip install --user pillow
4    pip install --user lxml
```

2. Install Google's Protobuf compiler. The installation process depends on the operating system you are using. Follow these instructions for your OS:

    a. On Ubuntu: `sudo apt-get install protobuf-compiler`

    b. On other Linux OSs:

    ```
    wget -O protobuf.zip \
    https://github.com/google/protobuf/releases/download/
    v3.0.0/protoc-3.0.0-linux-x86_64.zip
    unzip protobuf.zip
    ```

    Remember the directory location you have installed Protobuf in, as you will need to provide the full path to `bin/protoc` when building the TensorFlow code.

    c. On macOS: `brew install protobuf`

3. Clone the TensorFlow `models` project from GitHub:

    ```
    git clone --depth 1 https://github.com/tensorflow/models.git
    ```

    You can also download the `models` project from the TensorFlow official repository at `https://github.com/tensorflow/models.git`.
    As shown in Figure 6-32, I downloaded the TensorFlow `models` project in a directory called `chapter6`.

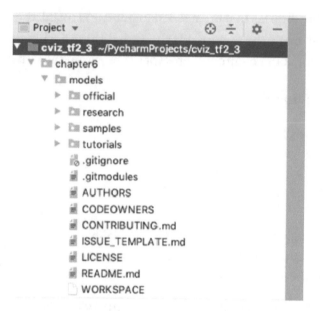

Figure 6-32. *Example directory structure consisting of TensorFlow* models *project*

4. Compile the models project using the Protobuf compiler. Run the following set of commands from the models/research directory:

```
$ cd models/research
$ protoc object_detection/protos/*.proto --python_out=.
```

If you installed Protobuf manually and unzipped it in a directory, provide the full path up to bin/protoc in the previous command.

5. Set the following environment variables. It's a standard practice to set these environment variables in ~/.bash_profile. Here are the instructions to do that:

a. Open your command prompt or terminal and type vi ~/.bash_profile. You can use any other editor, such as nano, to edit the .bash_profile file.

b. Add the following three lines at the end of .bash_profile. Make sure the paths match with the directory paths you have in your computer.

```
export PYTHONPATH=$PYTHONPATH:~/cviz_tf2_3/chapter6/
models/research/object_detection
```

```
export PYTHONPATH=$PYTHONPATH:~/cviz_tf2_3/chapter6/
models/research
export PYTHONPATH=$PYTHONPATH:~/cviz_tf2_3/chapter6/
models/research/slim
```

c. Save the file ~/.bash_profile after adding the previous line.

d. Close your terminal and relaunch it to effect the change. You will need to close your PyCharm IDE to have the environment variables update in your IDE. To test the setting, type the command echo $PYTHONPATH in your PyCharm terminal window. It should print the paths you just set up.

6. Build and install the research project that we just built using Protobuf. Execute the following commands from the models/research directory:

```
python setup.py build
python setup.py install
```

If these commands successfully run, you should output something like this:

```
Finished processing dependencies for object-detection==0.1
```

Our environment is now prepared to begin coding for image object detection. The model we'll be working with is the one we exported and downloaded from Colab. If you haven't completed this step yet, download the final model either from Google Colab or from Google Drive, assuming you stored your models there.

Code for Object Detection

In this section, we will write code that detects objects in an image and draws bounding boxes around them. To keep the code simple and easy to understand, we have divided it into the following parts:

Configuration and Initialization

In this section of the code, we initialize the model path, image input, and output directories. Listing 6-12 shows the first part of the code that includes the library imports and path setup.

Listing 6-12. Imports and Path Initialization Part of the Object Detection Code

```
Filename: Listing_6_9.py
1    import os
2    import pathlib
3    import random
4    import numpy as np
5    import tensorflow as tf
6    import cv2
7    # Import the object detection module.
8    from object_detection.utils import ops as utils_ops
9    from object_detection.utils import label_map_util
10
11   # to make gfile compatible with v2
12   #tf.gfile = tf.io.gfile
13
14   model_path = "ssd_model/final_model"
15   labels_path = "models/research/object_detection/data/pet_label_
     map.pbtxt"
16   image_dir = "images"
17   image_file_pattern = "*.jpg"
18   output_path="output_dir"
19
20   PATH_TO_IMAGES_DIR = pathlib.Path(image_dir)
21   IMAGE_PATHS = sorted(list(PATH_TO_IMAGES_DIR.glob(image_file_
     pattern)))
22
23   # List of the strings that is used to add the correct label for
     each box.
24   category_index = label_map_util.create_category_index_from_
     labelmap(labels_path, use_display_name=True)
25   class_num =len(category_index)
```

Line 1 through 6 are our usual imports. Lines 8 and 9 import the object detection APIs from the research module of the TensorFlow models project. Make sure the PYTHONPATH environment variable is correctly set (as explained earlier).

Line 12 initializes the gfile in the TensorFlow2 compatibility mode. The gfile provides I/O functionality in TensorFlow. Notice that this line is commented out since our model was trained using TensorFlow v2.x. Uncomment this line if the model was trained using TensorFlow v1.x.

Line 14 initializes the directory path where the object detection trained model is located.

Line 15 initializes the mapping file path. We set the same JSON-formatted file containing the class ID and class name mapping that we used for the training.

Line 16 is the input directory path containing images in which objects need to be detected.

Line 17 defines the pattern of file names in the input image path. If you want to load all files from the directory, use *.*.

Line 18 is the output directory path where the images with bounding boxes around the detected objects will be saved.

Lines 20 and 21 are to create iterable path objects that we will iterate through to read images one by one and detect objects in each of them.

Line 24 uses the label mapping file to create a category or class index.

Line 25 assigns the number of classes to the class_num variable.

In addition, we initialize a color table that we will use when drawing bounding boxes. Listing 6-13 shows the code.

Listing 6-13. Creating a Color Table Based on the Number of Object Classes

```
27   def get_color_table(class_num, seed=0):
28       random.seed(seed)
29       color_table = {}
30       for i in range(class_num):
31           color_table[i] = [random.randint(0, 255) for _ in range(3)]
32       return color_table
33
34   colortable = get_color_table(class_num)
35
```

Create `Model` Object by Loading the Trained Model

Listing 6-14 shows the function, `load_model()`, that takes the model path as input. Line 40 loads the saved model from the directory and creates a `model` object that is returned by this function. We will use this `model` object to predict the objects and bounding boxes.

Listing 6-14. Loading the Model from a Directory

```
36   # # Model preparation and loading the model from the disk
37   def load_model(model_path):
38
39       model_dir = pathlib.Path(model_path) / "saved_model"
40       model = tf.saved_model.load(str(model_dir))
41       #model = model.signatures['serving_default']
42       return model
43
```

Run the Prediction and Construct the Output in a Usable Form

We have written a function called `run_inference_for_single_image()` that takes two arguments: the `model` object and the image NumPy. This function returns a Python dictionary. The output dictionary contains the following key pairs:

> `detection_boxes`, which is a 2D array consisting of the four corners of bounding boxes.

> `detection_scores`, which is a 1D array of scores associated with each bounding box.

> `detection_classes`, which is a 1D array of integer representation of the object class-index associated with each bounding box.

> `num_detections`, which is a scalar that indicates the number of predicted object classes.

Listing 6-15 shows the implementation of the function `run_inference_for_single_image()`.

Let's examine the code listing line by line.

The TensorFlow `model` object takes a batch of image tensors to predict the object classes and bounding boxes around them. Line 48 converts the image NumPy into a tensor. Since we are processing one image at a time and the `model` object takes a batch, we need to convert the input image tensor into a batch of images. Line 50 does that. The `tf.newaxis` expression is used to increase the dimension of an existing array by 1, when used once. Thus, a 1D array will become a 2D array. A 2D array will become a 3D array. And so on.

Listing 6-15. Predicting Objects and Bounding Boxes and Organizing the Output

```
44    # Predict objects and bounding boxes and format the result
45    def run_inference_for_single_image(model, image):
46
47        # The input needs to be a tensor, convert it using `tf.convert_to_
          tensor`.
48        input_tensor = tf.convert_to_tensor(image)
49        # The model expects a batch of images, so add an axis with `tf.
          newaxis`.
50        input_tensor = input_tensor[tf.newaxis, ...]
51
52        # Run prediction from the model
53        output_dict = model(input_tensor)
54
55        # Input to model is a tensor, so the output is also a tensor
56        # Convert to numpy arrays, and take index [0] to remove the batch
          dimension.
57        # We're only interested in the first num_detections.
58        num_detections = int(output_dict.pop('num_detections'))
59        output_dict = {key: value[0, :num_detections].numpy()
60                          for key, value in output_dict.items()}
61        output_dict['num_detections'] = num_detections
62
63        # detection_classes should be ints.
```

```
64      output_dict['detection_classes'] = output_dict['detection_
        classes'].astype(np.int64)

65

66      # Handle models with masks:
67      if 'detection_masks' in output_dict:
68          # Reframe the bbox mask to the image size.
69          detection_masks_reframed = utils_ops.reframe_box_masks_to_
            image_masks(
70              output_dict['detection_masks'], output_dict['detection_
                boxes'],
71              image.shape[0], image.shape[1])
72          detection_masks_reframed = tf.cast(detection_masks_
            reframed > 0.5,
73                                              tf.uint8)
74          output_dict['detection_masks_reframed'] = detection_masks_
            reframed.numpy()

75

76      return output_dict
```

Line 53 is the one that does the actual object detection. The function model(input_tensor) predicts the object classes, bounding boxes, and associated scores. The model(input_tensor) function returns a dictionary that we will format in a usable form so that it contains the output corresponding to the input image only.

Since the model takes a batch of images, the function returns output for the batch. Because we have only one input image at a time, we are interested in the first result of this output dictionary (accessed by the 0th index). Line 59 extracts the first output and reassigns the output_dict variable.

Line 61 stores a number of detections in the dictionary so that we have this number handy when we work with the result.

Lines 66 through 74 are applicable only for a Mask R-CNN when masks need to be predicted. For all other predictors, these lines may be omitted.

Line 76 returns the output dictionary, which consists of coordinates of detected bounding boxes, object classes, scores, and number of detections. In the case of a Mask R-CNN, it also includes object masks.

Next, we will examine how output_dict is used to draw bounding boxes around detected objects in the images.

Write Code to Infer the Output, Draw Bounding Boxes Around Detected Objects, and Store the Result

The function infer_object() in Listing 6-16 is used to infer output_dict that was returned by the function run_inference_for_single_image(). The function infer_object() draws bounding boxes around each detected object in the image. It also labels the objects with class names and scores and finally saves the result to the output directory location. Listing 6-16 is the code segment that implements this function.

Listing 6-16. Drawing Bounding Boxes Around Detected Objects in Input Images

```
79   def infer_object(model, image_path):
80       # Read the image using openCV and create an image numpy
81       # The final output image with boxes and labels on it.
82       imagename = os.path.basename(image_path)
83
84       image_np = cv2.imread(os.path.abspath(image_path))
85       # Actual detection.
86       output_dict = run_inference_for_single_image(model, image_np)
87
88       # Visualization of the results of a detection.
89       for i in range(output_dict['detection_classes'].size):
90
91           box = output_dict['detection_boxes'][i]
92           classes = output_dict['detection_classes'][i]
93           scores = output_dict['detection_scores'][i]
94
95           if scores > 0.5:
96               h = image_np.shape[0]
97               w = image_np.shape[1]
98               classname = category_index[classes]['name']
99               classid =category_index[classes]['id']
100              #Draw bounding boxes
101              cv2.rectangle(image_np, (int(box[1] * w), int(box[0] * h)),
                     (int(box[3] * w), int(box[2] * h)), colortable[classid], 2)
```

```
102
103            #Write the class name on top of the bounding box
104            font = cv2.FONT_HERSHEY_COMPLEX_SMALL
105            size = cv2.getTextSize(str(classname) + ":" + str(scores),
               font, 0.75, 1)[0][0]
106
107            cv2.rectangle(image_np,(int(box[1] * w), int(box[0] *
               h-20)), ((int(box[1] * w)+size+5), int(box[0] * h)),
               colortable[classid],-1)
108            cv2.putText(image_np, str(classname) + ":" + str(scores),
109                    (int(box[1] * w), int(box[0] * h)-5), font, 0.75,
                       (0,0,0), 1, 1)
110        else:
111            break
112    # Save the result image with bounding boxes and class labels in
       file system
113    cv2.imwrite(output_path+"/"+imagename, image_np)
```

Line 79 defines the function infer_object() that takes two arguments: the model object and the path of the input image.

Line 82 simply gets the file name of the image that is used in line 113 and stores the resulting image with the same name to the output directory.

Line 84 reads the image using OpenCV and converts it into a NumPy array.

Line 86 calls the function run_inference_for_single_image() by passing to it the model object and the image NumPy. Recall that the function run_inference_ for_single_image() returns a dictionary containing the detected objects and bounding boxes.

The output dictionary may contain more than one object and bounding box. We need to loop through and draw bounding boxes around those objects for which the score is more than a threshold value. Lines 89 through 111 loops through each detected object class. The scores in the output dictionary are sorted in descending order. Therefore, when the score is less than the threshold value, the loop is exited.

Lines 91 through 93 simply extract the three important output arrays—bounding box coordinates, detected object class within this bounding box, and associated prediction score—and assign them to the corresponding variables.

In line 91, the variable box is an array containing the four corners of the bounding box as described here:

- box[0] is the y-coordinate, and box[1] is the x-coordinate of the top-left corner of the rectangular bounding box.

- box[2] and box[3] are the y- and x-coordinates of the bottom-right corner of the bounding box.

Line 95 checks to see whether the score is greater than a threshold. In this example, we used a threshold value of 0.5, but you can use a value suitable to your particular application. The bounding box will be drawn on the image only if the score is greater than the threshold value; otherwise, it will exit the for loop.

Recall that the images are resized before they are fed into the model for the training. The images are resized according to the height and width settings in the pipeline. config file that we used for the training. Therefore, the predicted bounding boxes are also scaled according to the resized images. Hence, we need to rescale the bounding boxes according to the original size of the input image used for detection. Multiplying the box coordinates with the image height and width scales the coordinates for the image size.

Line 101 draws rectangular bounding boxes using OpenCV's rectangle() function (review Chapter 2 for details about the rectangle() function). Notice that we used the colortable to dynamically get a different color for different classes.

Line 105 writes the predicted class name and corresponding score just above the bounding box. If you like, you can change the font style in line 104. In our example, the font color of the text and the borders of the bounding box are the same. You can use a different color by calling the colortable functions with different values. For example, add a constant to the class index and call the color table for the text color.

As mentioned earlier, the scores are sorted with the highest score at the top of the array. The first case of score after the threshold will break the loop to avoid unnecessary processing.

Line 113 saves the resulting image, with bounding boxes around detected objects, into the output directory.

Now that we have all the right settings and functions defined, we need to call them to trigger the detection process. Listing 6-17 shows how to trigger the detection.

Listing 6-17. Function Calls to Trigger the Detection Process

```
116   # Obtain the model object
117   detection_model = load_model(model_path)
118
119   # For each image, call the prediction
120   for image_path in IMAGE_PATHS:
121       infer_object(detection_model, image_path)
```

In Listing 6-17, line 117 calls the load_model() function by passing the path to the trained model. This function returns the model object that will be utilized in subsequent calls.

Line 120 iterates through each image file and calls infer_object() for each image. The function infer_object() is invoked for each image, and the final output with bounding boxes around the detected objects are saved in the output directory.

Putting It All Together

Let's put all the previous listings together to see the complete source code for object detection. Listing 6-18 is the fully working code.

Listing 6-18. Fully Working Code for Object Detection Using a Pretrained Model

```
Filename: Listing_6_18.py
1     import os
2     import pathlib
3     import random
4     import numpy as np
5     import tensorflow as tf
6     import cv2
7     # Import the object detection module.
8     from object_detection.utils import ops as utils_ops
9     from object_detection.utils import label_map_util
10
11    # to make gfile compatible with v2
```

```
12   #tf.gfile = tf.io.gfile
13
14   model_path = "ssd_model/final_model"
15   labels_path = "models/research/object_detection/data/pet_label_
     map.pbtxt"
16   image_dir = "images"
17   image_file_pattern = "*.jpg"
18   output_path="output_dir"
19
20   PATH_TO_IMAGES_DIR = pathlib.Path(image_dir)
21   IMAGE_PATHS = sorted(list(PATH_TO_IMAGES_DIR.glob(image_file_
     pattern)))
22
23   # List of the strings that is used to add correct label for each box.
24   category_index = label_map_util.create_category_index_from_
     labelmap(labels_path, use_display_name=True)
25   class_num =len(category_index)
26
27   def get_color_table(class_num, seed=0):
28       random.seed(seed)
29       color_table = {}
30       for i in range(class_num):
31           color_table[i] = [random.randint(0, 255) for _ in range(3)]
32       return color_table
33
34   colortable = get_color_table(class_num)
35
36   # # Model preparation and loading the model from the disk
37   def load_model(model_path):
38
39       model_dir = pathlib.Path(model_path) / "saved_model"
40       model = tf.saved_model.load(str(model_dir))
41       #model = model.signatures['serving_default']
42       return model
43
```

```
44   # Predict objects and bounding boxes and format the result
45   def run_inference_for_single_image(model, image):
46
47       # The input needs to be a tensor, convert it using `tf.convert_to_
         tensor`.
48       input_tensor = tf.convert_to_tensor(image)
49       # The model expects a batch of images, so add an axis with `tf.
         newaxis`.
50       input_tensor = input_tensor[tf.newaxis, ...]
51
52       # Run prediction from the model
53       output_dict = model(input_tensor)
54
55       # Input to model is a tensor, so the output is also a tensor
56       # Convert to numpy arrays, and take index [0] to remove the batch
         dimension.
57       # We're only interested in the first num_detections.
58       num_detections = int(output_dict.pop('num_detections'))
59       output_dict = {key: value[0, :num_detections].numpy()
60                      for key, value in output_dict.items()}
61       output_dict['num_detections'] = num_detections
62
63       # detection_classes should be ints.
64       output_dict['detection_classes'] = output_dict['detection_
         classes'].astype(np.int64)
65
66       # Handle models with masks:
67       if 'detection_masks' in output_dict:
68           # Reframe the bbox mask to the image size.
69           detection_masks_reframed = utils_ops.reframe_box_masks_to_
             image_masks(
70               output_dict['detection_masks'], output_dict['detection_
                 boxes'],
71               image.shape[0], image.shape[1])
```

```
72      detection_masks_reframed = tf.cast(detection_masks_
        reframed > 0.5,
73                                          tf.uint8)
74      output_dict['detection_masks_reframed'] = detection_masks_
        reframed.numpy()
75
76    return output_dict
77
78

79  def infer_object(model, image_path):
80      # Read the image using openCV and create an image numpy
81      # The final output image with boxes and labels on it.
82      imagename = os.path.basename(image_path)
83
84      image_np = cv2.imread(os.path.abspath(image_path))
85      # Actual detection.
86      output_dict = run_inference_for_single_image(model, image_np)
87
88      # Visualization of the results of a detection.
89      for i in range(output_dict['detection_classes'].size):
90
91          box = output_dict['detection_boxes'][i]
92          classes = output_dict['detection_classes'][i]
93          scores = output_dict['detection_scores'][i]
94          print(box, classes, scores)
95          if scores > 0.5:
96              h = image_np.shape[0]
97              w = image_np.shape[1]
98              classname = category_index[classes]['name']
99              classid =category_index[classes]['id']
100             #Draw bounding boxes
101             cv2.rectangle(image_np, (int(box[1] * w), int(box[0]
                * h)), (int(box[3] * w), int(box[2] * h)),
                colortable[classid-1], 2)

102
```

```
103                 #Write the class name on top of the bounding box
104                 font = cv2.FONT_HERSHEY_COMPLEX_SMALL
105                 size = cv2.getTextSize(str(classname) + ":" + str(scores),
                    font, 0.75, 1)[0][0]
106
107                 cv2.rectangle(image_np,(int(box[1] * w), int(box[0] *
                    h-20)), ((int(box[1] * w)+size+5), int(box[0] * h)),
                    colortable[classid-1],-1)
108                 cv2.putText(image_np, str(classname) + ":" + str(scores),
109                         (int(box[1] * w), int(box[0] * h)-5), font, 0.75,
                            (0,0,0), 1, 1)
110             else:
111                 break
112         # Save the result image with bounding boxes and class labels in
            file system
113         cv2.imwrite(output_path+"/"+imagename, image_np)
114         # cv2.imshow(imagename, image_np)
115
116 # Obtain the model object
117 detection_model = load_model(model_path)
118
119 # For each image, call the prediction
120 for image_path in IMAGE_PATHS:
121     infer_object(detection_model, image_path)
```

Figure 6-33 shows some sample outputs with the detected objects enclosed within bounding boxes.

Figure 6-33. *Example output images with detected animal faces and surrounding boxes*

Training a YOLOv7 Model for Object Detection

TensorFlow does not have an official implementation of the YOLOv7 model. The official YOLOv7 source code is written in PyTorch and made available as an open source release via the public GitHub repository at `https://github.com/WongKinYiu/yolov7`.

In this section, we will explore how to use the PyTorch-based public GitHub repository to train a custom YOLOv7 model, export the model in to ONNX format, and subsequently export it into TensorFlow and TensorFlow Lite (`tflite`) formats. We will utilize the `tflite` model for prediction and object detection in images.

Our objective is to train a YOLOv7 model to detect safety helmets and reflective jackets in images in various settings and environments.

This detection system will be valuable for sectors mandating the use of these safety items, such as construction, manufacturing, and mining. Employing this dataset for model training enables businesses to automate safety surveillance and mitigate the chances of accidents stemming from inadequate safety gear.

Dataset

We utilize an already labeled dataset consisting of images of individuals wearing safety helmets and reflective jackets. The dataset is available freely at the following URL:

```
https://www.kaggle.com/datasets/niravnaik/safety-helmet-and-reflective-
jacket/download?datasetVersionNumber=1
```

The dataset is also available at the following Google Drive link:

```
https://drive.google.com/file/d/1MdlK9lvwWXHQgRcsjZFcteA5oiFbOjBy/
view?usp=sharing
```

The dataset consists of 10,500 images, each annotated with bounding boxes around two classes of objects: safety helmets and reflective jackets. The dataset comprises pictures taken in various settings, including construction sites, factories, and outdoor workplaces. Within each image, there are approximately five marked objects, forming a comprehensive range of situations for the model's learning.

The dataset for YOLOv7 must be organized in the following folder structure:

1. Within a parent directory, there are three subdirectories: `train`, `test`, and `valid`.

2. Each of these subdirectories contains two additional subdirectories inside them: `images` and `labels`.

3. The `images` subdirectory contains image files (e.g., `.png`, `.jpg`, and `.jpeg`).

4. The `labels` subdirectory contains annotations in text files, one annotation text file per image. The file names in the `labels` directory must be the same as the image file names, except that it has the extension `.txt`. For example, if the image file name is `helmet_jacket_07350.jpg`, the annotation file name in the `labels` directory must be `helmet_jacket_07350.txt`.

5. The text file contains one or more lines of entries. Each line in the text file represents an object annotation. Multiple lines in the file means that the image contains multiple objects. Each line in the annotation text file must contain the annotated bounding box and object class in one single line in the following format:

<object-class> <x_center> <y_center> <width> <height>

where

<object-class> is the integer class index of the object, from 0 to
(num_class-1).

<x_center> and <y_center> are float values representing
the center of the bounding boxes relative to the image height
and width.

<width> and <height> are the width and height of bounding
boxes relative to the image height and width.

Note that the line entries in this file are separated by blank spaces
and not by commas or any other delimiters.

Figure 6-34 shows a sample directory structure for the YOLOv7 labeled dataset.

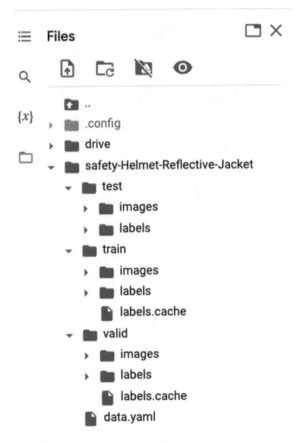

Figure 6-34. *YOLOv7 directory structure for training*

The parent directory `safety-Helmet-Reflective-Jacket` contains subdirectories: `train`, `test`, and `valid`. Within each, the `images` subdirectories contain image files used in training, testing, and validation and the `labels` subdirectories contain annotation text files. Figure 6-35 shows an example annotation with two objects in class 0 and class 1 and corresponding bounding boxes. Note that the coordinates of the bounding boxes are relative to the image size: the x-coordinate is divided by the image width, and the y-coordinate is divided by the image height. This makes the values of the bounding box coordinates range between 0 and 1.

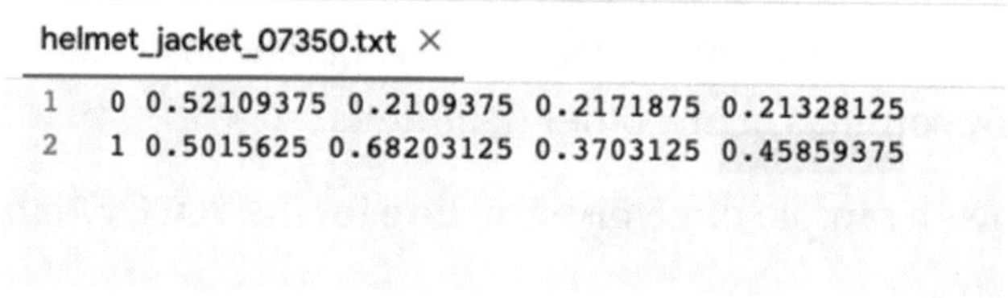

```
helmet_jacket_07350.txt  ×

1    0 0.52109375 0.2109375 0.2171875 0.21328125
2    1 0.5015625 0.68203125 0.3703125 0.45859375
```

Figure 6-35. *Annotations with bounding boxes and object classes*

Preparing Colab Environment

Start by creating a new project in Google Colab. Download the dataset described in the previous section to your local computer, and then upload it to the Colab environment. Again, uploading the dataset to a Google Drive directory is recommended to avoid losing the data upon the termination of your Colab session.

When you download the dataset, it will be in a compressed form (as a `.zip` file). Within a code block, execute the command shown in Listing 6-19 to uncompress the `.zip` file. After the command is successfully executed, the uncompressed data will be stored in the directory `/content/safety-Helmet-Reflective-Jacket`.

Listing 6-19. Command to Uncompress Dataset

```
!unzip '/content/drive/MyDrive/PPE/ppe.zip'
```

Creating the data.yaml File

The data.yaml file is used by the training function to locate the images and labels corresponding to the train, test, and validation sets. The data.yaml file already exists in the dataset directory /content/safety-Helmet-Reflective-Jacket. Edit this file to match with the entries shown in Figure 6-36.

```
data.yaml  ×
1    names: ['Safety-Helmet','Reflective-Jacket']
2    nc: 2
3    test: /content/safety-Helmet-Reflective-Jacket/test
4    train: /content/safety-Helmet-Reflective-Jacket/train
5    val: /content/safety-Helmet-Reflective-Jacket/valid
6
```

Figure 6-36. *data.yaml entries to indicate the locations of training, test, and validation sets*

The data.yaml file contains the following types of entries:

- names: The list of class names or objects names. The sequence of names in this list should align with the order employed during the annotation process.

- nc: The number of object classes. In our example, we have only two classes.

- test: The absolute path to the test directory containing images and labels subdirectories.

- train: The absolute path to the train directory containing images and labels subdirectories.

- val: The absolute path to the valid directory containing images and labels subdirectories.

Cloning YOLOv7 GitHub Repository

Within a separate code block, clone the official GitHub repository of YOLOv7 source code by running the command listed in Listing 6-20.

Listing 6-20. Command to Clone YOLOv7 GitHub Repository

```
!git clone https://github.com/WongKinYiu/yolov7.git /content/drive/MyDrive/
PPE/yolov7
```

This command will clone the GitHub repository within the Google Drive's PPE/
yolov7 directory. You can specify any directory of your choice. The advantage of using
the Google Drive directory is that the model we will train in this exercise will not get
deleted after the Colab session terminates.

Training YOLOv7 Model

As shown in Figure 6-37, the yolov7 directory (containing the cloned source code)
contains cfg and data subdirectories. The training subdirectory within the cfg
directory contains a few sample .yaml files. These .yaml files contain YOLOv7 neural
networks configurations. For the purpose of our exercise, we will utilize the yolov7.
yaml file.

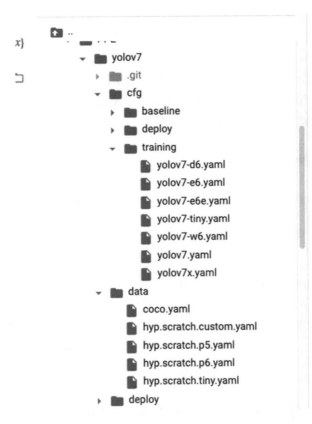

Figure 6-37. *Screenshot of* yolov7 *source's* cfg *and* data *directories*

Open the yolov7.yaml file and change the line containing the nc key to nc: 2, as our dataset contains only two types of objects: safety-helmet and reflective-jacket.

The section of the yolov7.yaml file showing the edited portion is shown in Figure 6-38.

```
yolov7.yaml  ×

1    # parameters
2    nc: 2|  # number of classes
3    depth_multiple: 1.0  # model depth multiple
4    width_multiple: 1.0  # layer channel multiple
5
```

Figure 6-38. *Portion of* yolov7.yaml *file showing* nc: 2 *to indicate two objects*

There are multiple options for training custom YOLOv7 models. In this section, we will explore some of the commonly used options.

Launching YOLOv7 Model Training

In the following sections, we will explore how to trigger the YOLO model training on machines equipped with either a single GPU or multiple GPUs. The process remains the same in both scenarios, with the only distinction being the configuration of the "device" parameter to match the hardware setup.

Training on a Single GPU

Listing 6-21 shows how to train a YOLOv7 model from scratch on a single GPU.

Listing 6-21. Training YOLOv7 Model from Scratch on a Single GPU

```
1    %%shell
2    cd /content/drive/MyDrive/PPE/yolov7
3    python train.py \
--epochs 10  \
--workers 8 \
--device 0 \
--batch-size 16 \
--data /content/safety-Helmet-Reflective-Jacket/data.yaml \
--img 640 640 \
--cfg cfg/training/yolov7.yaml \
--weights '' \
--name yolov7-ppe \
--hyp data/hyp.scratch.p5.yaml
```

Line 2 of Listing 6-21 changes the working directory to the yolov7 directory that we created at the time of cloning the YOLOv7 source from the repository. This directory contains a python script "train.py" that is used for training the model.

Line 3 calls the `train.py` script with the following arguments:

- `--epochs 10`: This indicates the training should run 10 epochs.

- `--workers 8`: This sets the number of data-loading worker processes to 8. These processes help load and preprocess training data efficiently.

- `--device 0`: This flag specifies which GPUs to use for training. In this case, a single GPU 0 is selected.

- `--batch-size 16`: This sets the batch size to 16 images per iteration.

- `--data /content/safety-Helmet-Reflective-Jacket/data.yaml`: This flag points to the YAML file containing dataset configuration details.

- `--img 640 640`: This defines the input image size as 640×640 pixels.

- `--cfg cfg/training/yolov7.yaml`: This specifies the model configuration file (YAML) for YOLOv7 architecture.

- `--weights ''`: This indicates that no pretrained weights are used for the model. Training will start from scratch. If the `weights` parameter is left blank, as in this example, it indicates training from scratch. Providing a file path to pr-existing weights indicates transfer learning.

- `--name yolov7-ppe`: This provides a name for the training session, which can be used to save checkpoints and logs. By default, the model is stored within the `runs/train` subdirectory of the project directory (`yolov7` in our example). In our example, the model will be stored in `/content/drive/MyDrive/PPE/yolov7/runs/train/yolov7-ppe` directory. If we run the training the second time, the model will get created in the directory `/content/drive/MyDrive/PPE/yolov7/runs/train/yolov7-ppe2`, and so on.

- `--hyp data/hyp.scratch.p5.yaml`: This points to the hyperparameters file in YAML format.

Training on Multiple GPUs

Training an object detection model demands substantial computational resources. When dealing with a large training dataset containing high-resolution images, it might be necessary to leverage multiple GPUs to accelerate the learning procedure.

By passing a few additional parameters into Listing 6-21, we can enhance the training process to utilize multiple GPUs. Listing 6-22 highlights these supplementary parameters in bold for clarity.

Listing 6-22. Training YOLOv7 Model from Scratch on a Single GPU

```
1    %%shell
2    cd /content/drive/MyDrive/PPE/yolov7
3    python \
```

```
-m torch.distributed.launch \
--nproc_per_node 4 \
--master_port 9527 \
train.py \
--epochs 100  \
--workers 8 \
--device 0,1,2,3 \
--sync-bn \
--batch-size 16 \
--data /content/safety-Helmet-Reflective-Jacket/data.yaml \
--img 640 640 \
--cfg cfg/training/yolov7.yaml \
--weights '' \
--name yolov7-ppe \
--hyp data/hyp.scratch.p5.yaml
```

The parameters and arguments to the training command are explained here:

- python -m torch.distributed.launch: This part of the command is invoking the Python interpreter with the torch.distributed.launch module, which facilitates distributed training across multiple nodes or GPUs.

- --nproc_per_node 4: This specifies the number of processes (GPUs) per node to be utilized for training. In this case, it's set to four GPUs.

- --master_port 9527: This parameter defines the communication port used by the master process for synchronization during distributed training.

- train.py: This is the name of the Python script responsible for initiating the training process.

- --workers 8: This sets the number of data-loading worker processes to 8. These processes help load and preprocess training data efficiently.

- `--device 0,1,2,3`: This flag specifies which GPUs to use for training. In this case, GPUs 0, 1, 2, and 3 are selected.

- `--sync-bn`: This indicates the usage of synchronized batch normalization, which can improve convergence during distributed training.

- `--batch-size 16`: This sets the batch size to 16 images per iteration.

- `--data /content/safety-Helmet-Reflective-Jacket/data.yaml`: This flag points to the YAML file containing dataset configuration details.

- `--img 640 640`: This defines the input image size as 640×640 pixels.

- `--cfg cfg/training/yolov7.yaml`: This specifies the model configuration file (YAML) for YOLOv7 architecture.

- `--weights ''`: This indicates that no pretrained weights are used for the model. Training will start from scratch.

- `--name yolov7-ppe`: This provides a name for the training session, which can be used to save checkpoints and logs.

- `--hyp data/hyp.scratch.p5.yaml`: This points to the hyperparameters file in YAML format.

Monitoring Training Progress

While the model is training, the Colab console displays the training log output. It displays various losses per epoch. For example, Figure 6-39 shows the log outputs of the last few epochs and at the end of the training.

```
Epoch     gpu_mem      box        obj       cls      total     labels  img_size
  7/9      11.5G    0.03545    0.01385   0.002048   0.05135       45       640: 100% 460/460 [07:32<00:00,  1.02it/s]
           Class     Images     Labels                 P          R      mAP@.5   mAP@.5:.95: 100% 50/50 [00:31<00:00,  1.58it/s]
            all       1575       5292               0.822       0.775     0.855       0.524

Epoch     gpu_mem      box        obj       cls      total     labels  img_size
  8/9      11.5G    0.03408    0.01382   0.001819   0.04972       32       640: 100% 460/460 [07:35<00:00,  1.01it/s]
           Class     Images     Labels                 P          R      mAP@.5   mAP@.5:.95: 100% 50/50 [00:31<00:00,  1.57it/s]
            all       1575       5292               0.841       0.78      0.872       0.551

Epoch     gpu_mem      box        obj       cls      total     labels  img_size
  9/9      11.5G    0.0333     0.01343   0.001733   0.04846       38       640: 100% 460/460 [07:34<00:00,  1.01it/s]
           Class     Images     Labels                 P          R      mAP@.5   mAP@.5:.95: 100% 50/50 [00:34<00:00,  1.43it/s]
            all       1575       5292               0.838       0.795     0.88        0.57
     Safety-Helmet     1575       2966               0.847       0.849     0.903       0.594
  Reflective-Jacket    1575       2326               0.83        0.741     0.858       0.546
10 epochs completed in 1.362 hours.

Optimizer stripped from runs/train/yolov7-ppe2/weights/last.pt, 74.8MB
Optimizer stripped from runs/train/yolov7-ppe2/weights/best.pt, 74.8MB
```

Figure 6-39. *Log output showing the training loss at each epoch and the end of the training*

At each epoch, the logout generates the following information (described here for epoch 8 in Figure 6-39):

- Epoch 8/9: This indicates that the training process is in its 8th epoch out of a total of 9 epochs.

- gpu_mem: This column shows the GPU memory usage during training. It states that 11.5GB of GPU memory are being used.

- box: This column represents the loss associated with bounding box prediction.

- obj: This column represents the loss associated with objectness prediction.

- cls: This column represents the loss associated with class prediction.

- total: This column represents the total loss, which is a combination of box loss, objectness loss, and class loss.

- labels: This column specifies the number of labeled objects in the training batch (32 in this epoch).

- img_size 640: The input image size used for training is 640×640 pixels.

- The progress bar [07:35<00:00, 1.01it/s] indicates the time remaining and the current processing speed for the current batch. The training is progressing at approximately one image per second.

- The section `Class Images Labels P R mAP@.5 mAP@.5:.95:` provides metrics related to object detection performance:

  - `all`: This refers to the aggregated metrics over all classes.

  - `Images`: The total number of images in the evaluation dataset (1575).

  - `Labels`: The total number of ground truth labels in the evaluation dataset (5292).

  - `P`: Precision, indicating the ratio of correctly predicted positive detections to all predicted positive detections.

  - `R`: Recall, indicating the ratio of correctly predicted positive detections to all ground truth positive instances.

  - `mAP@.5`: Mean average precision at IoU threshold 0.5, which measures the accuracy of detection.

  - `mAP@.5:.95`: Mean average precision at IoU thresholds ranging from 0.5 to 0.95, which provides a broader assessment of detection performance.

At the end of the training, the log output indicates the following (based on the example in Figure 6-39):

- `10 epochs completed in 1.362 hours`: This indicates that the training process has been run for a total of ten epochs, and it took approximately 1.362 hours to complete.

1. `Optimizer stripped from runs/train/yolov7-ppe2/weights/ last.pt, 74.8MB`: This message signifies that the optimizer state has been removed from the weights file corresponding to the last checkpoint of the training run. The weights file is located in the directory `runs/train/yolov7-ppe2/weights/last.pt`, and its size is 74.8 megabytes. Optimizer state is usually removed when saving checkpoints for sharing or deployment since it's not required for inference.

- `Optimizer stripped from runs/train/yolov7-ppe2/weights/ best.pt, 74.8MB`: Similarly, this message indicates that the optimizer state has been removed from the weights file corresponding to the best-performing checkpoint of the training run. The weights file is located in the directory runs/train/yolov7-ppe2/weights/best.pt, and its size is again 74.8 megabytes.

Monitoring Training Metrics Using TensorBoard

Listing 6-23 shows the command to launch TensorBoard by pointing it to the log directory. Ideally, you should launch TensorBoard from a different instance of Colab than the instance that is executing the training code. This will allow you to monitor the training progress while the training is running, and not after the training is completed.

Listing 6-23. Launching TensorBoard to Monitor the Training Metrics

```
1    %load_ext tensorboard
2    %tensorboard --logdir /content/drive/MyDrive/PPE/yolov7/runs/train
```

The TensorBoard web console will be presented directly within the Colab interface.

Inference or Object Detection Using the Training YOLOv7 Model

The final weights file is stored in the directory runs/train/yolov7-ppe/weights/best.pt. The command in Listing 6-24 shows how to use the trained model to detect objects within images. Listing 6-25 shows how to detect objects in a video file.

Listing 6-24. Launching TensorBoard to Monitor the Training Metrics of YOLO

```
1    %%shell
2    cd /content/drive/MyDrive/PPE/yolov7
3    python detect.py \
--project /content/detection \
--weights /content/drive/MyDrive/PPE/yolov7/runs/train/yolov7-ppe3/weights/
best.pt \
--conf 0.25 --img-size 640 \
--source /content/safety-Helmet-Reflective-ConstructionWorkers.png
```

Listing 6-25. Launching TensorBoard to Monitor the Training Metrics

```
1    %%shell
2    cd /content/drive/MyDrive/PPE/yolov7
3    python detect.py \
--project /content/detection \
--weights /content/drive/MyDrive/PPE/yolov7/runs/train/yolov7-ppe3/weights/
best.pt \
--conf 0.25 \
--img-size 640 \
--source /content/construction-site.mp4
```

In the preceding two examples, the argument `--project` takes a directory path where the prediction output is stored. If this argument is omitted, the output is stored in the project's `runs/detect/exp` directory.

A sample prediction output from the custom model, trained from scratch and with just ten epochs, is shown in Figure 6-40.

Figure 6-40. *Object detection output sample*

Exporting YOLOv7 Model to ONNX

ONNX, which stands for Open Neural Network Exchange, is an open and interoperable format for representing and exchanging deep learning models between various frameworks and tools.

The main idea behind ONNX is to provide a common standard for representing machine learning models, regardless of the framework they were trained in. This standardization makes it easier to deploy and use models across different platforms and frameworks, which is particularly valuable in scenarios where different parts of a workflow might involve different tools or environments.

ONNX supports a wide range of deep learning frameworks, including TensorFlow, PyTorch, Caffe, and more. With ONNX, we train a model in one framework and then export it to ONNX format. Once in the ONNX format, we can import the model into another supported framework, such as TensorFlow, for inference or further development without needing to retrain it.

In addition to the model format, ONNX also includes a runtime that allows us to load and run ONNX models on different devices, making it easier to deploy models on edge devices, mobile devices, or cloud servers.

To export the YOLOv7 model to ONNX, we need to install the onnx module from pypi. In Google Colab, open a new code block and simply use !pip install onnx to install the ONNX runtime.

The code block in Listing 6-26 exports the YOLOv7 model to ONNX and stores it in the same location where the best.pt model is located.

Listing 6-26. Exporting YOLOv7 Model to ONNX Format

```
1    %%shell
2    cd /content/drive/MyDrive/PPE/yolov7
3    python export.py \
--weights /content/drive/MyDrive/PPE/yolov7/runs/train/yolov7-ppe3/weights/
best.pt \
--grid --end2end --simplify \
--topk-all 100 --iou-thres 0.65 \
--conf-thres 0.35 \
--img-size 640 640 --max-wh 640
```

The exported model file is named as best.onnx and its location in an example directory structure is shown in Figure 6-41.

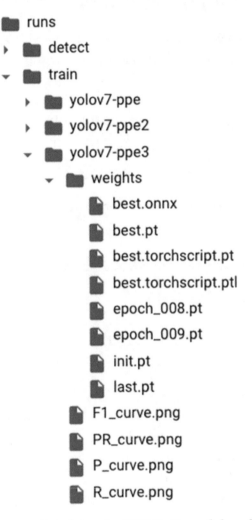

Figure 6-41. *Directory structure where the YOLOv7 model and exported ONNX model are located*

Converting the ONNX Model to TensorFlow and TensorFlow Lite Formats

We need to install the `onnx-tf` and `tensorflow-probability` modules in order to convert the ONNX model into TF and TFLite formats. Listing 6-27 shows the command to install these modules.

Listing 6-27. Installing Packages Needed for Conversion from ONNX to TF and TFLite

```
!pip install onnx-tf tensorflow-probability
```

Listing 6-28 shows the code segment that actually converts the ONNX model to TensorFlow format.

Listing 6-28. Code to Convert ONNX to TensorFlow

```
1    %%shell
2    onnx-tf convert \
-i /content/drive/MyDrive/PPE/yolov7/runs/train/yolov7-ppe3/weights/
best.onnx \
-o yolov7-tf
```

The argument `-i` takes the ONNX file path. The argument `-o` specifies the output location where the TensorFlow model will be saved. In Listing 6-28, the TF model will be stored in the directory `yolov7-tf` in the current working directory.

TensorFlow Lite is a lightweight version of the TensorFlow deep learning framework developed by Google. It's specifically designed for deploying machine learning models on resource-constrained devices, such as smartphones, embedded devices, microcontrollers, and other edge devices.

Listing 6-29 shows a Python code snippet that converts a TensorFlow model into TensorFlow Lite format.

Listing 6-29. Code to Convert TensorFlow to TensorFlow Lite

```
1    import tensorflow as tf
2    converter = tf.lite.TFLiteConverter.from_saved_model('/content/
     yolov7-tf/')
```

```
3    tflite_model = converter.convert()
4
5    with open('/content/tflite/yolov7-tiny.tflite', 'wb') as f:
6        f.write(tflite_model)
```

After executing this code, the YOLOv7 model will get converted into TensorFlow Lite format.

In the following section, we will explore how to predict object detection from the TensorFlow Lite model.

Predicting Using TensorFlow Lite Model

Listing 6-30 showcases the code used for detecting objects using a TensorFlow Lite model that was exported from the YOLO model.

Listing 6-30. Code to Predict Using TensorFlow Lite Model

```
FileName: listing_6_30
1    import cv2
2    import random
3    import numpy as np
4    from PIL import Image
5    import tensorflow as tf
6    from google.colab.patches import cv2_imshow
7    import numpy as np
8    # Load the TFLite model
9    interpreter = tf.lite.Interpreter(model_path="/content/tflite/yolov7-
     tiny.tflite")
10
11   def resize_image(im, new_shape=(640, 640), color=(114, 114, 114),
     auto=True, scaleup=True, stride=32):
12       # Resize and pad image while meeting stride-multiple constraints
13       shape = im.shape[:2]  # current shape [height, width]
14       if isinstance(new_shape, int):
15           new_shape = (new_shape, new_shape)
16
17       # Scale ratio (new / old)
18       r = min(new_shape[0] / shape[0], new_shape[1] / shape[1])
```

```
19        if not scaleup:  # only scale down, do not scale up (for better
          val mAP)
20            r = min(r, 1.0)
21
22        # Compute padding
23        new_unpad = int(round(shape[1] * r)), int(round(shape[0] * r))
24        dw, dh = new_shape[1] - new_unpad[0], new_shape[0] - new_
          unpad[1]  # wh padding
25
26        if auto:  # minimum rectangle
27            dw, dh = np.mod(dw, stride), np.mod(dh, stride)  # wh padding
28
29        dw /= 2  # divide padding into 2 sides
30        dh /= 2
31
32        if shape[::-1] != new_unpad:  # resize
33            im = cv2.resize(im, new_unpad, interpolation=cv2.INTER_LINEAR)
34        top, bottom = int(round(dh - 0.1)), int(round(dh + 0.1))
35        left, right = int(round(dw - 0.1)), int(round(dw + 0.1))
36        im = cv2.copyMakeBorder(im, top, bottom, left, right, cv2.BORDER_
          CONSTANT, value=color)  # add border
37        return im, r, (dw, dh)
38
39    #Name of the classes according to class indices.
40    names = ['safety-Helmet','Reflective-Jacket']
41
42    #Creating random colors for bounding box visualization.
43    colors = {name:[random.randint(0, 255) for _ in range(3)] for i,name
      in enumerate(names)}
44
45    #Load and preprocess the image.
46    img = cv2.imread('/content/CosntructionWorkers.png')
47    img = cv2.cvtColor(img, cv2.COLOR_BGR2RGB)
48
49    image = img.copy()
50    image, ratio, dwdh = resize_image(image, auto=False)
```

```
51    image = image.transpose((2, 0, 1))
52    image = np.expand_dims(image, 0)
53    image = np.ascontiguousarray(image)
54
55    im = image.astype(np.float32)
56    im /= 255
57
58    #Allocate tensors.
59    interpreter.allocate_tensors()
60    # Get input and output tensors.
61    input_details = interpreter.get_input_details()
62    output_details = interpreter.get_output_details()
63
64    # Test the model on random input data.
65    input_shape = input_details[0]['shape']
66    interpreter.set_tensor(input_details[0]['index'], im)
67
68    interpreter.invoke()
69
70    # The function `get_tensor()` returns a copy of the tensor data.
71    # Use `tensor()` in order to get a pointer to the tensor.
72    output_data = interpreter.get_tensor(output_details[0]['index'])
73
74    ori_images = [img.copy()]
75
76    for i,(batch_id,x0,y0,x1,y1,cls_id,score) in enumerate(output_data):
77        image = ori_images[int(batch_id)]
78        box = np.array([x0,y0,x1,y1])
79        box -= np.array(dwdh*2)
80        box /= ratio
81        box = box.round().astype(np.int32).tolist()
82        cls_id = int(cls_id)
83        score = round(float(score),3)
84        name = names[cls_id]
85        color = colors[name]
86        name += ' '+str(score)
```

```
87        cv2.rectangle(image,box[:2],box[2:],color,2)
88        cv2.putText(image,name,(box[0], box[1] - 2),cv2.FONT_HERSHEY_
          SIMPLEX,0.75,[225, 255, 255],thickness=2)
89
90    prediction = Image.fromarray(ori_images[0])
91
92    open_cv_image = np.array(prediction)
93    # Convert RGB to BGR
94    open_cv_image = open_cv_image[:, :, ::-1].copy()
95    cv2_imshow(open_cv_image)
```

Here's a breakdown of the code step by step:

- *Import required libraries*: Lines 1 to 7 import the necessary libraries such as cv2 (OpenCV), random, numpy, PIL, and tensorflow. It also imports the cv2_imshow function from Google Colab's patches for displaying images within a Colab notebook.

- *Load the TFLite model*: Line 9 creates an instance of the tf.lite. Interpreter class and loads the YOLOv7-tiny model from the given model path (/content/tflite/yolov7-tiny.tflite).

- resize_image(): The resize_image() function takes an input image and resizes it while ensuring it meets certain constraints required by YOLO. It calculates the necessary padding and scaling to match the desired dimensions and stride. This function is important because YOLO works with a specific input size and stride, and the image needs to be resized and padded accordingly.

- *Class names and colors*: The code defines a list of class names (names) and generates random colors for each class. These colors will be used for drawing bounding boxes and class labels on the detected objects.

- *Load and preprocess image*: The code loads an input image (CosntructionWorkers.png) using OpenCV, converts it from BGR to RGB format, and then resizes and preprocesses the image using the resize_image() function. The resulting image is reshaped, transposed, expanded in dimensions, and normalized to values between 0 and 1.

- *Allocate and set tensors*: The TFLite interpreter's tensors are allocated and input data is set using the preprocessed image. The interpreter is then invoked to run inference on the input image.

- *Retrieve output data*: The output data from the interpreter is retrieved using the get_tensor() function. This output data contains information about detected objects, including their bounding box coordinates, class IDs, and scores.

- *Visualize detected objects*: The code then processes the output data, extracts relevant information, and uses OpenCV to draw bounding boxes and labels on the original image. The boxes are scaled and positioned correctly based on the resizing and preprocessing steps performed earlier.

- *Display prediction*: Finally, the code converts the modified image (with bounding boxes and labels) to a format compatible with OpenCV and displays it using the cv2_imshow() function within the Colab notebook.

Summary

In this chapter, we have explored various object detection algorithms and compared their detection speed and accuracy. We have trained two detection models, namely SSD and YOLOv7, and have covered the entire process from data ingestion to saving prediction outputs.

Additionally, we identified how to leverage Google Colab to train object detection models in the cloud, taking advantage of powerful GPUs.

It's important to note that the focus of this chapter has primarily been on detecting objects in images, and we haven't worked on any examples involving videos. However, the process of object detection in videos is similar to that of object detection in images since videos are essentially composed of frames, which are individual images. Chapter 7 will be dedicated to exploring object detection in videos.

Furthermore, we will apply the concepts covered in this chapter to Chapters 9 and 10, where we will develop real-world use cases of computer vision utilizing deep learning techniques.

Practical Example: Object Tracking in Videos

The focus of this chapter is on two critical capabilities of computer vision: object detection and object tracking. In general, and in the context of a set of images, object detection provides the ability to identify one or more objects in an image, and object tracking provides the ability to track a detected object across a set of images. In previous chapters, we explored the technical aspects of training deep learning models to detect objects. In this chapter, we will delve into the practical application of our object detection knowledge within the realm of video.

Object tracking in a video, or simply *video tracking*, involves detecting and locating an object and tracking it over time. Video tracking is used not only to detect an object in different frames but also to track it across frames. When an object is first detected, its unique identity is extracted and then tracked in subsequent frames.

Object tracking has many applications in the real world, such as the following:

- Autonomous cars
- Security and surveillance
- Traffic control
- Augmented reality (AR)
- Crime detection and criminal tracking
- Medical imaging and more

In this chapter, we will explore how to implement video tracking and work through code examples. At the end of this chapter, you should have a fully functional video tracking system.

© Shamshad Ansari 2023
S. Ansari, *Building Computer Vision Applications Using Artificial Neural Networks*,
https://doi.org/10.1007/978-1-4842-9866-4_7

Our high-level plan of implementation is as follows:

1. *Video source*: We will use OpenCV to read live streams of video from a webcam or the built-in camera of a laptop. You can also read videos from a file or IP camera.

2. *Object detection model*: We will use an SSD model pretrained on the COCO dataset. You can train your own model for your specific use cases (review Chapter 6 for information on training the object detection model). Alternatively, you can use a YOLO model instead.

3. *Prediction*: We will predict object classes (detection) and their bounding boxes (localization) within each frame of the video (review Chapter 6 for information on detecting objects in images).

4. *Unique identity*: We will use a hashing algorithm to create a unique identity of each object. More details about the hashing algorithm are provided later in this chapter.

5. *Tracking*: We will use the Hamming distance algorithm (more on this later in this chapter) to track the previously detected objects.

6. *Display*: We will stream the output video for display in web browsers. To do so, we will use Flask, a lightweight web application microframework.

Preparing the Working Environment

Let's establish a directory structure so that it is easy to follow the code and work through the exercise. We will analyze code fragments of each of the six steps just described. At the end, we will put everything together to make the object tracking system complete and workable.

We have a directory called `video_tracking`. Inside this we have a subdirectory called `templates`, which has an HTML file called `index.html`. The subdirectory, `templates`, is the standard place where Flask looks for HTML pages. In the `video_tracking` directory, we have four Python files: `videoasync.py`, `object_tracker.py`, `tracker.py`, and `video_server.py`. Figure 7-1 shows this directory structure.

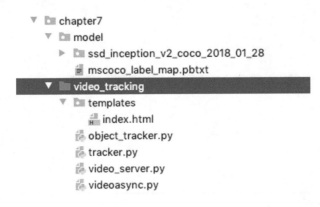

Figure 7-1. *Code directory structure*

We will import `videoasync` as a module in `object_tracker.py`. Therefore, the directory `video_tracking` must be recognized as a source directory in PyCharm. To make it a source directory in PyCharm, click the PyCharm menu option at the top left of the screen, then click Preferences, expand Project in the left panel, click Project Structure, select the `video_tracking` directory, and click Sources as the Mark As option (located at the top of the screen), as shown in Figure 7-2. Finally, click OK to close the window.

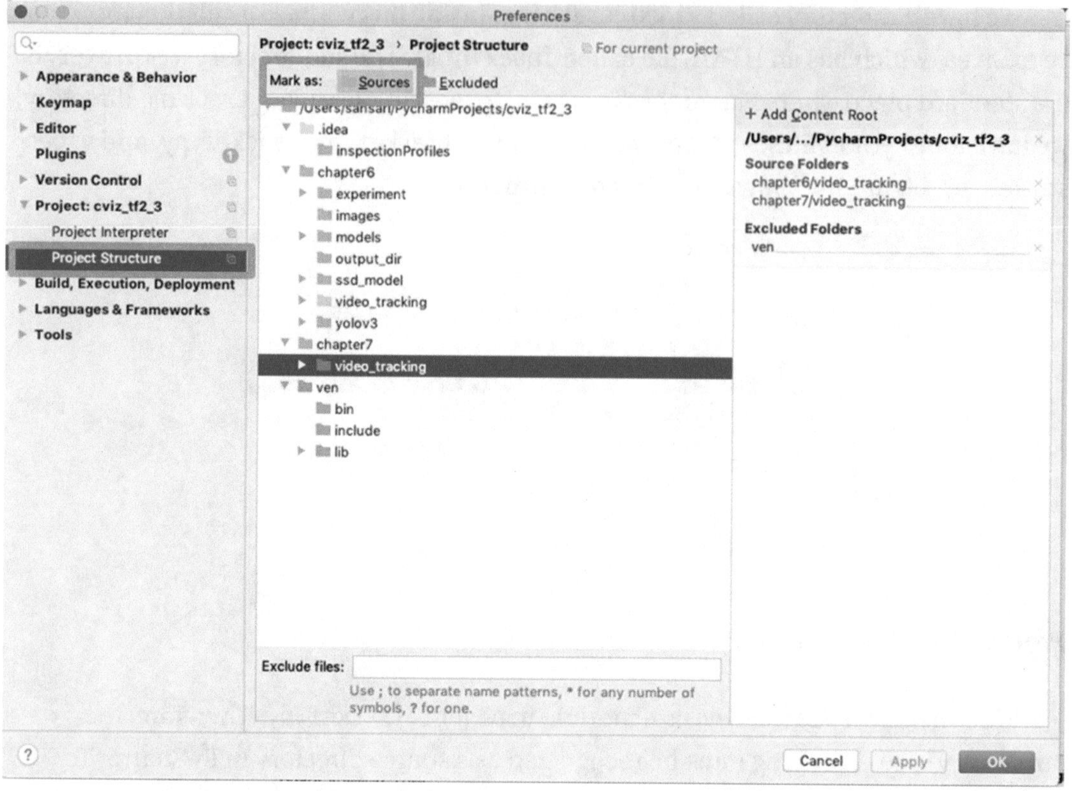

Figure 7-2. *Marking a directory as a source in PyCharm*

Reading a Video Stream

OpenCV provides convenient methods to connect to a video source and read images from the video frames. The images from these frames are internally converted by OpenCV into NumPy arrays. These NumPy arrays are further processed to detect and track objects in them. The detection process is compute-intensive, and it may not be able to keep up with the speed of reading frames. Therefore, reading the frames and performing detection operations in the main thread will exhibit slow performance, especially when dealing with high-definition (HD) videos. In Listing 7-1, we will implement multithreading for capturing frames. We will call it an *asynchronous reading* of video frames.

Listing 7-1. Implementation of Asynchronous Reading of Video Frames

```
1    # file: videoasync.py
2    import threading
3    import cv2
4
5    class VideoCaptureAsync:
6        def __init__(self, src=0):
7            self.src = src
8            self.cap = cv2.VideoCapture(self.src)
9            self.grabbed, self.frame = self.cap.read()
10           self.started = False
11           self.read_lock = threading.Lock()
12
13       def set(self, key, value):
14           self.cap.set(key, value)
15
16       def start(self):
17           if self.started:
18               print('[Warning] Asynchronous video capturing is already
                     started.')
19               return None
20           self.started = True
21           self.thread = threading.Thread(target=self.update, args=())
22           self.thread.start()
23           return self
24
25       def update(self):
26           while self.started:
27               grabbed, frame = self.cap.read()
28               with self.read_lock:
29                   self.grabbed = grabbed
30                   self.frame = frame
31
32       def read(self):
33           with self.read_lock:
```

```
34                    frame = self.frame.copy()
35                    grabbed = self.grabbed
36              return grabbed, frame
37
38        def stop(self):
39            self.started = False
40            # self.cap.release()
41            # cv2.destroyAllWindows()
42            self.thread.join()
43
44        def __exit__(self, exec_type, exc_value, traceback):
45            self.cap.release()
```

The file videoasync.py implements the class VideoCaptureAsync (line 5), which consists of a constructor and functions to start the thread, read frames, and stop the thread.

Line 6 defines a constructor that takes the video source as an argument. The default value of this source, src=0 (also called the *device index*), represents the input from the built-in camera on the laptop/computer. If you have a USB camera, set the value of this src accordingly. There is no standard way to find the device index if you have multiple cameras attached to your computer ports. One way could be to loop through from a starting index of 0 until you connect to the device. You can print the device properties to identify the device you want to connect to. For IP-based cameras, pass the IP address or the URL.

If your video source is a file, pass the path to the video file.

Line 8 uses OpenCV's VideoCapture() function and passes the source ID to connect to the video source. The VideoCapture object assigned to the self.cap variable is used for reading the frames.

Line 9 reads the first frame and occupies the connection to the video camera.

Line 10 is the flag that is used to manage the lock. Line 11 actually acquires the thread lock.

Lines 13 and 14 implement a function to set properties to the VideoCapture object, such as frame height, width, and frames per second (FPS).

Lines 16 through 23 implement the start() function to start the thread for asynchronous reading frames.

Lines 25 through 30 implement an update() function to read the frame and update the class-level frame variable. The update() function is internally used within the start() function, in line 21, to asynchronously read the video frames.

Lines 32 through 36 implement the read() function, which simply returns the frame updated in the update() function block. This also returns a Boolean to indicate whether the frame was successfully read.

Lines 38 through 40 implement the stop() function to stop the thread and return the control to the main thread. The join() function prevents the shutdown of the main thread until the child thread completes its execution.

Upon exit, the video source is released (line 45).

We will now write code to utilize the asynchronous video reading module. In the same directory, video_tracking, we will create a Python file called object_tracker.py that implements the following functionality.

Loading the Object Detection Model

We will use the same pretrained SSD model that we used in Chapter 6 for detecting objects in images. If you have trained a model based on your own images, you can use that model; simply provide the path to the model directory. Listing 7-2 shows how to load the trained model from the disk. Recall that this is the same function that we used in Chapter 6's Listing 6-11. We will load the model only once and use it for detecting objects in all frames.

Listing 7-2. load_model() Function to Load Trained Model from the Disk

```
43    # # Model preparation
44    def load_model(model_path):
45        model_dir = pathlib.Path(model_path) / "saved_model"
46        model = tf.saved_model.load(str(model_dir))
47        model = model.signatures['serving_default']
48        return model
49
50    model = load_model(model_path)
```

Detecting Objects in Video Frames

The code for detecting objects is almost the same as the code we used in Chapter 6 for the same purpose. The difference is that here we create an infinite loop inside which we read one image frame at a time and make a function call to `track_object()` for tracking objects within that frame. The `track_object()` function internally calls the same `run_inference_for_single_image()` function that we implemented in Chapter 6's Listing 6-15.

The output from the `run_inference_for_single_image()` function is a dictionary containing `detection_classes`, `detection_boxes`, and `detection_scores`. We will utilize these values to calculate the unique identity of each object and track their locations.

Listing 7-3 shows the `streamVideo()` function that implements the infinite loop to read streaming frames from the video source.

In Listing 7-3, line 115 starts the block of the `streamVideo()` function. Line 116 uses the `global` keyword with the thread lock.

Line 117 starts the infinite `while` loop. Inside this loop, the first line, line 118, reads the current video frame (image) by calling the `read()` function of the `VideoCaptureAsync` class. The `read()` function returns a tuple of a Boolean indicating whether the frame is read successfully, and a NumPy array of the image frame.

If the frame is successfully retrieved (line 119), acquire the lock (line 120) so that other threads do not modify the frame NumPy while the current thread's image is still being detected for objects.

Line 121 calls the `track_object()` function by passing the model object and frame NumPy. We will see later in Listing 7-13 what this `track_object()` function does. In line 123, the output NumPy array is converted into the compressed `.jpg` image so that it is lightweight and easily transferable over the network. We used `cv2.imencode()` to convert the NumPy array to an image. This function returns a tuple of a Boolean indicating whether the conversion is successful and returns the encoded image.

If the image conversion is not successful, skip that frame (line 125).

Finally, on line 127, it yields the byte-encoded image. The `yield` keyword returns a read-once iterator from the `while` loop.

Lines 130 through 137 are cleanup functions when either the program is terminated or the screen is killed by pressing Q to quit.

Listing 7-3. Implementing Infinite Loop for Reading Streams of Video Frames and Internally Calling an Object Tracking Function for Each Frame

```
114  # Function to implement infinite while loop to read video frames and
         generate the output    #for web browser
115  def streamVideo():
116      global lock
117      while (True):
118          retrieved, frame = cap.read()
119          if retrieved:
120              with lock:
121                  frame = track_object(model, frame)
122
123                  (flag, encodedImage) = cv2.imencode(".jpg", frame)
124                  if not flag:
125                      continue
126
127                  yield (b'--frame\r\n' b'Content-Type: image/jpeg\
                     r\n\r\n' +
128                      bytearray(encodedImage) + b'\r\n')
129
130          if cv2.waitKey(1) & 0xFF == ord('q'):
131              cap.stop()
132              cv2.destroyAllWindows()
133              break
134
135      # When everything done, release the capture
136      cap.stop()
137      cv2.destroyAllWindows()
```

Creating a Unique Identity for Objects Using dHash

We use perceptual hashing to create a unique identity of an object detected within the image. Difference hashing, or simply dHash, is one of the most commonly used algorithms to calculate a unique hash of an image. A dHash provides several advantages that makes it a suitable choice for identifying and comparing images. The following are some of the noteworthy benefits of using a dHash:

- The image hash does not change if the aspect ratio changes.

- Changes in brightness or contrast will either not change the image hash or change it slightly. This means the hashes remain close to others with varying contrasts.

- The computation of a dHash is extremely fast.

We do not use cryptographic hashes, such as MD-5 or SHA-1. The reason is that for these hashing algorithms, if there is a slight change in the image, the cryptographic hashes will be totally different. Even a single-pixel change will result in a completely different hash. Therefore, if two images are perceptually similar, their cryptographic hashes will be totally different. This makes it not a fit for the application when we have to compare two images.

The dHash algorithm is simple. The following are the steps to compute the dHash:

1. Convert the image or snippet of the image into grayscale. This makes computation much faster, and the dHash will not change much if there is a slight variation of color. In object detection, we crop the detected objects using the bounding boxes and convert the cropped image into grayscale.

2. Resize the grayscale image. To compute a 64-bit hash, the image is resized to 9×8 pixels, ignoring its aspect ratio. The aspect ratio is ignored to ensure that the resulting image hash will match similar images regardless of their initial spatial dimensions.

   Why 9×8 pixels? In a dHash, the algorithm computes the difference of gradients of adjacent pixels. The difference of nine rows with adjacent rows will yield only eight rows in the result, thus making the final output with 8×8 pixels, which will give us a 64-bit hash.

3. Build the hash by converting each pixel into either 0 or 1 by applying the "greater than" formula, as shown here:

$$\text{If } P[x{=}1] > P[x], \text{ then 1 else 0.}$$

The binary values are then converted into an integer hash.

Listing 7-4 shows the Python and OpenCV implementation of the dHash.

Listing 7-4. Calculating the dHash from an Image

```
32      def getCropped(self, image_np, xmin, ymin, xmax, ymax):
33          return image_np[ymin:ymax, xmin:xmax]
34
35      def resize(self, cropped_image, size=8):
36          resized = cv2.resize(cropped_image, (size+1, size))
37          return resized
38
39      def getHash(self, resized_image):
40          diff = resized_image[:, 1:] > resized_image[:, :-1]
41          # convert the difference image to a hash
42          dhash = sum([2 ** i for (i, v) in enumerate(diff.flatten()) if v])
43          return int(np.array(dhash, dtype="float64"))
```

Lines 32 and 33 implement the cropping function. We pass the NumPy arrays of the full image frame and the four coordinates of the bounding box that surrounds an object. The function crops the portion of the image that contains the detected object.

Lines 35 through 37 are for resizing the cropped image into a 9×8 size.

Lines 39 through 43 implement the calculation of the dHash. Line 40 finds the difference of adjacent pixels by applying the greater-than rule described earlier. Line 42 builds the numeric hash from the binary bit values. Line 43 converts the hash to integer and returns the ***dhash*** from the function.

Using the Hamming Distance to Determine Image Similarity

The Hamming distance is commonly used to compare two hashes. The Hamming distance measures the number of different bits in two hashes.

If the Hamming distance of two hashes is zero, it means the two hashes are identical. The lower the Hamming distance, the more similar the two hashes are.

Listing 7-5 shows how to calculate the Hamming distance between two hashes.

Listing 7-5. Calculation of the Hamming Distance

```
45      def hamming(self, hashA, hashB):
46          # compute and return the Hamming distance between the integers
47          return bin(int(hashA) ^ int(hashB)).count("1")
```

The function hamming() in line 45 takes two hashes as input and returns the number of bits that are different in these two input hashes.

Object Tracking

After an object is detected in an image, its unique identity is created by calculating the dHash of the cropped part of the image that contains the object. The object is tracked from one frame to the other by calculating the Hamming distance of the object's dHash. There are many use cases of tracking. In our example, we created two tracking functions to do the following:

1. Track the path of the object from the first occurrence of the object in a frame to all occurrences in the subsequent frames. This function tracks the center of the bounding boxes and draws a line or path connecting all these centers. Listing 7-6 shows this implementation. The function createHammingDict() takes the current object's dHash, its center of the bounding box, and the history of all objects and its centers. The function compares the dHash of the current object with all dHashes seen so far and uses the Hamming distance to find similar objects to track its movements or the path.

Listing 7-6. Tracking the Centers of Bounding Boxes of Detected Objects Between Multiple Frames

```
49      def createHammingDict(self, dhash, center, hamming_dict):
50          centers = []
51          matched = False
52          matched_hash = dhash
53          # matched_classid = classid
54
```

```python
55              if hamming_dict.__len__() > 0:
56                  if hamming_dict.get(dhash):
57                      matched = True
58
59                  else:
60                      for key in hamming_dict.keys():
61
62                          hd = self.hamming(dhash, key)
63
64                          if(hd < self.threshold):
65                              centers = hamming_dict.get(key)
66                              if len(centers) > self.max_track_frame:
67                                  centers.pop(0)
68                              centers.append(center)
69                              del hamming_dict[key]
70                              hamming_dict[dhash] = centers
71                              matched = True
72                              break
73
74              if not matched:
75                  centers.append(center)
76                  hamming_dict[dhash] = centers
77
78              return  hamming_dict
```

2. Get the unique identifiers of the objects and track the number of unique objects detected. Listing 7-7 implements a function called getObjectCounter() that counts the number of unique objects detected across frames. It compares the dHash of the current object with all dHashes computed so far across all previous frames.

Listing 7-7. Function to Track Count of Unique Objects Detected in
Video Frames

```
79
80      def getObjectCounter(self, dhash, hamming_dict):
81          matched = False
82          matched_hash = dhash
83          lowest_hamming_dist = self.threshold
84          object_counter = 0
85
86          if len(hamming_dict) > 0:
87              if dhash in hamming_dict:
88                  lowest_hamming_dist = 0
89                  matched_hash = dhash
90                  object_counter = hamming_dict.get(dhash)
91                  matched = True
92
93              else:
94                  for key in hamming_dict.keys():
95                      hd = self.hamming(dhash, key)
96                      if(hd < self.threshold):
97                          if hd < lowest_hamming_dist:
98                              lowest_hamming_dist = hd
99                              matched = True
100                             matched_hash = key
101                             object_counter = hamming_dict.get(key)
102         if not matched:
103             object_counter = len(hamming_dict)
104         if matched_hash in hamming_dict:
105             del hamming_dict[matched_hash]
106
107         hamming_dict[dhash] = object_counter
108         return  hamming_dict
109
```

Displaying a Live Video Stream in a Web Browser

We will publish our video tracking code to Flask, a lightweight web framework. This will allow us to view the live stream of the video, with tracked objects, in web browsers using a URL. You can use other frameworks, such as Django, to publish the video to be accessible from a web browser. I selected Flask for our example because it is lightweight, flexible, and easy to implement with just a few lines of code.

Let's explore how to use Flask in our current context. We will start with installing Flask to our virtualenv.

Installing Flask

We will use the `pip` command to install Flask. Make sure you activate your virtualenv and execute the command `pip install flask`, as shown here:

```
(cv_tf2) computername:~ username$ pip install flask
```

Flask Directory Structure

Refer to the directory structure in Figure 7-1. We have created a subdirectory called `templates` in the `video_tracking` directory. We will create an HTML file, `index.html`, that will contain the code to display streaming video. We will save `index.html` to the `templates` directory. The name of the directory must be `templates`, as Flask looks for this directory to find the HTML files.

HTML for Displaying a Video Stream

Listing 7-8 shows the HTML code that is saved in the `index.html` page. Line 7 is the most important line because it will display the live video stream. This is a standard `<img>` tag of HTML that is typically used to display an image in a web browser. The `{{...}}` portion of the code in line 7 is the Flask symbol that instructs Flask to load the image from a URL. When this HTML page is loaded, it will make a call to the `/video_feed` URL and fetch the image from there to display within the `<img>` tag.

Listing 7-8. HTML Code for Displaying the Video Stream

```
1    <html>
2     <head>
3       <title>Computer Vision</title>
4     </head>
5     <body>
6       <h1>Video Surveillance</h1>
7       <img src="{{ url_for('video_feed') }}" > </img>
8     </body>
9    </html>
```

Now we need some server-side code that will serve this HTML page. We also need a server-side implementation to serve images when the /video_feed URL is called.

We will implement these two functions in a separate Python file, video_server.py, that is saved in the video_tracking directory. Make sure that this video_server.py file and the templates directory are in the same parent directory.

Listing 7-9 shows a server-side implementation of Flask services. Line 2 imports Flask and its related packages. Line 3 imports our object_tracker package that has the implementation of object detection and tracking.

Line 4 creates a Flask application using the constructor app = Flask(__name__), which takes the current module as an argument. By calling the constructor, we instantiate the Flask web application framework and assign this to a variable called app. We will bind all server-side services to this app.

All Flask services are served through URLs, and we have to bind the URL or route to the service it will serve. The following are the two services we need to implement for our example:

- Service that will render index.html from the home URL, e.g., http://localhost:5019/

- Service that will serve a stream of video from /video_feed URL, e.g., http://localhost:5019/video_feed

Flask to Load the HTML Page

Line 6 of Listing 7-9 has a route binding of /, which indicates the home URL. When the home URL is called from a web browser, the function index() is called to serve the request (line 7). The index() function simply renders an HTML page from a template, index.html, that we created in Listing 7-8.

Flask to Serve the Video Stream

Line 11 of Listing 7-9 binds the /video_feed URL to the Python function video_feed(). This function, in turn, calls the streamVideo() function that we implemented for detecting and tracking objects in video. Line 15 creates the Response object from the video frames and sends a multipart HTTP response to the caller.

Listing 7-9. Flask Server-Side Code to Launch index.html and Serve Video Stream

```
1    # video_server.py
2    from flask import Flask, render_template, Response
3    import object_tracker as ot
4    app = Flask(__name__)
5
6    @app.route("/")
7    def index():
8        # return the rendered template
9        return render_template("index.html")
10
11   @app.route("/video_feed")
12   def video_feed():
13       # return the response generated along with the specific media
14       # type (mime type)
15       return Response(ot.streamVideo(),mimetype = "multipart/x-mixed-
         replace; boundary=frame")
16
17   if __name__ == '__main__':
18       app.run(host="localhost", port="5019", debug=True,
19                   threaded=True, use_reloader=False)
```

Running the Flask Server

Execute the video_server.py file from a terminal by typing the command python video_server.py from the video_tracking directory. Make sure you have your virtualenv activated.

```
(cv) computername:video_tracking username$ python video_server.py
```

This will start the Flask server and run on host="localhost" and port="5019" (line 18 of Listing 7-9). You should change the host and port for your production environment. Also, turn off the debug mode by setting debug=False in line 18.

When the server starts, point your web browser to the URL http://localhost:5019/ to see the live video streams with object tracking.

Putting It All Together

We have explored the building blocks of our video tracking system. Let's put them all together to have a fully functional system. Figure 7-3 shows the high-level sequence of function calls of our video tracking system.

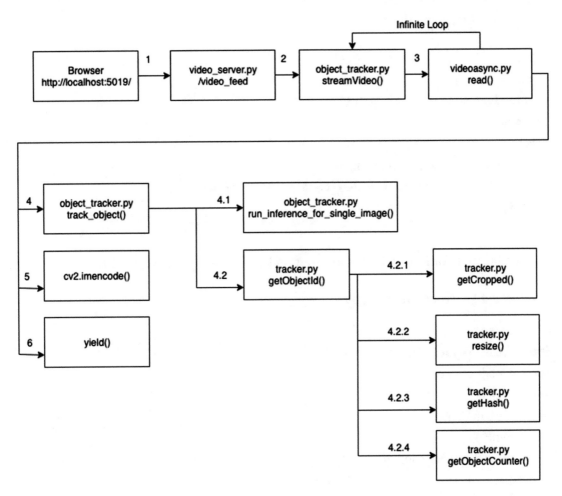

Figure 7-3. *Schematic of sequence of function calls of the video tracking system*

When the web browser is launched with the URL `http://localhost:5019/`, the Flask backend server serves the `index.html` page, which internally calls the URL `http://localhost:5019/video_feed` that invokes the server-side function `video_feed()`. The rest of the function calls, as shown in Figure 7-3, when completed, send the video frames with the detected objects with their tracking information to the web browser for display. Listings 7-10 through 7-14 provide the complete source code of the video tracking system.

The file path for Listing 7-10 is `video_tracking/templates/index.html`.

Listing 7-10. index.html

```html
<html>
 <head>
   <title>Computer Vision</title>
 </head>
 <body>
   <h1>Video Surveillance</h1>
   <img src="{{ url_for('video_feed') }}" > </img>
 </body>
</html>
```

The file path for Listing 7-11 is video_tracking/video_server.py.

Listing 7-11. video_server.py

```python
# video_server.py
from flask import Flask, render_template, Response
import object_tracker as ot
app = Flask(__name__)
@app.route("/")
def index():
   # return the rendered template
   return render_template("index.html")
@app.route("/video_feed")
def video_feed():
   # return the response generated along with the specific media
   # type (mime type)
   return Response(ot.streamVideo(),mimetype = "multipart/x-mixed-replace;
   boundary=frame")
if __name__ == '__main__':
   app.run(host="localhost", port="5019", debug=True,
       threaded=True, use_reloader=False)
```

The file path for Listing 7-12 is video_tracking/object_tracker.py.

Listing 7-12. `object_tracker.py`

```python
import os
import pathlib
import random
import numpy as np
import tensorflow as tf
import cv2
import threading
# Import the object detection module.
from object_detection.utils import ops as utils_ops
from object_detection.utils import label_map_util
from videoasync import VideoCaptureAsync
import tracker as hasher
lock = threading.Lock()
# to make gfile compatible with v2
tf.gfile = tf.io.gfile
model_path = "./../model/ssd_inception_v2_coco_2018_01_28"
labels_path = "./../model/mscoco_label_map.pbtxt"
# List of the strings that is used to add correct label for each box.
category_index = label_map_util.create_category_index_from_labelmap(labels_
path, use_display_name=True)
class_num =len(category_index)+100
object_ids = {}
hasher_object = hasher.ObjectHasher()
# Function to create color table for each object class
def get_color_table(class_num, seed=50):
    random.seed(seed)
    color_table = {}
    for i in range(class_num):
        color_table[i] = [random.randint(0, 255) for _ in range(3)]
    return color_table
colortable = get_color_table(class_num)
# Initialize and start the asynchronous video capture thread
cap = VideoCaptureAsync().start()
# # Model preparation
```

```
def load_model(model_path):
    model_dir = pathlib.Path(model_path) / "saved_model"
    model = tf.saved_model.load(str(model_dir))
    model = model.signatures['serving_default']
    return model
model = load_model(model_path)
# Predict objects and bounding boxes and format the result
def run_inference_for_single_image(model, image):
    # The input needs to be a tensor, convert it using `tf.convert_to_
tensor`.
    input_tensor = tf.convert_to_tensor(image)
    # The model expects a batch of images, so add an axis with `tf.newaxis`.
    input_tensor = input_tensor[tf.newaxis, ...]
    # Run prediction from the model
    output_dict = model(input_tensor)
    # Input to model is a tensor, so the output is also a tensor
    # Convert to NumPy arrays, and take index [0] to remove the batch
dimension.
    # We're only interested in the first num_detections.
    num_detections = int(output_dict.pop('num_detections'))
    output_dict = {key: value[0, :num_detections].numpy()
                   for key, value in output_dict.items()}
    output_dict['num_detections'] = num_detections
    # detection_classes should be ints.

output_dict['detection_classes'] = output_dict['detection_classes'].
astype(np.int64)
    return output_dict
# Function to draw bounding boxes and tracking information on the
image frame
def track_object(model, image_np):
    global object_ids, lock
    # Actual detection.
    output_dict = run_inference_for_single_image(model, image_np)
    # Visualization of the results of a detection.
    for i in range(output_dict['detection_classes'].size):
```

```
        box = output_dict['detection_boxes'][i]
        classes = output_dict['detection_classes'][i]
        scores = output_dict['detection_scores'][i]
        if scores > 0.5:
            h = image_np.shape[0]
            w = image_np.shape[1]
            classname = category_index[classes]['name']
            classid =category_index[classes]['id']
            # Draw bounding boxes
cv2.rectangle(image_np, (int(box[1] * w), int(box[0] * h)), (int(box[3] * w),
int(box[2] * h)), colortable[classid], 2)
            # Write the class name on top of the bounding box
            font = cv2.FONT_HERSHEY_COMPLEX_SMALL

hash, object_ids = hasher_object.getObjectId(image_np, int(box[1] * w),
int(box[0] * h), int(box[3] * w), int(box[2] * h), object_ids)

size = cv2.getTextSize(str(classname) + ":" + str(scores)+
"[Id: "+str(object_ids.get(hash))+"]", font, 0.75, 1)[0][0]

cv2.rectangle(image_np,(int(box[1] * w), int(box[0] * h-20)), ((int(box[1]
* w)+size+5), int(box[0] * h)), colortable[classid],-1)

cv2.putText(image_np, str(classname) + ":" + str(scores)+
"[Id: "+str(object_ids.get(hash))+"]",

(int(box[1] * w), int(box[0] * h)-5), font, 0.75, (0,0,0), 1, 1)

cv2.putText(image_np, "Number of objects detected: "+str(len(object_ids)),
                    (10,20), font, 0.75, (0, 0, 0), 1, 1)
        else:
            break
    return image_np
# Function to implement infinite while loop to read video frames and
generate the output for web browser
def streamVideo():
```

```python
    global lock
    while (True):
        retrieved, frame = cap.read()
        if retrieved:
            with lock:
                frame = track_object(model, frame)
                (flag, encodedImage) = cv2.imencode(".jpg", frame)
                if not flag:
                    continue
                yield (b'--frame\r\n' b'Content-Type: image/jpeg\r\n\r\n' +
                    bytearray(encodedImage) + b'\r\n')
        if cv2.waitKey(1) & 0xFF == ord('q'):
            cap.stop()
            cv2.destroyAllWindows()
            break
    # When everything done, release the capture
    cap.stop()
    cv2.destroyAllWindows()
```

The file path for Listing 7-13 is video_tracking/videoasync.py.

Listing 7-13. videoasync.py

```python
# file: videoasync.py
import threading
import cv2
class VideoCaptureAsync:
    def __init__(self, src=0):
        self.src = src
        self.cap = cv2.VideoCapture(self.src)
        self.grabbed, self.frame = self.cap.read()
        self.started = False
        self.read_lock = threading.Lock()
    def set(self, var1, var2):
        self.cap.set(var1, var2)
    def start(self):
        if self.started:
```

```python
        print('[Warning] Asynchronous video capturing is already
        started.')
        return None
    self.started = True
    self.thread = threading.Thread(target=self.update, args=())
    self.thread.start()
    return self
def update(self):
    while self.started:
        grabbed, frame = self.cap.read()
        with self.read_lock:
            self.grabbed = grabbed
            self.frame = frame
def read(self):
    with self.read_lock:
        frame = self.frame.copy()
        grabbed = self.grabbed
    return grabbed, frame
def stop(self):
    self.started = False
    # self.cap.release()
    # cv2.destroyAllWindows()
    self.thread.join()
def __exit__(self, exec_type, exc_value, traceback):
    self.cap.release()
```

The file path for Listing 7-14 is video_tracking/tracker.py.

Listing 7-14. tracker.py

```python
# tracker.py
import numpy as np
import cv2
class ObjectHasher:

def __init__(self, threshold=20, size=8, max_track_frame=10, radius_
tracker=5):
```

```python
        self.threshold = 20
        self.size = 8
        self.max_track_frame = 10
        self.radius_tracker = 5
    def getCenter(self, xmin, ymin, xmax, ymax):
        x_center = int((xmin + xmax)/2)
        y_center = int((ymin+ymax)/2)
        return (x_center, y_center)

def getObjectId(self, image_np, xmin, ymin, xmax, ymax, hamming_dict={}):

croppedImage = self.getCropped(image_np,int(xmin*0.8), int(ymin*0.8),
int(xmax*0.8), int(ymax*0.8))
        croppedImage = cv2.cvtColor(croppedImage, cv2.COLOR_BGR2GRAY)
        resizedImage = self.resize(croppedImage, self.size)
        hash = self.getHash(resizedImage)
        center = self.getCenter(xmin*0.8, ymin*0.8, xmax*0.8, ymax*0.8)
        # hamming_dict = self.createHammingDict(hash, center, hamming_dict)
        hamming_dict = self.getObjectCounter(hash, hamming_dict)
        return hash, hamming_dict
    def getCropped(self, image_np, xmin, ymin, xmax, ymax):
        return image_np[ymin:ymax, xmin:xmax]
    def resize(self, cropped_image, size=8):
        resized = cv2.resize(cropped_image, (size+1, size))
        return resized
    def getHash(self, resized_image):
        diff = resized_image[:, 1:] > resized_image[:, :-1]
        # convert the difference image to a hash
        dhash = sum([2 ** i for (i, v) in enumerate(diff.flatten()) if v])
        return int(np.array(dhash, dtype="float64"))
    def hamming(self, hashA, hashB):
        # compute and return the Hamming distance between the integers
        return bin(int(hashA) ^ int(hashB)).count("1")
    def createHammingDict(self, dhash, center, hamming_dict):
        centers = []
        matched = False
```

```python
            matched_hash = dhash
            # matched_classid = classid
            if hamming_dict.__len__() > 0:
                if hamming_dict.get(dhash):
                    matched = True
                else:
                    for key in hamming_dict.keys():
                        hd = self.hamming(dhash, key)
                        if(hd < self.threshold):
                            centers = hamming_dict.get(key)
                            if len(centers) > self.max_track_frame:
                                centers.pop(0)
                            centers.append(center)
                            del hamming_dict[key]
                            hamming_dict[dhash] = centers
                            matched = True
                            break
        if not matched:
            centers.append(center)
            hamming_dict[dhash] = centers
        return  hamming_dict
    def getObjectCounter(self, dhash, hamming_dict):
        matched = False
        matched_hash = dhash
        lowest_hamming_dist = self.threshold
        object_counter = 0
        if len(hamming_dict) > 0:
            if dhash in hamming_dict:
                lowest_hamming_dist = 0
                matched_hash = dhash
                object_counter = hamming_dict.get(dhash)
                matched = True
            else:
                for key in hamming_dict.keys():
                    hd = self.hamming(dhash, key)
```

```python
            if(hd < self.threshold):
                if hd < lowest_hamming_dist:
                    lowest_hamming_dist = hd
                    matched = True
                    matched_hash = key
                    object_counter = hamming_dict.get(key)
    if not matched:
        object_counter = len(hamming_dict)
    if matched_hash in hamming_dict:
        del hamming_dict[matched_hash]
    hamming_dict[dhash] = object_counter
    return  hamming_dict
def drawTrackingPoints(self, image_np, centers, color=(0,0,255)):

image_np = cv2.line(image_np, centers[0], centers[len(centers) - 1], color)
    return image_np
```

Run the Flask server by executing the command python video_server.py from a terminal. To see the live stream of video, launch your web browser and point to the URL http://localhost:5019.

Summary

In this chapter, we developed a fully operational video tracking system by leveraging a pretrained SSD model. Additionally, we gained insights into the difference hashing (dHash) algorithm and harnessed the Hamming distance to gauge image similarity. Our implementation was integrated into the Flask microweb framework, enabling us to present real-time video tracking directly in a web browser.

CHAPTER 8

Practical Example: Face Recognition

Face recognition is a computer vision problem to detect and identify human faces in an image or video. The first step of facial recognition is to detect and locate the position of the face in the input image. This is a typical object detection task that we explored in the previous chapters. After the face is detected, a feature set, also called a *facial footprint* or *face embedding*, is created from various key points on the face. A human face has 80 nodal points or distinguishing landmarks that are used to create the feature set (USPTO Patent Number US7634662B2, `https://patents.google.com/patent/US7634662B2/`). The face embedding is then compared against a database to establish the identity of the face.

There are many applications of facial recognition in the real world, such as the following:

- As a password for access control to high-security areas

- In airport customs and border protection

- In identifying genetic disorders

- As a way to predict the age and gender of individuals (e.g., used in controlling age-based access, such as alcohol purchases)

- In law enforcement (e.g., police find potential crime suspects and witnesses by scanning millions of photos)

- In organizing digital photo albums (e.g., photos on social media)

401

In this chapter, we will explore FaceNet, a popular face recognition algorithm developed by Google engineers. We will examine how to train a FaceNet-based neural network to develop a face recognition model. At the end, we will write code to develop a fully functional face recognition system that can detect faces in real time from a video stream.

FaceNet

FaceNet was invented by three Google engineers, Florian Schroff, Dmitry Kalenichenko, and James Philbin. They published their work in 2015 in a paper titled "FaceNet: A Unified Embedding for Face Recognition and Clustering" (`https://arxiv.org/pdf/1503.03832.pdf`).

FaceNet is a unified system that provides the following capabilities:

- Face verification (is this the same person?)

- Recognition (who is this person?)

- Clustering (are there similar faces?)

FaceNet is a deep neural network that does the following:

- Computes a 128D compact feature vector, called *face embedding*, from the input images. Recall from Chapter 4 that a feature vector contains information that describes an object's significant characteristics. The 128D feature vector, which is a list of 128 real-valued numbers, represents output that attempts to quantify the face.

- Learns by optimizing a triplet loss function (described later in this chapter).

FaceNet Neural Network Architecture

Figure 8-1 shows the FaceNet architecture.

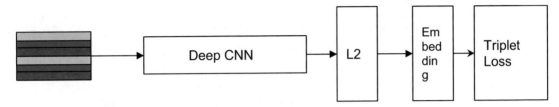

Figure 8-1. *FaceNet neural network architecture*

The components of a FaceNet network are described in the following sections.

Input Images

The training set consists of thumbnails of faces cropped from the images. Other than translation and scaling, no other alignments to the face crops are needed.

Deep CNN

FaceNet was trained using deep convolutional neural networks using SGD with backpropagation and an AdaGrad optimizer. The initial learning rate was taken as 0.05 and decreased with iterations to finalize the model. The training was performed on a CPU-based cluster for 1,000 to 2,000 hours.

The FaceNet paper describes two different architectures of deep convolutional neural networks having different trade-offs. The first architecture was inspired by Zeiler and Fergus, and the second is the inception from Google. The two architectures differ mainly in two aspects: the number of parameters and the floating-point operations per second (FLOPS). FLOPS is a standard measure of computer performance that requires floating-point computations.

The Zeiler and Fergus CNN architecture consists of 22 layers and trains on 140 million parameters at 1.6 billion FLOPS per image. This CNN architecture is referred to as NN1 that has an input size of 220×220.

Table 8-1 shows the network configuration based on Zeiler and Fergus that is used in FaceNet.

Table 8-1. *Deep CNN Based on Zeiler and Fergus Network Architecture (Source: Schroff et al, https://arxiv.org/pdf/1503.03832.pdf)*

layer	size-in	size-out	kernel	param	FLPS
conv1	$220 \times 220 \times 3$	$110 \times 110 \times 64$	$7 \times 7 \times 3, 2$	9K	115M
pool1	$110 \times 110 \times 64$	$55 \times 55 \times 64$	$3 \times 3 \times 64, 2$	0	
rnorm1	$55 \times 55 \times 64$	$55 \times 55 \times 64$		0	
conv2a	$55 \times 55 \times 64$	$55 \times 55 \times 64$	$1 \times 1 \times 64, 1$	4K	13M
conv2	$55 \times 55 \times 64$	$55 \times 55 \times 192$	$3 \times 3 \times 64, 1$	111K	335M
rnorm2	$55 \times 55 \times 192$	$55 \times 55 \times 192$		0	
pool2	$55 \times 55 \times 192$	$28 \times 28 \times 192$	$3 \times 3 \times 192, 2$	0	
conv3a	$28 \times 28 \times 192$	$28 \times 28 \times 192$	$1 \times 1 \times 192, 1$	37K	29M
conv3	$28 \times 28 \times 192$	$28 \times 28 \times 384$	$3 \times 3 \times 192, 1$	664K	521M
pool3	$28 \times 28 \times 384$	$14 \times 14 \times 384$	$3 \times 3 \times 384, 2$	0	
conv4a	$14 \times 14 \times 384$	$14 \times 14 \times 384$	$1 \times 1 \times 384, 1$	148K	29M
conv4	$14 \times 14 \times 384$	$14 \times 14 \times 256$	$3 \times 3 \times 384, 1$	885K	173M
conv5a	$14 \times 14 \times 256$	$14 \times 14 \times 256$	$1 \times 1 \times 256, 1$	66K	13M
conv5	$14 \times 14 \times 256$	$14 \times 14 \times 256$	$3 \times 3 \times 256, 1$	590K	116M
conv6a	$14 \times 14 \times 256$	$14 \times 14 \times 256$	$1 \times 1 \times 256, 1$	66K	13M
conv6	$14 \times 14 \times 256$	$14 \times 14 \times 256$	$3 \times 3 \times 256, 1$	590K	116M
pool4	$14 \times 14 \times 256$	$7 \times 7 \times 256$	$3 \times 3 \times 256, 2$	0	
concat	$7 \times 7 \times 256$	$7 \times 7 \times 256$		0	
fc1	$7 \times 7 \times 256$	$1 \times 32 \times 128$	maxout p=2	103M	103M
fc2	$1 \times 32 \times 128$	$1 \times 32 \times 128$	maxout p=2	34M	34M
fc7128	$1 \times 32 \times 128$	$1 \times 1 \times 128$		524K	0.5M
L2	$1 \times 1 \times 128$	$1 \times 1 \times 128$		0	
total				140M	1.6B

The second type of network is the inception model based on GoogLeNet. This model has 20× fewer parameters (around 6.6 million to 7.5 million) and 5× fewer FLOPS (around 500 million to 1.6 billion).

There are a few variants of the inception model based on the input size. They are briefly described here:

- *NN2*: This is an inception model that takes images of size 224×224 and trains on 7.5 million parameters at 1.6 billion FLOPS per image.

Table 8-2 shows the NN2 inception model used in FaceNet.

Table 8-2. *Inception Model Architecture Based on GoogLeNet (Source: Schroff et al,* $https://arxiv.org/pdf/1503.03832.pdf$*)*

type	output size	depth	#1×1	#3×3 reduce	#3×3	#5×5 reduce	#5×5	pool proj (p)	params	FLOPS
conv1 (7×7×3, 2)	112×112×64	1							9K	119M
max pool + norm	56×56×64	0						m 3×3, 2		
inception (2)	56×56×192	2		64	192				115K	360M
norm + max pool	28×28×192	0						m 3×3, 2		
inception (3a)	28×28×256	2	64	96	128	16	32	m, 32p	164K	128M
inception (3b)	28×28×320	2	64	96	128	32	64	L_2, 64p	228K	179M
inception (3c)	14×14×640	2	0	128	256,2	32	64,2	m 3×3,2	398K	108M
inception (4a)	14×14×640	2	256	96	192	32	64	L_2, 128p	545K	107M
inception (4b)	14×14×640	2	224	112	224	32	64	L_2, 128p	595K	117M
inception (4c)	14×14×640	2	192	128	256	32	64	L_2, 128p	654K	128M
inception (4d)	14×14×640	2	160	144	288	32	64	L_2, 128p	722K	142M
inception (4e)	7×7×1024	2	0	160	256,2	64	128,2	m 3×3,2	717K	56M
inception (5a)	7×7×1024	2	384	192	384	48	128	L_2, 128p	1.6M	78M
inception (5b)	7×7×1024	2	384	192	384	48	128	m, 128p	1.6M	78M
avg pool	1×1×1024	0								
fully conn	1×1×128	1							131K	0.1M
L2 normalization	1×1×128	0								
total									7.5M	1.6B

- *NN3*: This is identical in architecture compared to NN2 except that it uses a 160×160 input size resulting in a smaller network size.

- *NN4*: This network has a 96×96 input size resulting in drastically reduced parameters that require only 285 million FLOPS per image (compared to 1.6 billion on NN1 and NN2). Because of the reduced size and lower FLOPS requiring less CPU time, NN4 is suitable for mobile devices.

- *NNS1*: This is also called a "mini" inception due to its smaller size. It has an input size of 165×165 and 26 million parameters that require only 220 million FLOPS per image.

- *NNS2*: This is called a "tiny" inception. It has an input size of 140×116 and 4.3 million parameters that require 20 million FLOPS.

NN4, NNS1, and NNS2 are suitable for mobile devices because of the smaller number of parameters requiring low CPU FLOPS per image.

It is important to mention that the model accuracy is higher with larger FLOPS. In general, a network with lower FLOPS runs faster and consumes less memory but results in lower accuracy.

Figure 8-2 shows a plot of FLOPS versus accuracy with different types of CNN architectures.

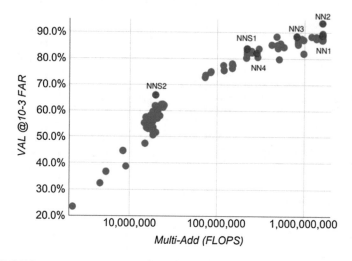

Figure 8-2. *FLOPS versus accuracy (Source: FaceNet,* `https://arxiv.org/pdf/1503.03832.pdf`*)*

Face Embedding

The face embeddings of sizes 1×1×128 are generated from the L2 normalization layer of the deep CNN (as shown in Figure 8-1 and Tables 8-1 and 8-2).

After the embeddings are calculated, the face verification (or finding similar faces) is performed by calculating the Euclidean distances between the embeddings and finding similar faces based on the following:

- The faces of the same person have smaller distances.

- The faces of different people have larger distances.

The face recognition is performed by the standard K-nearest neighbors (K-NN) classification.

The clustering is done using algorithms like K-means or agglomerative clustering techniques.

Triplet Loss Function

The loss function used in FaceNet is known as the *triplet loss function*.

The embeddings of the same faces are called *positives* and of different faces are *negatives*. The face being analyzed is called the *anchor*. To calculate the loss, a triplet consisting of an anchor, a positive, and a negative embedding is formed, and their Euclidean distances are analyzed. The learning objective of FaceNet is to minimize the distance between an anchor and a positive and maximize the distance between the anchor and a negative.

Figure 8-3 illustrates the triplet loss function and the learning process.

Figure 8-3. *The triplet loss minimizes the distance between an anchor and a positive, both of which have the same identity, and it maximizes the distance between the anchor and a negative of a different identity (Source: FaceNet,* `https://arxiv.org/pdf/1503.03832.pdf`*.)*

Each face image is a feature vector, representing a d-dimensional Euclidean hypersphere, and represented by a function $\|f(x)\|_2 = 1$.

Assume the face image x_i^a (anchor) is closer to the face x_i^p (hard positive) of the same person than x_i^n (hard negative) faces of different people. Further, assume that there are N triplets in the training set. The triplet loss function is represented by the following equation:

$$\sum_{i}^{N}\left[\left\|f\left(x_i^a\right)-f\left(x_i^p\right)\right\|_2^2-\left\|f\left(x_i^a\right)-f\left(x_i^n\right)\right\|_2^2+\alpha\right]$$

where α is a margin of distance between positive and negative embeddings.

If we consider every possible combination of triplets, there will be lots of triplets, and the previous function may take a lot of time to converge. Also, not every triplet contributes to the model learning. Therefore, we need a method to select the right triplets so that our model training is efficient and the accuracy is optimum.

Triplet Selection

Ideally, we should select triplets in such a way that $\left\| f\left(x_i^a\right) - f\left(x_i^p\right) \right\|_2^2$ is minimum and $\left\| f\left(x_i^a\right) - f\left(x_i^n\right) \right\|_2^2$ is maximum. But calculating this min and max across all datasets may be infeasible. Therefore, we need a method to efficiently calculate the min and max of distances. This may be done offline and then fed to the algorithm or be determined online using some algorithms.

In an online method, we divide the embeddings into mini-batches. Each mini-batch contains a small set of positives and some randomly selected negatives. The inventors of FaceNet used mini-batches consisting of 40 positives and randomly selected negatives embeddings. The min and max distances are calculated for each mini-batch to create triplets.

In the next sections, we will train our own model based on FaceNet and build a system for real-time face recognition.

Training a Face Recognition Model

One of the most popular TensorFlow implementations of FaceNet is by David Sandberg. This is an open source version and freely available under the MIT License at GitHub at `https://github.com/davidsandberg/facenet`. I have forked the original GitHub repository and committed a slightly modified version to my GitHub repository located at `https://github.com/ansarisam/facenet`. I did not modify the core neural network and triplet loss function implementations. My modified version of FaceNet, forked from David Sandberg's repository, uses OpenCV for reading and manipulating images. I also upgraded some of the library functions of TensorFlow.

In the following example, we will use Google Colab to train our face detection model. It is important to note that a face detection model is compute-intensive and may take several days to learn, even on GPUs. Therefore, Colab is not an ideal platform to train a long-running model, because you will lose all the data and settings after the Colab session expires. You should consider using a cloud-based GPU environment for training a production-quality face recognition model. Chapter 10 will show you how to scale your model training on the cloud. For now, let's use Colab for the purposes of learning.

Before we start, create a new Colab project and give it a meaningful name, such as FaceNet Training.

Checking Out FaceNet from GitHub

Check out the source code of the TensorFlow implementation of FaceNet. In Colab, add a code cell by clicking the +Code icon. Write the command to clone the GitHub repository, as shown in Listing 8-1. Click the Execute button to run the command. After the successful execution, you should see the directory facenet in your Colab file browser panel.

Listing 8-1. Cloning the GitHub Repository of TensorFlow Implementation of FaceNet

```
1    %%shell
2    git clone https://github.com/ansarisam/facenet.git
```

If you clone the official GitHub repository, you will need to make the following changes to make it run with the TensorFlow 2.x environment. The official FaceNet was implemented in TensorFlow 1.x and does not currently have support for TensorFlow 2.x. In order to run the code provided in this chapter, update the source code as follows:

1. Install the tf_slim module using the following command:

   ```
   !pip install tf_slim
   ```

2. Make the following changes to the facenet/src/models/ inception_resnet_v1.py file:

a. Comment the import tensorflow as tf and add the following lines:

   ```
   import tensorflow.compat.v1 as tf
   tf.disable_v2_behavior()
   ```

b. Comment the import tensorflow.contrib.slim as slim and add the following import line:

   ```
   import tf_slim as slim
   ```

3. In the facenet/src/train_tripletloss.py file, comment the import tensorflow as tf and add the following lines:

   ```
   import tensorflow.compat.v1 as tf
   tf.disable_v2_behavior()
   ```

4. In the facenet/src/facenet.py file, comment the import
    tensorflow as tf and add the following lines:

```
import tensorflow.compat.v1 as tf
tf.disable_v2_behavior()
```

The repository https://github.com/ansarisam/facenet.git is already updated
with the preceding changes and does not need any editing to execute the code listings
provided in this chapter.

Dataset

We will use the VGGFace2 dataset for training our face recognition model. VGGFace2
is a large-scale image dataset for face recognition, provided by Visual Geometry Group,
https://www.robots.ox.ac.uk/~vgg/data/vgg_face2/.

The VGGFace2 dataset consists of 3.3 million faces of more than 9,000 people
(referred to as *identities*). The data sample has 362 images (on an average) per
identity. The dataset is described in the paper at http://www.robots.ox.ac.uk/~vgg/
publications/2018/Cao18/cao18.pdf published in 2018 by Q. Cao, L. Shen, W. Xie,
O. M. Parkhi, and A. Zisserman.

The size of the training set is 35GB, and the test set is 1.9GB. The datasets are
available as compressed (zipped) files. The face images are organized in subdirectories.
The name of each subdirectory is the identity class ID in the format n< classID >.
Figure 8-4 shows a sample directory structure containing training images.

Figure 8-4. *Subdirectories containing images*

A separate metadata file in CSV format is provided. The header of this metadata file is as follows:

Identity ID, name, sample number, train/test flag and gender

Here is a brief description:

- `Identity ID` maps to the subdirectory name.

- `name` is the name of the person whose face image is included.

- `sample number` represents the number of images in the subdirectory.

- `train/test` flag indicates whether the identity is in the training set or test set. The training set is represented by flag 1 and the test set as 0.

- `gender` is the gender of the person.

It is important to note that the size of this dataset is too large to fit in the free version of Google Colab or Google Drive.

However, you could use just a subset of the data (maybe a few hundred identities) for the purpose of learning.

Of course, you can use your own images if you want to build a custom face recognition model. All you need to do is to save images of the same person in one directory, with each person having their own directory, and match the directory structure to look like Figure 8-4. Make sure your directory names and image file names do not have any blank spaces.

Downloading VGGFace2 Data

To download the images, you will need to register at `http://zeus.robots.ox.ac.uk/vgg_face2/signup/`. After the registration, log in to download the data directly from `https://www.robots.ox.ac.uk/~vgg/data/vgg_face2/`, save the compressed training and test files to your local drive, and then upload them to Colab.

If you prefer to download the images directly in Colab, you can use the code in Listing 8-2. Run the program with the correct URLs to download both the training and test sets.

Listing 8-2. Python Code to Download VGGFace2 Images (Source: `https://github.com/MistLiao/jgitlib/blob/master/download.py`)

```
1    import sys
2    import getpass
3    import requests
4
5    VGG_FACE_URL = "http://zeus.robots.ox.ac.uk/vgg_face2/login/"
6    IMAGE_URL = "http://zeus.robots.ox.ac.uk/vgg_face2/get_
     file?fname=vggface2_train.tar.gz"
```

```
7   TEST_IMAGE_URL="http://zeus.robots.ox.ac.uk/vgg_face2/get_
    file?fname=vggface2_test.tar.gz"
8
9   print('Please enter your VGG Face 2 credentials:')
10  user_string = input('    User: ')
11  password_string = getpass.getpass(prompt='    Password: ')
12
13  credential = {
14      'username': user_string,
15      'password': password_string
16  }
17
18  session = requests.session()
19  r = session.get(VGG_FACE_URL)
20
21  if 'csrftoken' in session.cookies:
22      csrftoken = session.cookies['csrftoken']
23  elif 'csrf' in session.cookies:
24      csrftoken = session.cookies['csrf']
25  else:
26      raise ValueError("Unable to locate CSRF token.")
27
28  credential['csrfmiddlewaretoken'] = csrftoken
29
30  r = session.post(VGG_FACE_URL, data=credential)
31
32  imagefiles = IMAGE_URL.split('=')[-1]
33
34  with open(imagefiles, "wb") as files:
35      print(f"Downloading the file: `{imagefiles}`")
36      r = session.get(IMAGE_URL, data=credential, stream=True)
37      bytes_written = 0
38      for data in r.iter_content(chunk_size=400096):
39          files.write(data)
40          bytes_written += len(data)
```

```
41          MegaBytes = bytes_written / (1024 * 1024)
42          sys.stdout.write(f"\r{MegaBytes:0.2f} MiB downloaded...")
43          sys.stdout.flush()
44
45  print("\n Images are successfully downloaded. Exiting the process.")
```

After you download the training and test sets, uncompress them to get the training and test directories and subdirectories as per the structure shown in Figure 8-4. To uncompress, you can execute the commands in Listing 8-3.

Listing 8-3. Commands to Uncompress Files

```
1   %%shell
2   tar xvzf vggface2_train.tar.gz
3   tar xvzf vggface2_test.tar.gz
```

Data Preparation

The training set for FaceNet should be images of the face portion only. Therefore, we need to crop the images to extract the faces, align them, and resize them, if needed. We will use an algorithm called *multitask cascaded convolutional networks (MTCNN)* that has proven to outperform many face detection benchmarks while retaining real-time performance.

The FaceNet source we cloned from the GitHub repository has a TensorFlow implementation of MTCNN. The implementation of this model is outside the scope of this book. We will use the Python program align_dataset_mtcnn.py available in the align module to get the bounding boxes of all the faces detected in the training and test sets. This program will retain the directory structure and save the cropped images in the same directory hierarchy, as shown in Figure 8-4.

Listing 8-4 shows the script to perform the face cropping and alignment. Note that this code works with the https://github.com/ansarisam/facenet.git repository only.

Listing 8-4. Code for Face Detection Using MTCNN, Cropping and Alignment

```
1   %%shell
2   export PYTHONPATH=$PYTHONPATH:/content/facenet
3   export PYTHONPATH=$PYTHONPATH:/content/facenet/src
```

```
4    for N in {1..10}; do \
5    python /content/facenet/src/align/align_dataset_mtcnn.py \
/content/drive/MyDrive/chapter8/train \
/content/drive/MyDrive/chapter8/train_aligned \
--image_size 182 \
--margin 44 \
--random_order \
--gpu_memory_fraction 0.10 \
& done
```

In Listing 8-4, line 1 activates the shell.

Lines 2 and 3 set the PYTHONPATH environment variable to the facenet and facenet/src directories. If you are using a virtual machine or physical machine and have direct access to the operating system, you should consider setting the environment variable in the ~/.bash_profile file.

To speed up the face detection and alignment process, we have created ten parallel processes (line 4), and for each process we are using 10 percent of the GPU memory (line 5). If your dataset is smaller and you want to process the MTCNN in a single process, set the value of N to 1.

Line 5 calls the file align_dataset_mtcnn.py and passes the following arguments:

- The first argument, /content/drive/MyDrive/chapter8/train, is the directory path where training images are located.

- The second argument, /content/drive/MyDrive/chapter8/train_aligned, is the directory path where the aligned images will be stored.

- The third argument, --image_size, is the size of the cropped images. We set this to 182×182 pixels.

- The argument --margin, which is set to 44, creates a margin around all four sides of the cropped images.

- The next parameter, --random_order, if present, will select images in random order by the parallel processes.

- The last argument, --gpu_memory_fraction, is used to tell the algorithm what fraction of the GPU memory to use for each of the parallel processes.

The cropped image size in the previous script is 182×182 pixels. The input to the Inception-ResNet-v1 is only 160×160. This gives an additional margin for random crops. The use of the additional margin 44 is used to add any contextual information to the model. This additional margin of 44 should be tuned based on your particular situations, and the cropping performance should be assessed.

Execute the previous script to start the cropping and alignment processes. Note that this is a compute-intensive process and may take several hours to complete.

Repeat the previous process for the test images.

If you would like to use the MTCNN library and write a custom Python code to crop the face portion in an image and perform the alignment, the code in Listing 8-5 will be helpful. First, install the mtcnn library using the command:

```
!pip install mtcnn
```

Listing 8-5. Script to Crop the Face Portion of an Image

```
1    import numpy as np
2    import pandas as pd
3    import cv2
4    from mtcnn.mtcnn import MTCNN
5    from matplotlib import pyplot as plt
6    from keras.models import load_model
7    from PIL import Image
8    import os
9    # extract a single face from a given photograph
10   def extract_face(filename, required_size=(160, 160)):
11       # load image from file
12       image = Image.open(filename)
13       # convert to RGB, if needed
14       image = image.convert('RGB')
15       # convert to array
16       pixels = np.asarray(image)
17       # create the detector, using default weights
18       detector = MTCNN()
19       # detect faces in the image
20       results = detector.detect_faces(pixels)
```

```
21        # extract the bounding box from the first face
22        x1, y1, width, height = results[0]['box']
23        # deal with negative pixel index
24        x1, y1 = abs(x1), abs(y1)
25        x2, y2 = x1 + width, y1 + height
26        # extract the face
27        face = pixels[y1:y2, x1:x2]
28        # resize pixels to the model size
29        image = Image.fromarray(face)
30        image = image.resize(required_size)
31        face_array = np.asarray(image)
32        return face_array
33
34   inputpath = "/content/drive/MyDrive/chapter8/train"
35   outputpath = "/content/drive/MyDrive/chapter8/train_aligned"
36   for dir in os.listdir(inputpath):
37     if os.path.isdir(os.path.join(inputpath, dir)):
38       os.makedirs(os.path.join(outputpath, dir))
39       for image in os.listdir(os.path.join(inputpath, dir)):
40           pixels = extract_face(os.path.join(inputpath, dir, image))
41           cv2.imwrite(os.path.join(outputpath, dir, image), pixels)
42     else:
43       print(dir, " is not a dir")
```

Here is the breakdown of Listing 8-5:

1. Import necessary libraries:

    a. numpy as np: A library for numerical computations in Python

    b. pandas as pd: A library for data manipulation and analysis

    c. cv2: The OpenCV library for computer vision tasks

    d. MTCNN from mtcnn.mtcnn: An implementation of the Multi-Task Cascaded Convolutional Networks for face detection

    e. pyplot from matplotlib: A library for creating plots and visualizations

     f.   load_model from keras.models: Loads a pretrained Keras model

     g.   Image from PIL: A module from the Python Imaging Library for image manipulation

     h.   os: A module for interacting with the operating system

2.   Define a function `extract_face(filename, required_size)`:

     a.   This function takes an image file path (`filename`) and an optional argument for the required face size (`required_size`).

     b.   It loads the image using PIL, converts it to RGB if needed, converts it to a NumPy array, and then uses MTCNN to detect faces in the image.

     c.   It extracts the bounding box coordinates of the first detected face.

     d.   It extracts the face region from the image based on the bounding box, resizes it to the required size, and returns the resulting face array.

3.   Define input and output paths for image processing.

4.   Loop through the directories in the input path:

     a.   For each subdirectory (dir) within the input path, the code creates a corresponding subdirectory within the output path.

     b.   It then processes each image within the subdirectory:

        – It calls the `extract_face()` function to extract the face from the image.

        – It saves the extracted face as an image in the corresponding subdirectory within the output path using OpenCV's `cv2.imwrite()` function.

     c.   If the item in the input path is not a directory, it prints a message indicating that it's not a directory.

Model Training

Listing 8-6 is a script for training a facial recognition model using the FaceNet framework. FaceNet learns to map faces to a high-dimensional space where similar faces are closer together. This script uses the triplet loss as its training objective.

Listing 8-6. Script to Train the FaceNet Model with the Triplet Loss Function

```
1    !export PYTHONPATH=$PYTHONPATH:/content/facenet/src
2    !python facenet/src/train_tripletloss.py \
--logs_base_dir logs/facenet/ \
--models_base_dir /content/drive/MyDrive/chapter8/facenet_model4/ \
--data_dir /content/drive/MyDrive/chapter8/train_aligned/ \
--image_size 160 \
--model_def models.inception_resnet_v1 \
--optimizer ADAGRAD \
--learning_rate 0.01 \
--weight_decay 1e-4 \
--max_nrof_epochs 10 \
--epoch_size 200 \
--people_per_batch 45 \
--images_per_person 20 \
```

Let's break down the code in Listing 8-6 step by step.

Line 1 adds the /content/facenet/src directory to the Python path. This is necessary so that the subsequent python command can find and import the necessary modules from this directory.

Line 2 executes the train_tripletloss.py script located in the facenet/src directory using the python interpreter. It's passing the following command-line arguments to the script to configure and control the training process:

- --logs_base_dir logs/facenet/: Specifies the base directory where training logs will be stored.

- --models_base_dir /content/drive/MyDrive/chapter8/facenet_ model4/: Specifies the base directory where trained models will be saved.

- `--data_dir /content/drive/MyDrive/chapter8/train_aligned/`: Specifies the directory containing the training data, which are aligned face images.

- `--image_size 160`: Sets the target size for the input face images during training.

- `--model_def models.inception_resnet_v1`: Specifies the architecture of the neural network model. In this case, it's using the Inception ResNet v1 architecture.

- `--optimizer ADAGRAD`: Specifies the optimizer to use during training. ADAGRAD is the chosen optimization algorithm.

- `--learning_rate 0.01`: Sets the learning rate for the optimizer.

- `--weight_decay 1e-4`: Specifies the weight decay coefficient for the optimizer.

- `--max_nrof_epochs 10`: Defines the maximum number of training epochs.

- `--epoch_size 200`: Specifies the number of batches in an epoch.

- `--people_per_batch 45`: Sets the number of different people (classes) in each batch.

- `--images_per_person 20`: Specifies the number of images per person in a batch.

Execute the training by clicking the Run button in Colab. Depending on the training size and the training parameters, it may take several hours or even days to complete the model.

After the model is successfully trained, the checkpoints are saved in the directory `--model_base_dir` that we configured in Listing 8-6.

Evaluation

While the model is running, the losses for each epoch and each batch will print to the console. This should give you an idea of how the model is learning. Ideally, the losses should be decreasing and should become stable at a very low value, close to zero. Figure 8-5 shows a sample output while the training is going on.

```
Epoch:  [0][138/2000]   Time 0.623   Loss 1.521
Epoch:  [0][139/2000]   Time 0.631   Loss 1.553
Epoch:  [0][140/2000]   Time 0.624   Loss 1.560
Epoch:  [0][141/2000]   Time 0.586   Loss 1.622
Epoch:  [0][142/2000]   Time 0.625   Loss 1.623
Epoch:  [0][143/2000]   Time 0.628   Loss 1.534
Epoch:  [0][144/2000]   Time 0.631   Loss 1.556
Epoch:  [0][145/2000]   Time 0.673   Loss 1.469
Epoch:  [0][146/2000]   Time 0.653   Loss 1.605
Epoch:  [0][147/2000]   Time 0.637   Loss 1.572
Epoch:  [0][148/2000]   Time 0.626   Loss 1.631
Epoch:  [0][149/2000]   Time 0.638   Loss 1.573
Epoch:  [0][150/2000]   Time 0.624   Loss 1.605
```

Figure 8-5. *Colab console output while the training is in progress, showing the loss per batch per epoch*

You can also evaluate the model performance using TensorBoard. Launch the TensorBoard dashboard using the command in Listing 8-7.

Listing 8-7. Launching TensorBoard by Pointing to the `logs` Directory

```
1    %tensorflow_version 2.x
2    %load_ext tensorboard
3    %tensorboard --logdir /content/logs/facenet
```

Developing a Real-Time Face Recognition System

A face recognition system requires three important items:

- A face detection model

- A classification model

- An image or video source

Face Detection Model

We explored how to train a face detection model in the previous section. We can either use the model that we built or use an available pretrained model that fits our requirements. Table 8-3 lists pretrained models that are publicly available for free, along with the location from which they can be downloaded. These models are graciously provided by David Sandberg.

Table 8-3. *Face Recognition Pretrained Models Provided by David Sandberg*

Model Name	Training Dataset	Download Location
20180408-102900	CASIA-WebFace	https://drive.google.com/open?id=1R77HmFADxe87GmoLwzfgMu_HYOIhcyBz
20180402-114759	VGGFace2	https://drive.google.com/open?id=1EXPBSXwTaqrSCOOhUdXNmKSh9qJUQ55-

The models were evaluated against the Labeled Faces in the Wild (LFW) dataset, available at `http://vis-www.cs.umass.edu/lfw/`. Table 8-4 shows the model architecture and accuracy.

Table 8-4. *Evaluation Results with Accuracy of FaceNet Models Trained on the CASIA-WebFace and VGGFace2 Datasets (Information Provided by David Sandberg)*

Model Name	LFW accuracy	Training Dataset	Architecture
20180408-102900	0.9905	CASIA-WebFace	Inception ResNet v1
20180402-114759	0.9965	VGGFace2	Inception ResNet v1

For our example, we will use the VGGFace2 model.

Classifier for Face Recognition

We will build a model to recognize faces (who the person is). We will train the model to recognize George W. Bush, Barack Obama, and Donald Trump, the three most recent U.S. presidents preceding Joe Biden.

To keep this simple, we will download a few images of each of the three former presidents and organize them in subdirectories that will look like Figure 8-6.

Figure 8-6. *Input image directory structure*

For the purpose of this exercise, we will develop a face detector model on our person computer/laptop. In Chapter 10, we will explore how to use GPUs on Cloud to build large scale CV models. We will develop the face detector on our personal computer/ laptop. Before we train our classifier, we need to clone the FaceNet GitHub repository. Execute the following command:

```
git clone https://github.com/ansarisam/facenet.git
```

After the FaceNet source is cloned, set PYTHONPATH to facenet/src and add it to the environment variable:

```
export PYTHONPATH=$PYTHONPATH:/home/user/facenet/src
```

The path to the src directory must be the actual directory path in your computer.

Face Alignment

In this section, we will perform the face alignment of the images. We will use the same MTCNN model as we did in the previous section. Since we have a small set of images, we will use a single process to align these faces. Listing 8-8 shows the script for face alignment.

Listing 8-8. Script for Face Alignment Using MTCNN

```
1    python facenet/src/align/align_dataset_mtcnn.py \
2    ~/presidents/ \
3    ~/presidents_aligned \
4    --image_size 182 \
5    --margin 44
```

Note   On Mac-based computers, the image directories may have a hidden file called .DS_Store. Make sure you delete this file from all subdirectories that contain your input images. Also, ensure that the subdirectories contain the images only and no other files.

Execute the script from Listing 8-8 to crop and align the faces. Figure 8-7 shows sample output.

Figure 8-7. *Cropped and aligned faces of three U.S. presidents*

Classifier Training

With this minimal setup, we are ready to train the classifier. Listing 8-9 shows the script that launches the classifier training.

Listing 8-9. Script to Launch the Face Classifier Training

```
1     python facenet/src/classifier.py TRAIN \
~/presidents_aligned \
~/20180402-114759/20180402-114759.pb \
~/presidents_aligned/face_classifier.pkl \
--batch_size 1000 \
--min_nrof_images_per_class 40 \
--nrof_train_images_per_class 35 \
--use_split_dataset
```

Here is an explanation of the code in Listing 8-9:

- `python facenet/src/classifier.py TRAIN`: Initiates the execution of the `classifier.py` script and specifies that the script should be in TRAIN mode, indicating that the purpose is to train a classifier.

- `~/presidents_aligned`: Specifies the directory containing the aligned face images of the individuals to be recognized by the classifier. This directory contains subdirectories, each of which corresponds to a different individual and contains their face images.

- `~/20180402-114759/20180402-114759.pb`: Specifies the path to a pretrained FaceNet model in TensorFlow's Protocol Buffer (protobuf) format (`.pb` file). This model will be used to extract facial feature embeddings from the input face images.

- `~/presidents_aligned/face_classifier.pkl`: Specifies the path where the trained face classifier will be saved as a serialized Python object (a pickled file). The classifier will be trained to distinguish between the different individuals using the feature embeddings generated by the FaceNet model.

- `--batch_size 1000`: Sets the batch size for training. The training images will be processed in batches of 1,000 images at a time.

- `--min_nrof_images_per_class 40`: Specifies the minimum number of images required for each class (individual) in the training data. Classes with fewer than this number of images will be excluded from training.

- --nrof_train_images_per_class 35: Sets the number of images per class (individual) that will be used for training. This value limits the number of images per class that are used for training, potentially to create a balanced dataset.

- --use_split_dataset: Indicates that the training data should be split into training and validation sets. The validation set is used for monitoring the training progress and preventing overfitting.

After the classifier model is successfully executed, the trained classifier is stored in the file ~/presidents_aligned/face_classifier.pkl.

Face Recognition in a Video Stream

In Listing 7-1, we used OpenCV's convenient function cv2.VideoCapture() to read video frames from either the built-in camera of the computer or a USB or IP camera. The argument 0 to the VideoCapture() function is typically used to read frames from the built-in camera. In this section, we will explore how to use YouTube as our video source.

To read YouTube videos, we will use a Python library called pafy, which internally uses the youtube_dl library. Install these libraries using PIP in your development environment. Simply execute the commands in Listing 8-10 to install pafy.

Listing 8-10. Commands to Install YouTube-Related Libraries

```
pip install pafy
pip install youtube_dl
```

The FaceNet repository that we cloned for this exercise provides the source code, real_time_face_recognition.py in the contributed module, for recognizing faces in a video. Listing 8-11 shows how to use the Python API to detect and recognize faces from a video.

Listing 8-11. Script to Call Real-Time Face Recognition API

```
1    python real_time_face_recognition.py \
2    --source youtube \
3    --url https://www.youtube.com/watch?v=ZYkxVbYxy-c \
4    --facenet_model_checkpoint ~/20180402-114759/20180402-114759.pb \
5    --classfier_model ~/presidents_aligned/face_classifier.pkl
```

In Listing 8-11, line 1 calls `real_time_face_recognition.py` and passes the following arguments:

- Line 2 sets the value of the argument `--source`, which in this case is youtube. If you skip this argument, it will default to the built-in camera with the computer. You can explicitly pass the argument `webcam` to read frames from the built-in camera.

- Line 3 is to pass the YouTube video URL. This argument is not needed in the case of the camera source.

- Line 4 provides the path to the pretrained FaceNet model. You can supply the path to either the checkpoint directory or the frozen `*.pb` model.

- Line 5 provides the file path of the classifier model that we trained in the previous section, such as the classifier model for recognizing the faces of three U.S. presidents.

When you execute Listing 8-11, it will read the YouTube video frames and display the recognized faces with bounding boxes. Figure 8-8 shows a sample recognition.

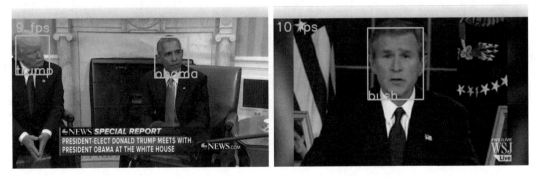

Figure 8-8. *Sample screenshots taken from videos with faces recognized (the input source of the video is YouTube)*

Summary

Face detection is an interesting computer vision problem that involves detecting classifying facial embeddings to identify, in an image, who the person is. In this chapter, we explored FaceNet, a popular face recognition algorithm based on ResNet. We analyzed the technique to crop the face portion of the image using the MTCNN algorithm. We also trained our own classifier and worked through an example to classify faces of three U.S. presidents. Finally, we ingested streams of videos from YouTube and implemented a real-time face recognition system.

CHAPTER 9

Industrial Application: Real-Time Defect Detection in Industrial Manufacturing

Computer vision has numerous applications in industrial manufacturing, one of which is its role in automating visual inspections to ensure quality control and assurance, which is the focus of this chapter.

Most manufacturing companies train their employees to conduct manual visual inspections. This method not only involves a subjective assessment process that can vary in accuracy based on the inspector's experience and judgment, but also is labor intensive.

When issues such as machine calibration problems, adverse environmental conditions, or equipment malfunctions arise, if they are not detected immediately, they can potentially lead to the entire production batch being defective. Delayed detection, which is more likely with manual inspection, may result in the entire batch (which could number in the hundreds or thousands) being discarded, a potentially huge expense (in addition to the cost of replacing them). At the very least, the batch will be suspect, requiring manual post-production inspection and the cost associated with that.

In summary, the manual inspection procedure is both slow and imprecise, leading to high costs. By comparison, a computer vision–based visual inspection system has the capability to instantly identify surface defects by analyzing continuous streams of video frames. This system can promptly issue real-time alerts upon detecting any defects or a sequence of defects, allowing for immediate production halts to prevent potential losses.

© Shamshad Ansari 2023
S. Ansari, *Building Computer Vision Applications Using Artificial Neural Networks*,
https://doi.org/10.1007/978-1-4842-9866-4_9

In this chapter, we will develop a deep learning–based computer vision system to detect surface defects, such as patches, scratches, pitted surfaces, and crazings. We will work with a dataset containing labeled images of hot-rolled steel strips. We will first transform the dataset, train an SSD model, and utilize the model to build a defect detector. We will also cover how to label our own images for any object detection task.

Real-Time Surface Defect Detection System

In this section, we will first examine the dataset that we will use for training and testing a surface defect detection model. We will transform the images and annotations into TFRecord files and train an SSD model on Google Colab. We will apply the object detection concepts presented in Chapter 6.

Dataset

We will utilize a dataset provided by K. Song and Y. Yan at Northeastern University (NEU). The dataset consists of six types of surface defects of hot-rolled steel strips. These defects are labeled as follows:

- Rolled-in scale (RS), which typically occurs when the mill scale is rolled into metal during the rolling process

- Patches (Pa), which may be irregular surface patches

- Crazing (Cr), which is a network of cracks on the surface

- Pitted surface (PS), which consists of a number of small shallow holes

- Inclusion (In), which is compound materials embedded inside steel

- Scratches (Sc)

Figure 9-1 shows labeled images of steel surfaces with these six defects.

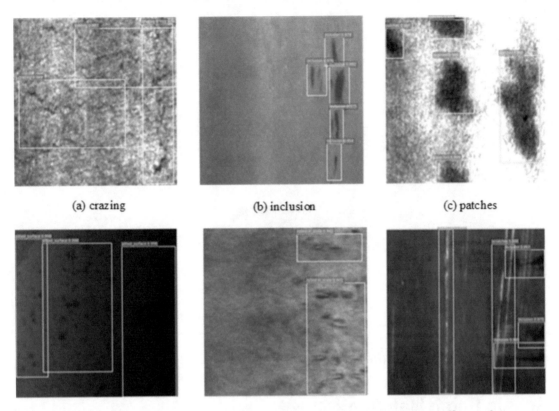

(a) crazing   (b) inclusion   (c) patches

(d) pitted surface   (e) rolled-in scale   (f) scratches

Figure 9-1. *Sample of labeled images of surfaces having six different types of defects (Source: http://faculty.neu.edu.cn/yunhyan/NEU_surface_defect_database.html)*

The dataset includes 1,800 grayscale images with 300 samples of each of the defect classes.

The dataset is available for free download for education and research purposes at https://drive.google.com/file/d/1qrdZlaDi272eA79bOuCwwqPrm2Q_WI3k/view. Download the dataset from this link and uncompress it. The uncompressed dataset is organized in the directory structure shown in Figure 9-2. The images are in the subdirectory IMAGES. The ANNOTATIONS subdirectory contains XML files of annotations of bounding boxes and the defect class in PASCAL VOC annotation format.

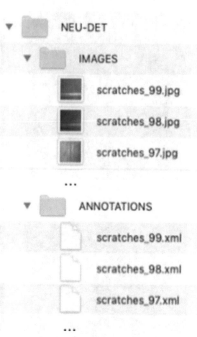

Figure 9-2. *NEU-DET dataset directory structure*

Google Colab Notebook

First, create a new notebook on Google Colab and give it a name such as Surface Defect Detection v1.0.

Since the NEU dataset is located on Google Drive, you can directly copy it to your private Google Drive. On Colab, we will mount the private Google Drive, uncompress the dataset, and set up the development environment (see Listing 9-1). Please review Chapter 6 to refresh your understanding of the implementation.

Listing 9-1. Mounting Google Drive, Downloading, Building, and Installing TensorFlow Models

```
Code Block 1: Mount Google Drive
1    # Code block 1: Mount Google Drive
2    from google.colab import drive
3    drive.mount('/content/drive')
Code Block 2: Uncompressing NEU dataset
4    # Code block 2: uncompress NEU data
```

```
5     %%shell
6     ls /content/drive/'MyDrive'/NEU-DET.zip
7     unzip /content/drive/'MyDrive'/NEU-DET.zip
```

Code Block 3: Installing necessary dependencies

```
8     !pip install numpy==1.23
9     !apt install -y protobuf-compiler
10    !pip install pillow==9.5
```

Code Block 4: Installing TensorFlow model project

```
11    %%shell
12    git clone --depth 1 https://github.com/tensorflow/models.git
13    cd models/research/
14    protoc object_detection/protos/*.proto --python_out=.
15    cd /content
16    git clone https://github.com/cocodataset/cocoapi.git
17    cd cocoapi/PythonAPI
18    make
19    cp -r pycocotools /content/models/research
20    cd
21    cd /content/models/research/
22    cp object_detection/packages/tf2/setup.py .
23    python -m pip install .
```

Data Transformation

We will transform the NEU dataset into TFRecord format (review the SSD model training section of Chapter 6). Listing 9-2 is TensorFlow-based code to transform images and annotations into TFRecord.

Listing 9-2. Transforming Images and Annotations in PASCAL VOC Format into TFRecord

```
File name: generic_xml_to_tf_record.py
1     import hashlib
2     import io
3     import logging
4     import os
```

```
5
6    from lxml import etree
7    import PIL.Image
8    import tensorflow as tf
9
10   from object_detection.utils import dataset_util
11   from object_detection.utils import label_map_util
12   import random
13
14   flags = tf.app.flags
15   flags.DEFINE_string('data_dir', '', 'Root directory to raw PASCAL VOC
     dataset.')
16
17   flags.DEFINE_string('annotations_dir', 'annotations',
18                       '(Relative) path to annotations directory.')
19   flags.DEFINE_string('image_dir', 'images',
20                       '(Relative) path to images directory.')
21
22   flags.DEFINE_string('output_path', '', 'Path to output TFRecord')
23   flags.DEFINE_string('label_map_path', 'data/pascal_label_map.pbtxt',
24                       'Path to label map proto')
25   flags.DEFINE_boolean('ignore_difficult_instances', False, 'Whether to
     ignore '
26                       'difficult instances')
27   FLAGS = flags.FLAGS
28
29   # This function generates a list of images for training and
     validation.
30   def create_trainval_list(data_dir):
31       trainval_filename = os.path.abspath(os.path.join(data_
         dir,"trainval.txt"))
32       trainval = open(os.path.abspath(trainval_filename), "w")
33       files = os.listdir(os.path.join(data_dir, FLAGS.image_dir))
34       for f in files:
35           absfile =os.path.abspath(os.path.join(data_dir, FLAGS.image_
             dir, f))
```

```
36          trainval.write(absfile+"\n")
37          print(absfile)
38      trainval.close()
39
40
41  def dict_to_tf_example(data,
42                         dataset_directory,
43                         label_map_dict,
44                         ignore_difficult_instances=False,
45                         image_subdirectory=FLAGS.image_dir):
46    """Convert XML derived dict to tf.Example proto.
47
48    Notice that this function normalizes the bounding box coordinates
    provided
49    by the raw data.
50
51    Args:
52      data: dict holding PASCAL XML fields for a single image
53      dataset_directory: Path to root directory holding PASCAL dataset
54      label_map_dict: A map from string label names to integers ids.
55      ignore_difficult_instances: Whether to skip difficult
      instances in the
56        dataset  (default: False).
57      image_subdirectory: String specifying subdirectory within the
58        PASCAL dataset directory holding the actual image data.
59
60    Returns:
61      example: The converted tf.Example.
62
63    Raises:
64      ValueError: if the image pointed to by data['filename'] is not a
      valid JPEG
65    """
66    filename = data['filename']
67
```

```
68    if filename.find(".jpg") < 0:
69        filename = filename+".jpg"
70    img_path = os.path.join("",image_subdirectory, filename)
71    full_path = os.path.join(dataset_directory, img_path)
72
73    with tf.gfile.GFile(full_path, 'rb') as fid:
74      encoded_jpg = fid.read()
75    encoded_jpg_io = io.BytesIO(encoded_jpg)
76    image = PIL.Image.open(encoded_jpg_io)
77    if image.format != 'JPEG':
78      raise ValueError('Image format not JPEG')
79    key = hashlib.sha256(encoded_jpg).hexdigest()
80
81    width = int(data['size']['width'])
82    height = int(data['size']['height'])
83
84    xmin = []
85    ymin = []
86    xmax = []
87    ymax = []
88    classes = []
89    classes_text = []
90    truncated = []
91    poses = []
92    difficult_obj = []
93    if 'object' in data:
94      for obj in data['object']:
95        difficult = bool(int(obj['difficult']))
96        if ignore_difficult_instances and difficult:
97          continue
98
99        difficult_obj.append(int(difficult))
100
101        xmin.append(float(obj['bndbox']['xmin']) / width)
102        ymin.append(float(obj['bndbox']['ymin']) / height)
```

```
103     xmax.append(float(obj['bndbox']['xmax']) / width)
104     ymax.append(float(obj['bndbox']['ymax']) / height)
105     classes_text.append(obj['name'].encode('utf8'))
106     classes.append(label_map_dict[obj['name']])
107     truncated.append(int(obj['truncated']))
108     poses.append(obj['pose'].encode('utf8'))
109
110 example = tf.train.Example(features=tf.train.Features(feature={
111     'image/height': dataset_util.int64_feature(height),
112     'image/width': dataset_util.int64_feature(width),
113     'image/filename': dataset_util.bytes_feature(
114         data['filename'].encode('utf8')),
115     'image/source_id': dataset_util.bytes_feature(
116         data['filename'].encode('utf8')),
117     'image/key/sha256': dataset_util.bytes_feature(key.
        encode('utf8')),
118     'image/encoded': dataset_util.bytes_feature(encoded_jpg),
119     'image/format': dataset_util.bytes_feature('jpeg'.
        encode('utf8')),
120     'image/object/bbox/xmin': dataset_util.float_list_feature(xmin),
121     'image/object/bbox/xmax': dataset_util.float_list_feature(xmax),
122     'image/object/bbox/ymin': dataset_util.float_list_feature(ymin),
123     'image/object/bbox/ymax': dataset_util.float_list_feature(ymax),
124     'image/object/class/text': dataset_util.bytes_list_
        feature(classes_text),
125     'image/object/class/label': dataset_util.int64_list_
        feature(classes),
126     'image/object/difficult': dataset_util.int64_list_
        feature(difficult_obj),
127     'image/object/truncated': dataset_util.int64_list_
        feature(truncated),
128     'image/object/view': dataset_util.bytes_list_feature(poses),
129 }))
130 return example
131
```

```
132  def create_tf(examples_list, annotations_dir, label_map_dict,
     dataset_type):
133      writer = None
134      if not os.path.exists(FLAGS.output_path+"/"+dataset_type):
135          os.mkdir(FLAGS.output_path+"/"+dataset_type)
136
137      j = 0
138      for idx, example in enumerate(examples_list):
139
140          if idx % 100 == 0:
141              logging.info('On image %d of %d', idx, len(examples_list))
142              print((FLAGS.output_path + "/tf_training_" + str(j) +
                     ".record"))
143              writer = tf.python_io.TFRecordWriter(FLAGS.output_path +
                     "/"+dataset_type+"/tf_training_" + str(j) + ".record")
144              j = j + 1
145
146          path = os.path.join(annotations_dir, os.path.basename(example).
                 replace(".jpg", '.xml'))
147
148          with tf.gfile.GFile(path, 'r') as fid:
149              xml_str = fid.read()
150          xml = etree.fromstring(xml_str)
151          data = dataset_util.recursive_parse_xml_to_dict(xml)
                 ['annotation']
152
153          tf_example = dict_to_tf_example(data, FLAGS.data_dir, label_
                 map_dict,
154                          FLAGS.ignore_difficult_instances)
155          writer.write(tf_example.SerializeToString())
156
157  def main(_):
158
159      data_dir = FLAGS.data_dir
160      create_trainval_list(data_dir)
```

```
161
162    label_map_dict = label_map_util.get_label_map_dict(FLAGS.label_
       map_path)
163
164    examples_path = os.path.join(data_dir,'trainval.txt')
165    annotations_dir = os.path.join(data_dir, FLAGS.annotations_dir)
166    examples_list = dataset_util.read_examples_list(examples_path)
167
168    random.seed(42)
169    random.shuffle(examples_list)
170    num_examples = len(examples_list)
171    num_train = int(0.7 * num_examples)
172    train_examples = examples_list[:num_train]
173    val_examples = examples_list[num_train:]
174
175    create_tf(train_examples, annotations_dir, label_map_dict, "train")
176    create_tf(val_examples, annotations_dir, label_map_dict, "val")
177
178  if __name__ == '__main__':
179     tf.app.run()
```

Listing 9-2 does the following:

1. First, call the function `create_trainval_list()` to create a text file containing a list of absolute paths of all images from the `IMAGES` subdirectory.

2. Split the list of image paths into a 70:30 ratio to generate separate lists of images for the training and validation sets.

3. For each image in the training set, create a TFRecord using the function `dict_to_tf_example()`. The TFRecord contains the bytes of the image, bounding boxes, the annotated class name, and several other metadata about the image. The TFRecord is serialized and written to a file. Multiple TFRecord files are created, and the number of files depends on the total number of images and the number of images to be included in each TFRecord file.

4. Similarly, create TFRecords for each of the validation images and serialize them to files.

5. Save the training and validation sets into two separate subdirectories—`train` and `val`—inside the `output` directory.

We need a mapping file that maps the class index with the class name. This file contains JSON content and typically has the extension `.pbtxt`. We have six defect classes, and we can manually write the label mapping file as shown here:

```
File name: steel_label_map.pbtxt
    item {
      id: 1
      name: 'rolled-in_scale'
    }
    item {
      id: 2
      name: 'patches'
    }
    item {
      id: 3
      name: 'crazing'
    }
    item {
      id: 4
      name: 'pitted_surface'
    }
    item {
      id: 5
      name: 'inclusion'
    }
    item {
      id: 6
      name: 'scratches'
    }
```

Upload the `steel_label_map.pbtxt` file to your Colab environment to the `/content` directory (or any other directory you want, but be sure to provide the correct path in Listing 9-3).

Listing 9-3. Executing `generic_xml_to_tf_record.py` That Creates TFRecord Files

```
1    %%shell
2    python /content/generic_xml_to_tf_record.py \
 --label_map_path=/content/steel_label_map.pbtxt \
--data_dir=/content/NEU-DET \
--output_path=/content/NEU-DET/out \
--annotations_dir=ANNOTATIONS \
--image_dir=IMAGES
```

The script in Listing 9-3 executes `generic_xml_to_tf_record.py` by providing these parameters:

- `--label_map_path`: The path to the `steel_label_map.pbtxt`.

- `--data_dir`: The root directory where images and annotations directories are located.

- `--output_path`: The path where you want to save the generated TFRecord files. Ensure that this directory exists. If not, create this directory before executing this script.

- `--annotations_dir`: The name of the subdirectory where the annotation XML files are located.

- `--image_dir`: The name of the subdirectory where images are located.

Run the script in Listing 9-3 to create TFRecord files in the output directory. You will see two subdirectories—`train` and `val`—where TFRecords for training and validation are saved.

Again, the output directory must exist before you execute the code in Listing 9-3.

Training the SSD Model

We are now ready with the right input set in TFRecord format to train our SSD model. The training step is exactly the same as we followed in Chapter 6. First download a pretrained SSD model for a transfer learning based on the training and validation set we created earlier. You can download a suitable model for the transfer learning from the following URL: `https://github.com/tensorflow/models/blob/master/research/object_detection/g3doc/tf2_detection_zoo.md`.

Listing 9-4 shows the same code that we used in Chapter 6 (Listing 6-5).

Listing 9-4. Downloading a Pretrained Object Detection Model

```
1    %%shell
2    mkdir pre-trained-model
3    cd pre-trained-model
4    wget http://download.tensorflow.org/models/object_detection/
     tf2/20200711/ssd_resnet50_v1_fpn_640x640_coco17_tpu-8.tar.gz
5    tar -xvf ssd_resnet50_v1_fpn_640x640_coco17_tpu-8.tar.gz
```

We will now edit the `pipeline.config` file, as explained in the section "Configuring the Object Detection Pipeline" of Chapter 6. Listing 9-5 shows the sections of the `pipeline.config` file edited as per the current configuration.

Listing 9-5. Section of `pipeline.config` That Must to Be Edited to Point to the Appropriate Directory Structure

```
model {
  ssd {
    num_classes: 6
    image_resizer {
      fixed_shape_resizer {
        height: 640
        width: 640
      }
    }
```

```
......
      batch_norm {
        decay: 0.999700009823
        center: true
        scale: true
        epsilon: 0.0010000000475
        train: true
      }
    }
          override_base_feature_extractor_hyperparams: true
  }
  .....
  matcher {
    argmax_matcher {
      matched_threshold: 0.7
      unmatched_threshold: 0.3
      ignore_thresholds: false
      negatives_lower_than_unmatched: true
      force_match_for_each_row: true
    }
  }
  ......
  fine_tune_checkpoint: "/content/pre-trained-model/ssd_resnet50_v1_
  fpn_640x640_coco17_tpu-8/checkpoint/ckpt-0"
  from_detection_checkpoint: true
  num_steps: 100000
}
train_input_reader {
  label_map_path: "/content/steel_label_map.pbtxt"
  tf_record_input_reader {
    input_path: "/content/NEU-DET/out/train/*.record"
  }
}
```

```
eval_config {
  num_examples: 8000
  max_evals: 10
  use_moving_averages: false
}
eval_input_reader {
  label_map_path: "/content/steel_label_map.pbtxt"
  shuffle: false
  num_readers: 1
  tf_record_input_reader {
    input_path: "/content/NEU-DET/out/val/*.record"
  }
}
```

As shown in Listing 9-5, we must edit the sections highlighted in bold in Listing 9-5.

```
num_classes: 6
fine_tune_checkpoint: path to pre-trained model checkpoint
label_map_path: path to .pbtxt file
input_path: path to the training TFRecord files
label_map_path: path to the .pbtxt file
input_path: path to the validation TFRecord files
```

Edit the pipeline.config file and upload it to the Colab environment.

Execute the model training using the script shown in Listing 9-6. Review the discussion of Listing 6-6 in Chapter 6 to refresh the concepts.

Listing 9-6. Executing the Model Training

```
1    %%shell
2    export PYTHONPATH=$PYTHONPATH:/content/models/research
3    export PYTHONPATH=$PYTHONPATH:/content/models/research/slim
4    cd /content/models/research
5    PIPELINE_CONFIG_PATH=/content/computer_vision/pre-trained-model/ssd_
     resnet50_v1_fpn_640x640_coco17_tpu-8/pipeline.config
6    MODEL_DIR=/content/computer_vision/surface_defect_detection_model/
7    NUM_TRAIN_STEPS=100000
```

```
8    SAMPLE_1_OF_N_EVAL_EXAMPLES=1
9    python object_detection/model_main_tf2.py \
--pipeline_config_path=${PIPELINE_CONFIG_PATH} \
--model_dir=${MODEL_DIR} \
--num_train_steps=${NUM_TRAIN_STEPS} \
--sample_1_of_n_eval_examples=${SAMPLE_1_OF_N_EVAL_EXAMPLES} \
--alsologtostderr
```

While the model is learning, the logs are printed on the Colab console. Make a note of the loss per epoch and tune the model's hyperparameters, if needed.

Model Evaluation

Listing 9-7 is the code that we need to execute to run the model evaluation.

Listing 9-7. Executing the Model Evaluation

```
1    %%shell
2    python /content/models/research/object_detection/model_main_tf2.py \
--model_dir=/content/computer_vision/surface_defect_detection_model \
--pipeline_config_path=/content/computer_vision/pre-trained-model/ssd_
resnet50_v1_fpn_640x640_coco17_tpu-8/pipeline.config \
--checkpoint_dir=/content/computer_vision/ surface_defect_detection_model
```

Launch the TensorBoard dashboard to evaluate the model quality. Listing 9-8 shows how to launch the TensorBoard dashboard.

Listing 9-8. Launching the TensorBoard Dashboard

```
1    %tensorflow_version 2.x
2    %load_ext tensorboard
3    %tensorboard --logdir /drive/'MyDrive'/NEU-DET-models/
```

Figure 9-3 shows sample training output of TensorBoard.

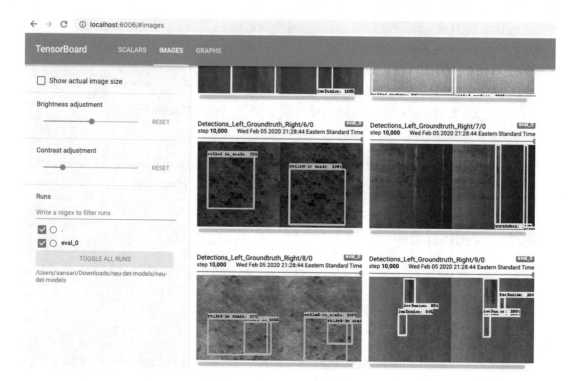

Figure 9-3. *TensorBoard output display of surface defect detection model training*

Exporting the Model

After the training has successfully completed, the checkpoints are saved in the directory specified in line 6 of Listing 9-6.

To utilize the model for real-time detection, we need to export the TensorFlow graph. Review the section "Exporting the TensorFlow Graph" of Chapter 6 for details on this.

Listing 9-9 shows how to export the SSD model that we just trained.

Listing 9-9. Exporting the Model to the TensorFlow Graph

```
1  %%shell
2  export PYTHONPATH=$PYTHONPATH:/content/models/research
3  export PYTHONPATH=$PYTHONPATH:/content/models/research/slim
4  cd /content/models/research
```

```
5  python /content/models/research/object_detection/exporter_main_
   v2.py --input_type image_tensor --pipeline_config_path /content/
   computer_vision/pre-trained-model/ssd_resnet50_v1_fpn_640x640_coco17_
   tpu-8/pipeline.config --trained_checkpoint_dir /content/computer_vision/
   surface_defect_detection_model --output_directory /content/computer_
   vision/surface_defect_detection_model/final_model
```

After exporting the model, you should save it to Google Drive. Download the final model from Google Drive to your local computer. We can use this model to detect surface defects from video frames in real time. Review the concepts presented in Chapter 7.

Prediction

If you have set up your working environment as described in the section "Detecting Objects Using Trained Models" of Chapter 6, you should have everything needed to predict surface defects in an image. Simply change the variables in Listing 9-10 and execute the Python code shown in Listing 6-18. Change the bold highlighted paths based on what you have setup in your computer.

Listing 9-10. Variable Initialization Portion of Code from Listing 6-18

```
model_path = "/Users/sansari/Downloads/neu-det-models/final_model"
labels_path = "/Users/sansari/Downloads/steel_label_map.pbtxt"
image_dir = "/Users/sansari/Downloads/NEU-DET/test/IMAGES"
image_file_pattern = "*.jpg"
output_path="/Users/sansari/Downloads/surface_defects_out"
```

Figure 9-4 shows some sample output of predictions for different classes of defects.

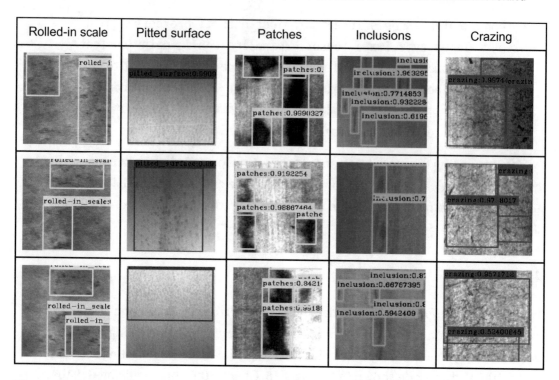

Figure 9-4. *Sample prediction output of defective surfaces with bounding boxes*

Real-Time Defect Detector

Follow the instructions provided in Chapter 7 and deploy the detection system that will read video images from the camera and detect surface defects in real time. If you have multiple cameras connected to the same device, use the appropriate value for the argument x in the function `cv2.VideoCapture(x)`. By default, x=0 reads video from the built-in camera of the computer. The values of x=1, x=2, etc., will read videos attached to computer ports. For an IP-based camera, the value of x should be the IP address.

Image Annotations

In all previous examples, we used images that were already annotated and labeled. In this section, we will explore how to annotate images for object detection or face recognition.

There are several open source and commercial tools for image labeling. We will explore the Microsoft Visual Object Tagging Tool (VoTT), which is an open source annotation and labeling tool for image and video assets. The source code of VoTT is available at `https://github.com/microsoft/VoTT`.

Installing VoTT

VoTT requires NodeJS and NPM.

To install NodeJS, download the executable binaries for your operating system from the official website at `https://nodejs.org/en/download/`. For example, download and install the Windows Installer (`.msi`) to install NodeJS on Windows OS, download and install the macOS Installer (`.pkg`) to install it on a Mac, or choose Linux Binaries (x64) for Linux.

NPM is installed with NodeJS. To check whether NodeJS and NPM are installed on your computer, execute the following commands in your terminal window:

```
node -v
npm -v
```

VoTT installers for different OSs are maintained at GitHub (`https://github.com/Microsoft/VoTT/releases`). Download the installer for your OS. At the time of writing this book, the latest VoTT is version 2.1.1, which can be downloaded from these locations:

- Windows: `https://github.com/microsoft/VoTT/releases/download/v2.1.0/vott-2.1.0-win32.exe`

- Mac: `https://github.com/microsoft/VoTT/releases/download/v2.1.0/vott-2.1.0-darwin.dmg`

- Linux: `https://github.com/microsoft/VoTT/releases/download/v2.1.0/vott-2.1.0-linux.snap`

Install VoTT on your computer by running the downloaded executable.

To run VoTT from the source, execute the following commands on your terminal:

```
git clone https://github.com/Microsoft/VoTT.git
 cd VoTT
 npm ci
 npm start
```

449

Running VoTT with the `npm start` command will launch both the electron version and the browser version. The major difference between the two versions is that the browser version cannot access the local file system, while the electron version can. Since our images are on the local file system, we will explore the electron version of VoTT.

When you launch the VoTT user interface, you will see the home screen in which to either create a new project, open a local project, or open a cloud project.

To annotate images, we will follow the steps in the next sections.

Create Connections

We will create two connections: one for input and the other for output. The input connection is to the directory where unlabeled images are stored. The output connection is to the directory where the annotations are stored.

Currently, VoTT supports connection to the following:

- Azure Blob Storage

- Bing Image Search

- Local file system

We will create a connection to the local file system. To create a new connection, click the New Connections icon in the left navigation bar to launch the connection screen. Click the plus icon corresponding to the label CONNECTIONS, located in the top-left panel. See Figure 9-5.

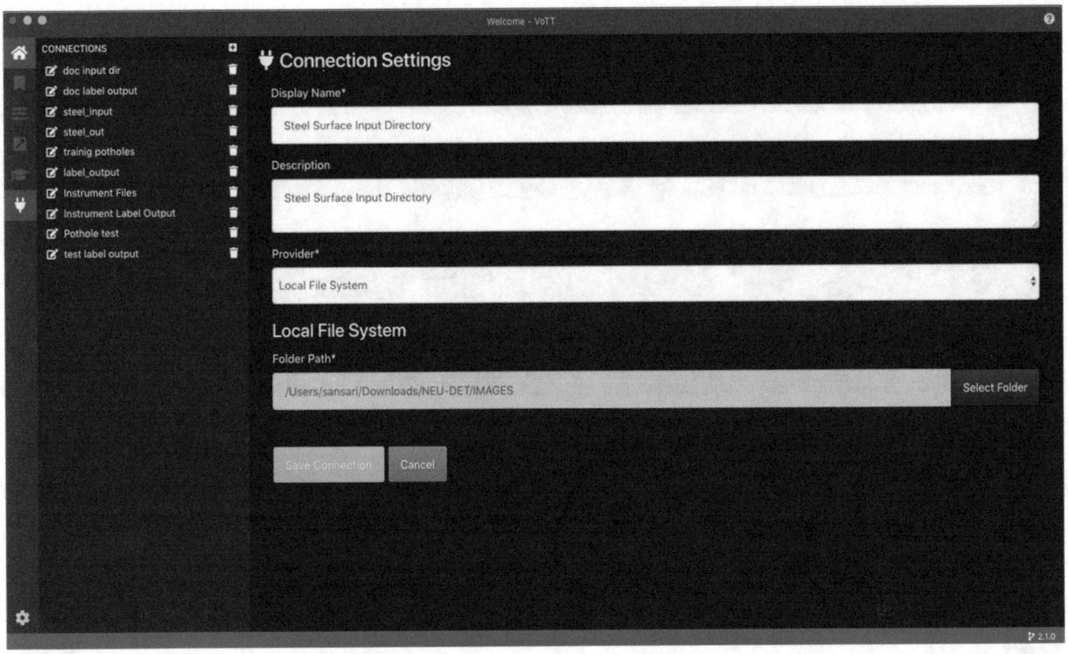

Figure 9-5. *Creating a new connection*

Select Local File System for the Provider field. Click Select Folder to open the local file system directory structure. Select the directory that contains input images that need to be labeled. Click the Save Connection button.

Similarly, create another connection for storing the output.

Create a New Project

The tasks of image annotations and labeling are managed under a project. To create a project, click the home icon and then New Project to open the Project Settings page. See Figure 9-6.

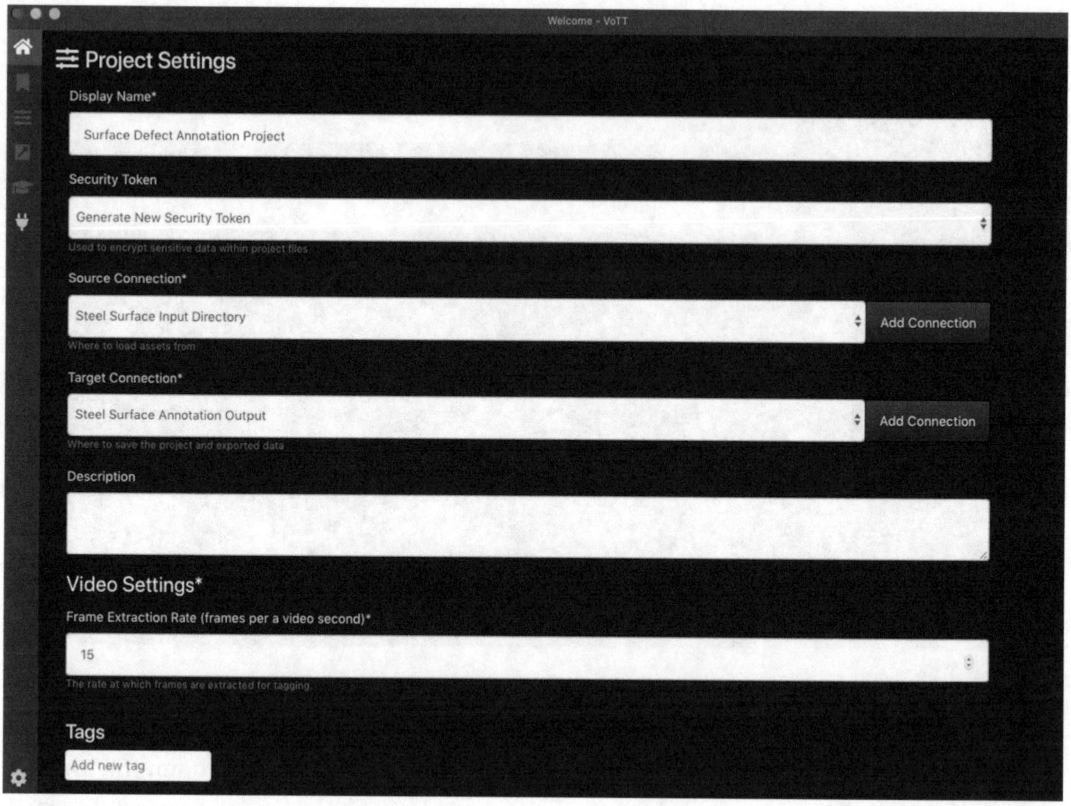

Figure 9-6. *Project Settings page to create a new project*

The two important fields on the Project Settings page are Source Connection and Target Connection. Select the appropriate connections that we created in the previous step for input and output directories. Click the Save Project button.

Create Class Labels

After saving the project settings, the screen transitions to the main labeling page. To create the class labels, click the (+) icon corresponding to the label TAGS located in the top-right corner of the panel on the right (as shown in Figure 9-7). Create all the class labels, such as crazing, patch, inclusion, etc.

Figure 9-7. *Creating class labels*

Label the Images

Select an image thumbnail from the left panel, and the image will open in the main tagging area. Draw rectangles or polygons around the defective areas of the image, and select the appropriate tag to annotate the image. See Figure 9-8.

Figure 9-8. *Drawing rectangles around the defective areas and selecting the class tag to annotate the image*

Similarly, annotate all images one by one.

Export Labels

VoTT supports the following formats for export:

- Azure Custom Vision Service

- Microsoft Cognitive Toolkit (CNTK)

- TensorFlow (Pascal VOC and TFRecords)

- VoTT (generic JSON schema)

- Comma-separated values (CSV)

We will configure the settings to export our annotations in the TensorFlow TFRecord file format.

To configure, click the export icon located in the left navigation bar. The export icon looks like a slanting arrow pointing upward. The Export Settings page opens. For the Provider field, select TensorFlow Records and click the Save Export Settings button (see Figure 9-9).

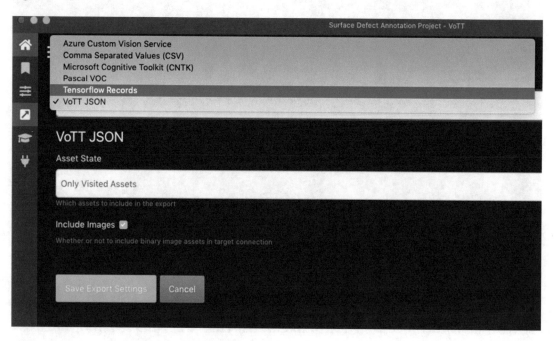

Figure 9-9. Export Settings page

Go back to the project page (click the Tag Editor icon). Click the ◰ icon located in the top toolbar to export the annotation to a TensorFlow Records file.

Check the output folder of the local file system. You will notice that a directory with the name containing `TFRecords-export` has been created in the output directory.

Exporting to the TFRecord format also generates a `tf_label_map.pbtxt` file that contains the class and index mapping.

For up-to-date information and instructions on image labeling, visit the official GitHub page of the VoTT project maintained by Microsoft: `https://github.com/microsoft/VoTT`.

Summary

In this section, we created a system dedicated to identifying surface defects. We trained an SSD model by using a preexisting dataset containing labeled images of hot-rolled steel strips, encompassing six distinct defect categories. We then used this trained model to predict surface defects found in both images and videos. Additionally, we delved into the utilization of an image annotation tool named VoTT, which facilitates the annotation of images and the subsequent export of labels into the TFRecord format.

CHAPTER 10

Computer Vision Modeling on the Cloud

Training state-of-the-art convolutional neural networks can require significant computer resources. It may take several hours or days to train a computer vision model, depending on the number of training samples, network configuration, and available hardware resources. A single GPU may not be feasible to train a complex network involving large numbers of training images. The models need to be trained on multiple GPUs. Only a limited number of GPUs can be installed on a single machine. A single machine with multiple GPUs may not be sufficient for training on a large number of images. It will be faster if the model is trained on multiple machines, each having multiple GPUs.

It is difficult to estimate the number of GPUs and machines needed to train a model in a certain time frame. In most practical cases, it is not known up front how many machines are needed for the modeling and how long the training will run. Also, modeling is not done frequently. A model that predicts with a high degree of accuracy may not need to be retrained for several days, weeks, months, or as long as it gives accurate results. Therefore, any hardware procured for the modeling may remain idle until the model is retrained.

Modeling on the cloud is a good way to scale the training across multiple machines and GPUs. Most cloud providers offer virtual machines, compute resources, and storage on a pay-as-you-go model. This means you will be charged only for the cloud resources used during the period when the model is learning. After the model is successfully trained, you can export the model to your application server and use it there for prediction. At this point, you can delete all cloud resources that are no longer required, which will reduce costs.

© Shamshad Ansari 2023
S. Ansari, *Building Computer Vision Applications Using Artificial Neural Networks*,
https://doi.org/10.1007/978-1-4842-9866-4_10

TensorFlow provides APIs to train machine learning models on multiple CPUs and GPUs installed on either a single machine or multiple machines.

In this chapter, we will explore distributed modeling and train computer vision models at scale on the cloud.

Our learning objectives in this chapter are as follows:

- To explore the TensorFlow APIs for distributed training

- To set up distributed TensorFlow clusters involving multiple virtual machines and GPUs on the three popular cloud providers: Amazon Web Services (AWS), Google Cloud Platform (GCP), and Microsoft Azure

- To train computer vision models on distributed clusters on the cloud

TensorFlow Distributed Training

This section discusses the core concepts of distributed training provided by TensorFlow.

What Is Distributed Training?

The state-of-the-art neural network for computer vision computes millions of parameters from a large number of images. The training is time-consuming if all of the computations are performed on a single CPU or GPU. In addition, the entire training dataset is required to be loaded in memory, which may exceed the memory of a single machine.

In distributed training, computations are performed concurrently on multiple CPUs or GPUs, and the results are combined to create the final model. Ideally, the computation should scale linearly with the number of GPUs or CPUs. In other words, if it takes H hours to train a model on one GPU, it should take H/N hours to train the model on N number of GPUs.

There are two commonly used methods to implement parallelism in distributed training: data parallelism and model parallelism. TensorFlow provides APIs to distribute the training by splitting models over multiple devices (CPUs, GPUs, or computers).

Data Parallelism

Large training datasets can be divided into smaller mini-batches, which then can be distributed across multiple computers in a cluster architecture. Stochastic gradient descent (SGD) can independently and in parallel compute weights on individual computers that have a small batch of data. The results can be combined from the individual computers to a central computer to get the final and optimized weights.

SGD can also optimize weights by using parallel processing in a single computer with multiple CPUs or GPUs. The distributed and parallel operations to compute optimized weights by using the SGD algorithm helps converge it faster.

Figure 10-1 shows a pictorial view of data parallelism.

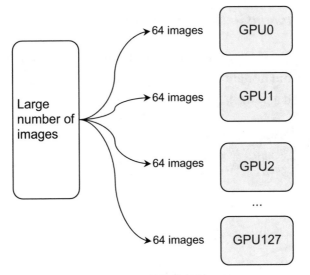

Batch size = 64 x 128 = 8,192 images

Figure 10-1. *Data parallelism and batch size calculation*

Data parallelism can be achieved in the following two ways:

- *Synchronous*: In this case, all nodes train over different chunks of input data and aggregate gradients at each step. The synchronization of gradients is done by an all-reduce method, as illustrated in Figure 10-2.

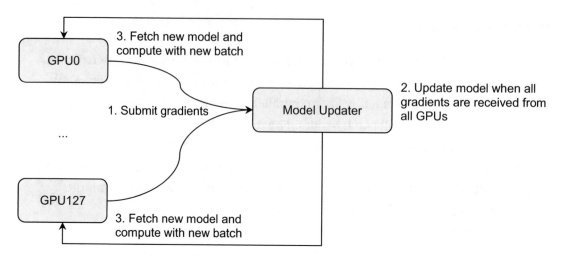

Figure 10-2. *Synchronous data parallelism*

- *Asynchronous*: In this case, all nodes independently train over the input data and update variables asynchronously through a dedicated server called a *parameter server*, as shown in Figure 10-3.

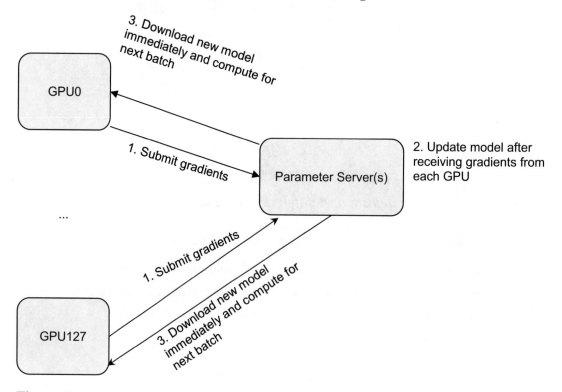

Figure 10-3. *Asynchronous data parallelism using parameter servers*

Model Parallelism

Deep neural networks, such as Darknet, compute billions of parameters. It is a challenge to load the entire network in the memory of a single CPU or GPU, even when the batch size is small. *Model parallelism* is a method in which the model is broken into different parts, with each part performing operations on the same set of data in different CPUs, GPUs, or nodes of the physical computer hardware. The same data batch is copied to all nodes in the cluster, but the nodes get different parts of the model. These model parts operate on its input dataset concurrently on different nodes.

When the parts of the model run in parallel, their shared parameters need to be synchronized. This approach of parallelism works the best in the case of multiple CPUs or GPUs on the same machine because the devices are connected by a high-speed bus.

We will now explore how TensorFlow distributes the training across multiple GPUs or machines.

TensorFlow Distribution Strategy

TensorFlow provides a high-level API to distribute the training across multiple GPUs or multiple nodes. The API is exposed via the `tf.distribute.Strategy` class. With just a few additional lines and minor code changes, we can distribute the neural networks that we have explored in all prior examples.

We can use `tf.distribute.Strategy` with Keras to distribute networks built by using the Keras API. We can also use this to distribute custom training loops. In general, any computation in TensorFlow can be distributed using this API.

TensorFlow supports the following types of distribution strategies.

MirroredStrategy

`MirroredStrategy` supports synchronous distributed training on multiple GPUs on one machine. All variables of the model are mirrored across all GPUs. These variables collectively are called `MirroredVariables`. The computations for the training are performed in parallel on each GPU. The variables are synchronized with each other by applying identical updates.

The `MirroredVariables` are updated across all devices by using all-reduce algorithms. An *all-reduce algorithm* aggregates tensors across all the devices by adding them up and making them available on each device. Figure 10-2 illustrates an example of an all-reduce algorithm. These algorithms are efficient and do not have much communication overhead for synchronization.

There are several all-reduce algorithms. TensorFlow uses NVIDIA NCCL as the default all-reduce algorithm in `MirroredStrategy`.

We will explore how to use `MirroredStrategy` to distribute the training of a deep neural network. To keep it simple and easy to understand, let's modify the code from Listing 5-2 and make it distributed. Refer to lines 11, 19, and 24 of Listing 5-2. Here is what these lines of code look like:

Line 11, Listing 5-2: `model = tf.keras.models.Sequential([...])`

Line 19, Listing 5-2: `model.compile(...)`

Line 24, Listing 5-2: `history = model.fit(...)`

The following are the steps to parallelize the training of Listing 5-2:

1. Create an instance of `MirroredStrategy`.

2. Move the creation and compilation of the model (lines 11 and 19 of Listing 5-2) inside the `scope()` method of the `MirroredStrategy` object.

3. Fit the model (line 24, without any change).

All other lines of Listing 5-2 remain unchanged.

Listing 10-1 shows this concept.

Listing 10-1. Synchronous Distributed Training Using `MirroredStrategy`

```
1      strategy = tf.distribute.MirroredStrategy()
2       with strategy.scope():
3          model = tf.keras.Sequential([...])
4          model.compile(...)
5      model.fit(...)
```

Thus, with just two additional lines of code and minor adjustments, we can distribute the training to multiple GPUs on a single machine.

As shown in Listing 10-1, within the scope() method of the MirroredStrategy object, we create the computation that we want to run in a distributed and parallel fashion. The MirroredStrategy object takes care of replicating the model's training on the available GPUs, aggregating gradients, and more.

Each batch of the input is divided equally among the replicas. For example, if the input batch size is 16 and we use MirroredStrategy with two GPUs, each GPU will get eight input examples in each step. We should tune the batch size appropriately to effectively utilize the computing power of the GPUs.

The tf.distribute.MirroredStrategy() method creates the default object that uses all available GPUs that are visible to TensorFlow. If you want to use only some of the GPUs of the machine, simply do the following:

```
strategy = tf.distribute.MirroredStrategy(devices=["/gpu:0", "/gpu:1"])
```

Here's an exercise for you: modify the code example shown in Listing 5-4 and train the digit recognition model in distributed mode using MirroredStrategy.

CentralStorageStrategy

CentralStorageStrategy places the model variables on the CPU and replicates the computations across all local GPUs on one machine. Except for the placement of variables on the CPU rather than replicating them on GPUs, the CentralStorageStrategy is similar to the MirroredStrategy.

At the time of writing this book, CentralStorageStrategy is experimental and likely to change in the future. To distribute the training under CentralStorageStrategy, simply replace line 1 of Listing 10-1 with the following:

```
strategy = tf.distribute.experimental.CentralStorageStrategy()
```

MultiWorkerMirroredStrategy

MultiWorkerMirroredStrategy is similar to MirroredStrategy. It distributes the training across multiple machines, each having one or more GPUs. It copies all variables in the model on each device across all machines. These machines where computations are performed are referred to as *workers*.

To keep the variables in sync across all workers, `MultiWorkerMirroredStrategy` uses `CollectiveOps` as the all-reduce communication method. A *collective op* is a single op in the TensorFlow graph. It can automatically choose an all-reduce algorithm in the TensorFlow runtime according to hardware, network topology, and tensor sizes.

To distribute the training across multiple workers under `MultiWorkerMirroredStrategy`, simply replace line 1 of Listing 10-1 with the following:

```
strategy = tf.distribute.experimental.MultiWorkerMirroredStrategy()
```

This creates the default `MultiWorkerMirroredStrategy` with `CollectiveCommunication.AUTO` as the default for `CollectiveOps`. You can choose one of the following two implementations of `CollectiveOps`:

- `CollectiveCommunication.RING` implements ring-based collectives using gRPC as the communication layer. gRPC is an open source implementation of Remote Procedure Call developed by Google. To use this, call the previous instantiation as follows:

  strategy = tf.distribute.experimental.MultiWorkerMirrored Strategy(tf.distribute.experimental.Collective Communication.RING)

- `CollectiveCommunication.NCCL` uses NVIDIA NCCL to implement collectives. Here is a usage example:

  strategy = tf.distribute.experimental.MultiWorkerMirrored Strategy(tf.distribute.experimental.Collective Communication.NCCL)

Cluster Configuration

TensorFlow makes it easy to distribute the training across multiple workers. But how does it know about the cluster configuration? Before we run our code that uses `MultiWorkerMirroredStrategy` to distribute the training, we must set the `TF_CONFIG` environment variable on all the workers that are going to participate in the model training. `TF_CONFIG` is described later in this section.

Dataset Sharding

How is data made available to workers? When we use model.fit(x=train_datasets, epochs=3, steps_per_epoch=5), we pass the training set directly to the fit() function. The dataset is sharded by TensorFlow automatically in a multiworker training.

Fault Tolerance

If any of the workers fails, the entire cluster will fail. There is no built-in failure recovery mechanism in TensorFlow. However, tf.distribute.Strategy with Keras provides a fault tolerance mechanism by saving the training checkpoints. If any worker fails, all the other workers will wait for the failed worker to restart. Since the checkpoints are saved, the training will start from the point where it stopped as soon as the failed worker comes back up.

To make the distributed cluster fault tolerant, you must save the training checkpoints (review Chapter 5 to see how checkpoints are saved using callbacks).

TPUStrategy

Tensor processing units (TPUs) are specialized application-specific integrated circuits (ASICs) designed by Google to dramatically accelerate the machine learning workloads. TPUs are available on Cloud TPU and Google Colab.

In terms of implementation, TPUStrategy is the same as MirroredStrategy except that the model variables are mirrored to TPUs. Listing 10-2 shows how to instantiate TPUStrategy.

Listing 10-2. Instantiation of TPUStrategy

```
1    cluster_resolver = tf.distribute.cluster_resolver.TPUClusterResolver(
     tpu=tpu_address)
2    tf.config.experimental_connect_to_cluster(cluster_resolver)
3    tf.tpu.experimental.initialize_tpu_system(cluster_resolver)
4    tpu_strategy = tf.distribute.experimental.TPUStrategy(cluster_
     resolver)
```

In line 1, specify the TPU address by passing it to the argument tpu=tpu_address.

ParameterServerStrategy

In ParameterServerStrategy, the model variables are placed on a dedicated machine, called the *parameter server*. In this case, some machines are designated as workers and others as parameter servers. Computations are replicated across all GPUs of all workers while the variables are updated in the parameter server.

The implementation of ParameterServerStrategy is the same as MultiWorkerMirroredStrategy. We must set the TF_CONFIG environment variable on each of the participating machines. TF_CONFIG is explained next.

To distribute the training under ParameterServerStrategy, simply replace line 1 of Listing 10-1 with the following:

```
strategy = tf.distribute.experimental.ParameterServerStrategy()
```

OneDeviceStrategy

Sometimes we want to test our distributed code on a single device (GPU) before moving it to a fully distributed system involving multiple devices. OneDeviceStrategy is designed for this purpose. When we use this strategy, the model variables are placed on a specified device.

To use this strategy, simply replace line 1 of Listing 10-1 with the following:

```
strategy = tf.distribute.OneDeviceStrategy(device="/gpu:0")
```

This strategy is only for testing the code. Switch to other strategies before training your model on a fully distributed environment.

It is important to note that all the previous strategies, except MirroredStrategy, for distributed training are experimental at this time.

TF_CONFIG: TensorFlow Cluster Configuration

A TensorFlow cluster for distributed training consists of one or more machines, called *workers*. The computations of the model training are performed in each worker. There is one specialized worker, called the *master* or *chief worker*, that has extra responsibilities in addition to being a normal worker. The additional responsibilities of the chief worker include saving the checkpoints and writing summary files for TensorBoard.

The TensorFlow cluster may also include dedicated machines for parameter servers. The parameter server is mandatory in the case of ParameterServerStrategy.

The TensorFlow cluster configuration is specified by a TF_CONFIG environment variable. We must set this environment variable on all the machines on the cluster.

The format of TF_CONFIG is a JSON file consisting of two components: cluster and task.

The cluster component provides information about the workers and parameter servers that participate in the model training. This is a dictionary list of the workers' hostnames and communication ports (e.g., localhost:1234).

The task component specifies the role of the worker for the current task. It is customary to specify the first worker, with index 0 in the worker list, as the master or chief worker.

Table 10-1 describes the key-value pairs of TF_CONFIG.

Table 10-1. *TF_CONFIG Format Description*

Key	Description	Example

cluster	A dictionary containing the keys worker, chief, and ps. Each of these keys is a list of hostname:port of all machines involved in the training.	cluster: {
		worker:["host1:12345","host2:2345"]
		}

task	Specifies the task a particular machine will perform. This has the following keys: type: This specifies the worker type and takes a string for worker, chief, or ps. index: The zero-based index of the task. Most distributed training jobs have a single master task, one or more parameter servers, and one or more workers. trial: This is used when hyperparameter tuning is performed. This value sets the number of trials to train. This helps to identify which trial is currently running. This takes a string value containing the trial number, starting from 1.	task: { type: chief, index:0} This indicates that host1:1234 is the chief node.
job	The job parameters you used when you initiated the job. This is optional and may be ignored in most cases.	

An Example TF_CONFIG

Assume that we have a cluster of three machines that we want to use for distributed training. The hostnames of these machines are host1.local, host2.local, and host3.local. Assume that they all communicate via port 8900.

Also, assume the following roles for each machine:

```
worker: host1.local (chief worker)
worker: host2.local (normal worker)
ps: host3.local (parameter server)
```

The TF_CONFIG environment variable that needs to be set on all three machines is shown in Table 10-2.

Table 10-2. *Example TF_CONFIG Environment Variable in Three-Node Cluster That Has Two Workers and One Parameter Server*

master	worker	ps
'cluster': {	'cluster': {	'cluster': {
'worker': ["host1.local:8900", "host2.local:8900"], "ps": ["host3.local:8900"] },	'worker': ["host1.local:8900", "host2.local:8900"], "ps": ["host3.local:8900"] },	'worker': ["host1.local:8900", "host2.local:8900"], "ps": ["host3.local:8900"] },
'task': {'type': worker, 'index': 0} }	'task': {'type': worker, 'index': 1} }	'task': {'type': ps, 'index': 0} }

Example Code of Distributed Training with a Parameter Server

Listing 10-3, a modified version of Listing 5-2, shows a simple implementation of ParameterServerStrategy to distribute training to multiple workers. We will explore how to execute this code on the cloud.

Listing 10-3. Distributing Training Across Multiple Workers Using
ParameterServerStrategy

File name: distributed_training_ps.py

```
1    import argparse
2    import tensorflow as tf
3    from tensorflow_core.python.lib.io import file_io
4
5    #Disable eager execution
6    tf.compat.v1.disable_eager_execution()
7
8    # Instantiate the distribution strategy -- ParameterServerStrategy.
     This needs to be in the beginning of the code.
9    strategy = tf.distribute.experimental.ParameterServerStrategy()
10
11   # Parse the command line arguments
12   parser = argparse.ArgumentParser()
13   parser.add_argument(
14       "--input_path",
15       type=str,
16       default="",
17       help="Directory path to the input file. Could you be cloud
         storage"
18   )
19   parser.add_argument(
20       "--output_path",
21       type=str,
22       default="",
23       help="Directory path to the input file. Could you be cloud
         storage"
24   )
25   FLAGS, unparsed = parser.parse_known_args()
26
27   # Load MNIST data using built-in datasets' download function
28   mnist = tf.keras.datasets.mnist
29   (x_train, y_train), (x_test, y_test) = mnist.load_data()
```

```
30
31   # Normalize the pixel values by dividing each pixel by 255
32   x_train, x_test = x_train / 255.0, x_test / 255.0
33
34   BUFFER_SIZE = len(x_train)
35   BATCH_SIZE_PER_REPLICA = 16
36   GLOBAL_BATCH_SIZE = BATCH_SIZE_PER_REPLICA * 2
37   EPOCHS = 10
38   STEPS_PER_EPOCH = int(BUFFER_SIZE/EPOCHS)
39
40   train_dataset = tf.data.Dataset.from_tensor_slices((x_train, y_
     train)).repeat().shuffle(BUFFER_SIZE).batch(GLOBAL_BATCH_SIZE,drop_
     remainder=True)
41   test_dataset = tf.data.Dataset.from_tensor_slices((x_test, y_test)).
     batch(GLOBAL_BATCH_SIZE)
42
43
44   with strategy.scope():
45       # Build the ANN with 4-layers
46       model = tf.keras.models.Sequential([
47       tf.keras.layers.Flatten(input_shape=(28, 28)),
48       tf.keras.layers.Dense(128, activation='relu'),
49       tf.keras.layers.Dense(60, activation='relu'),
50       tf.keras.layers.Dense(10, activation='softmax')])
51
52       # Compile the model and set optimizer, loss function, and metrics
53       model.compile(optimizer='adam',
54                 loss='sparse_categorical_crossentropy',
55                 metrics=['accuracy'])
56
57   # Save checkpoints to the output location -- most probably on a cloud
     storage, such as GCS
58   callback = tf.keras.callbacks.ModelCheckpoint(filepath=FLAGS.
     output_path)
59   # Finally, train or fit the model
```

```
60   history = model.fit(train_dataset, epochs=EPOCHS, steps_per_
     epoch=STEPS_PER_EPOCH, callbacks=[callback], use_multiprocessing=True)
61   # history = model.fit(train_dataset, epochs=EPOCHS, steps_per_
     epoch=STEPS_PER_EPOCH,  use_multiprocessing=True)
62   # Save the model to the cloud storage
63   model.save("model.h5")
64   with file_io.FileIO('model.h5', mode='r') as input_f:
65       with file_io.FileIO(FLAGS.output_path+ '/model.h5', mode='wb+') as
         output_f:
66           output_f.write(input_f.read())
```

The code in Listing 10-3 can be divided into four logical parts:

- *Read and parse the command-line arguments (lines 11–25)*: It accepts two arguments: the training data input path and the output path for saving checkpoints and the final model.

- *Load the input images and create the training and test sets (lines 27–41)*: It is important to note that ParameterServerStrategy does not support last partial batch handling, passing the steps_per_ epoch argument to model.fit() when the dataset is imbalanced on multiple workers. Notice the calculation of steps_per_epoch in Line 38.

- *Create and compile the Keras model within the scope of* ParameterServerStrategy *(line 9 and lines 44–55)*: Here are a few important points to consider:

  - Create the instance of ParameterServerStrategy or MultiWorkerMirroredStrategy at the beginning of the program and put the code that may create ops after the strategy is instantiated.

  - The portion of the code that needs to be distributed must be wrapped within the scope of the strategy.

  - Line 44 defines the scope() block within which we wrap the model definition and compilation.

  – Lines 46 through 50 create the model within the strategy scope.

  – Lines 53 through 55 compile the model within the strategy scope.

- *Train the model and save the checkpoints and final model (lines 57–66):*

  – Line 58 creates the model checkpoint object that is passed to the model's `fit()` function to save the checkpoints while the model trains.

  – Line 60 triggers the model training by calling the `fit()` function. As explained earlier, the `train_dataset` passed to the `fit()` function is automatically distributed by the distribution strategy (`ParameterServerStrategy` in this case).

  – Line 63 saves the complete model in the local directory.

  – Lines 64 through 66 copy the local model to cloud storage, such as Google Cloud Storage (GCS) or Amazon S3.

  – Lines 57 through 66 are outside the scope of the strategy.

We now have the model training code that can be distributed across multiple workers and trained in parallel mode using a parameter server. Next, we will run this training on the cloud using the architecture shown in Figure 10-4.

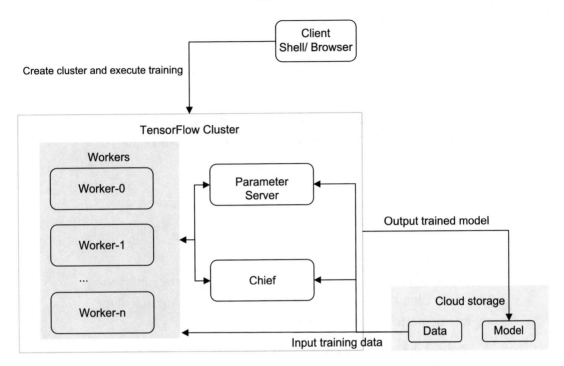

Figure 10-4. *TensorFlow cluster architecture with the chief, workers, and parameter server on the cloud VMs (the data and the model are on the scalable storage system)*

Steps for Running Distributed Training on the Cloud

We will deploy a TensorFlow cluster on the cloud, based on the architecture shown in Figure 10-4, and do the following steps to execute the training:

1. *Create a TensorFlow cluster.*

    a. *Parameter server, chief, and worker nodes*: All three cloud providers—AWS, GCP, Azure—provide a browser-based shell and graphical user interface (GUI) to create and manage virtual machines. We can create either GPU-based VMs or CPU-based VMs depending on the data size and the neural network's complexity.

2. *Install TensorFlow and all the prerequisite libraries on all VMs*: Review Chapter 1 for the instructions on installing prerequisites. For running the code in Listing 10-3, we will install TensorFlow only.

3. *Create the cloud storage directory (also called a bucket)*: Depending upon the cloud provider, we will create one of the following:

   - AWS S3 bucket

   - Google Cloud Storage (GCS) bucket

   - Azure container

4. *Upload the Python code and execute the training on each machine*: Using the Cloud Shell or any other SSH client, log in to each of the nodes and perform the following:

   a. Upload the Python package containing the dependencies and model training code (of Listing 10-3) to each of the nodes. Upload the code via scp or any other file transfer protocol. Since our code is committed in GitHub, we can clone the repository and download the code across all nodes.

   On each machine, clone the GitHub repository as shown in Listing 10-4.

Listing 10-4. Cloning the GitHub Repository

```
git clone https://github.com/ansarisam/dist-tf-modeling.git
```

   b. Set the machine role–specific TF_CONFIG environment variable on each machine and execute the Python code for distributed training, as shown in Listing 10-5.

Listing 10-5. Executing Distributed Training

```
export TF_CONFIG=$CONFIG;python distributed_training_ps.py --input_path
gs://cv_training_data --output_path gs://cv_distributed_model/output
```

It is not efficient to manually execute the command in Listing 10-5 on each of the nodes, especially when there is a large number of workers. We can write scripts to automate the launch of distributed training on a large cluster. The GitHub repository shown in Listing 10-4 has a Python script that can be used for automation. To understand how this works, we will follow the manual steps and launch the training on each VM one by one.

Distributed Training on Google Cloud

Google Cloud Platform (GCP) is a suite of cloud computing services that runs on the same infrastructure that Google uses internally for its end-user products, such as Google Search and YouTube.

We will use two GCP services for the purpose of running distributed training. These two services are Google Cloud Storage (GCS) for saving checkpoints and trained models and Compute Engine for virtual machines (VMs).

Let's get started!

Signing Up for GCP Access

If you already have a GCP account, skip this section. If not, create a GCP account at `https://cloud.google.com`. Google offers a $300 credit for education and learning. We will use this free account for our exercise in this section. You must enable billing for business and production deployment.

After creating an account, sign in to the Google Cloud Console, at `https://console.cloud.google.com`. A successful sign-in will take you to the GCP Dashboard, which looks like Figure 10-5.

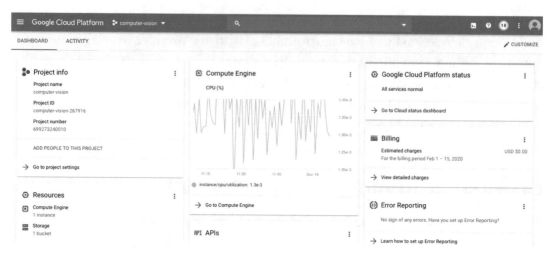

Figure 10-5. *Google Cloud Platform Dashboard*

Creating a Google Cloud Storage Bucket

GCS is a highly durable object storage on Google Cloud. It can scale to store exabytes of data. A GCS bucket is analogous to a directory in a file system. We can create the GCS bucket in one of the following two ways.

Creating the GCS Bucket from the Web UI

To create a bucket using the web UI, follow these steps:

1. Log in to the Google Cloud Console, at `https://cloud.google.com`. From the left-side navigation menu, click Storage and then Browser to launch the storage browser page (see Figure 10-6).

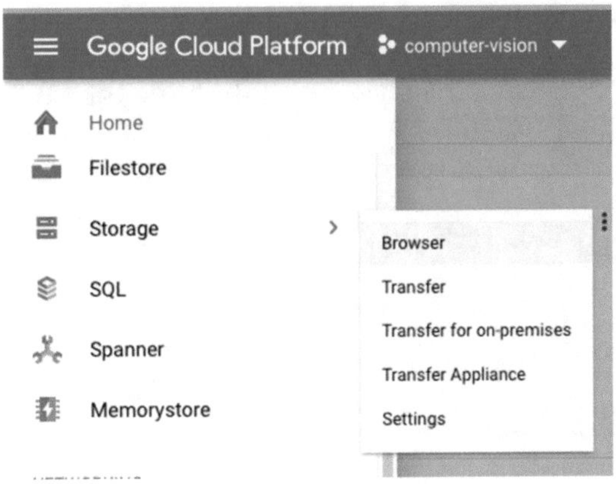

Figure 10-6. *Storage menu*

2. Click the Create Bucket button at the top of the page.

3. On the next page, fill in the bucket name (e.g., cv_model) and click Continue (see Figure 10-7).

4. Select Region for the location type, select the appropriate location such as "us-east4 (Northern Virginia)," and then click Continue.

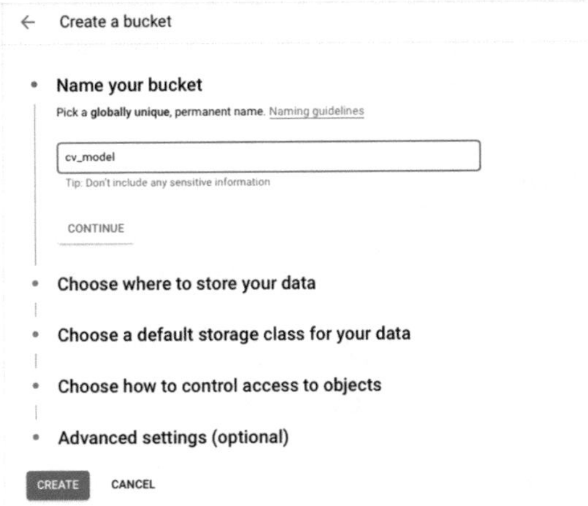

Figure 10-7. *Form to create a bucket*

5. Select Standard for the default storage class, and click Continue.

6. Select Uniform for the access control and then click Continue.

7. Click the Create button to create the bucket.

8. On the next page, click the Overview tab to see the bucket details (see Figure 10-8).

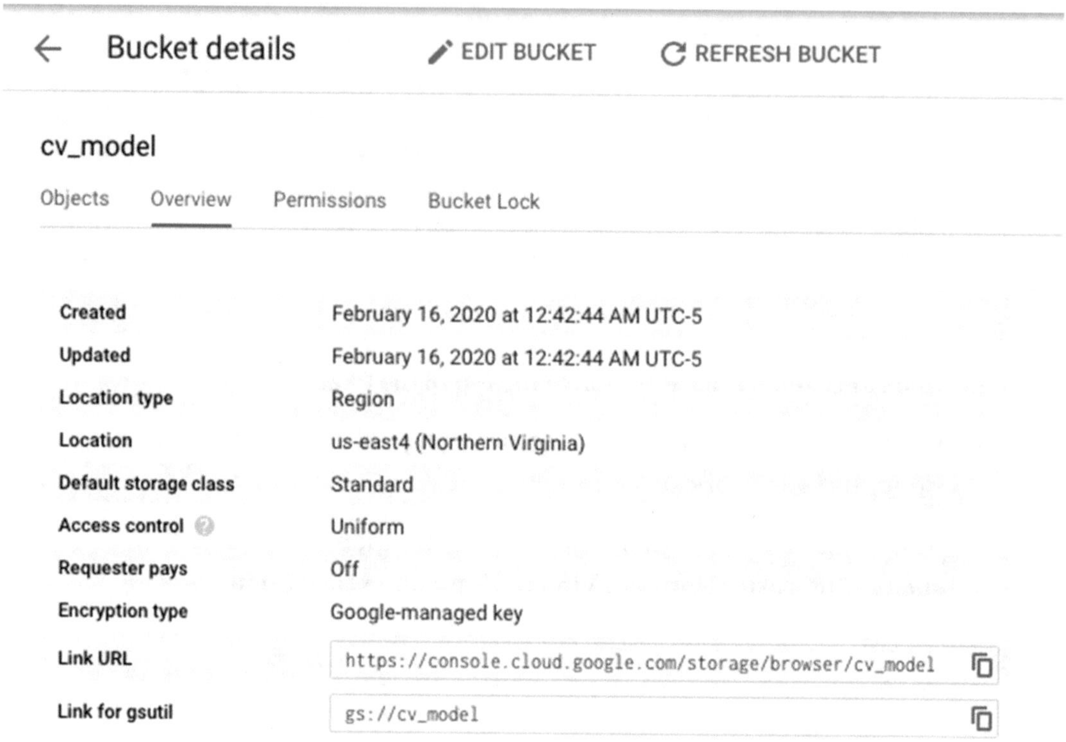

← **Bucket details**   ✎ EDIT BUCKET   ↻ REFRESH BUCKET

cv_model

Objects   Overview   Permissions   Bucket Lock

Created	February 16, 2020 at 12:42:44 AM UTC-5
Updated	February 16, 2020 at 12:42:44 AM UTC-5
Location type	Region
Location	us-east4 (Northern Virginia)
Default storage class	Standard
Access control ⓘ	Uniform
Requester pays	Off
Encryption type	Google-managed key
Link URL	`https://console.cloud.google.com/storage/browser/cv_model`
Link for gsutil	`gs://cv_model`

Figure 10-8. *Bucket detail page*

Creating the GCS Bucket from the Cloud Shell

If you have already created the bucket using the web UI, you do not need to follow these steps. It is easy to create the bucket using the command line.

1. Activate the Cloud Shell by clicking the 〉_ icon located in the top-right corner. The Cloud Shell will open at the bottom of the screen (within the same browser window).

2. Execute the command in Listing 10-6 in the Cloud Shell to create the bucket.

Listing 10-6. *gsutil Command to Create GCS Bucket*

```
gsutil mb -c regional -l us-east4 gs://cv_model
```

Provide the appropriate region and bucket name in the command in Listing 10-6. If you have created the bucket using the web UI, make sure to use a different bucket name.

gsutil is a Python application that lets us access cloud storage from the command line.

Figure 10-9 shows the gsutil command execution in the Cloud Shell.

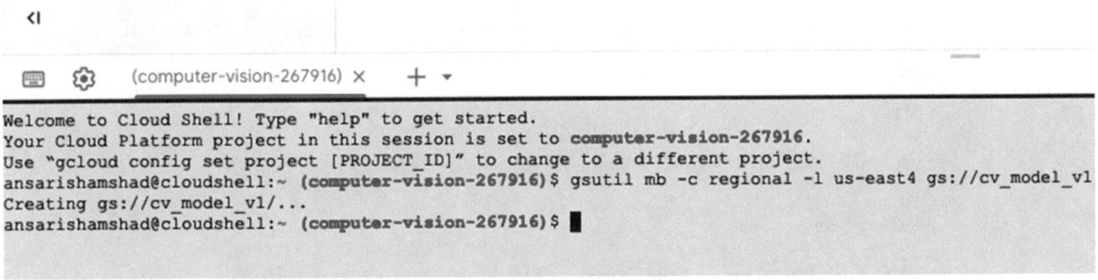

Figure 10-9. *gsutil command to create a bucket using the Cloud Shell*

Launching GCP Virtual Machines

We will launch the following types of virtual machines for our exercise:

- *One GPU-based VM*: Parameter server
- *One GPU-based VM*: Chief node
- *Two GPU-based VMs*: Worker nodes

The VMs will be launched in the same region where our GCS bucket is located (us-east4 in the previous example).

To launch the VMs, follow these steps:

1. In the main navigation menu, click Compute Engine and then VM Instances to launch the page that displays a list of the VMs previously launched.

2. Click Create to launch the web form that we need to fill in to create the instance. Figure 10-10 and Figure 10-11 show the instance creation form.

479

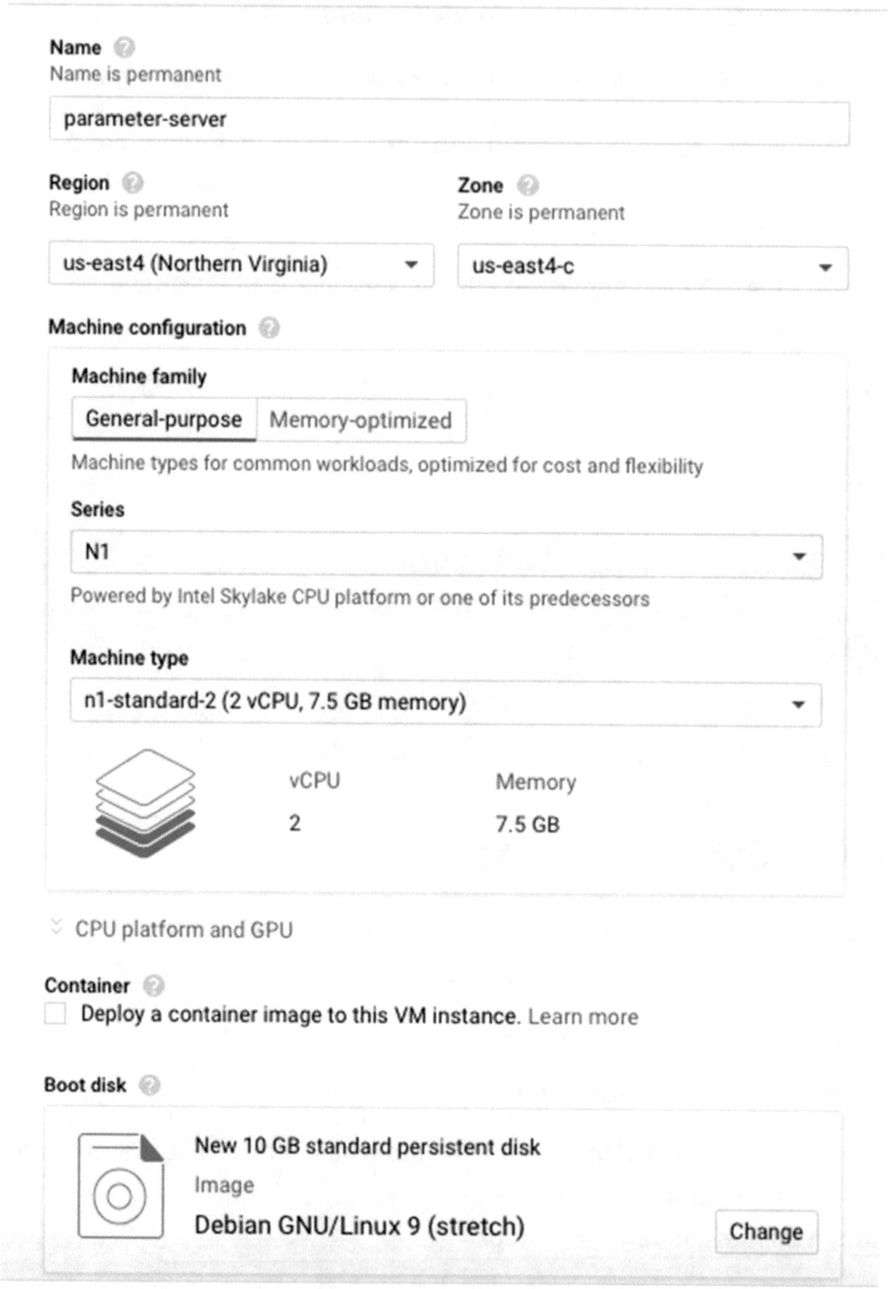

Figure 10-10. *Form (top portion) to provide information to create a VM*

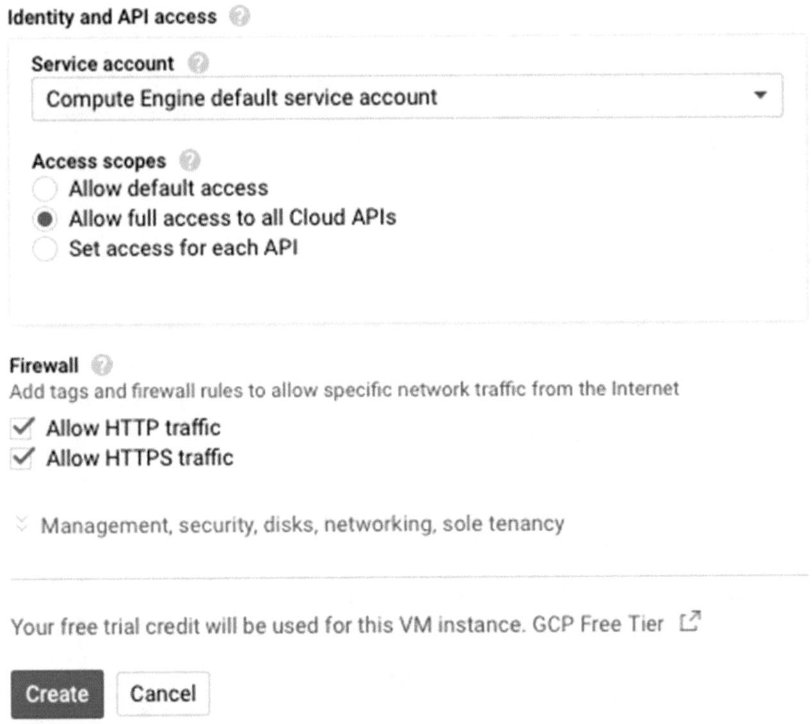

Identity and API access ⓘ

> **Service account** ⓘ
>
> Compute Engine default service account   ▾
>
> **Access scopes** ⓘ
> ○ Allow default access
> ● Allow full access to all Cloud APIs
> ○ Set access for each API

Firewall ⓘ
Add tags and firewall rules to allow specific network traffic from the Internet

☑ Allow HTTP traffic
☑ Allow HTTPS traffic

⌄ Management, security, disks, networking, sole tenancy

Your free trial credit will be used for this VM instance. GCP Free Tier ⤴

[Create] [Cancel]

Figure 10-11. *Bottom portion of the instance creation form*

3. We will create four GPU-based VMs to create the cluster. In the
    instance creation form, click the Change button next to the Image
    under Boot Disk (as shown in Figure 10-12).

Boot disk ⓘ

> New 10 GB standard persistent disk
> Image
> **Debian GNU/Linux 9 (stretch)**   [Change]

Figure 10-12. *Clicking the Change button to launch the Boot Disk selection page*

On the next screen (as shown in Figure 10-13), select Deep
Learning on Linux for the operating system and Deep Learning
Image: TensorFlow 1.15.0 m45 for the version.

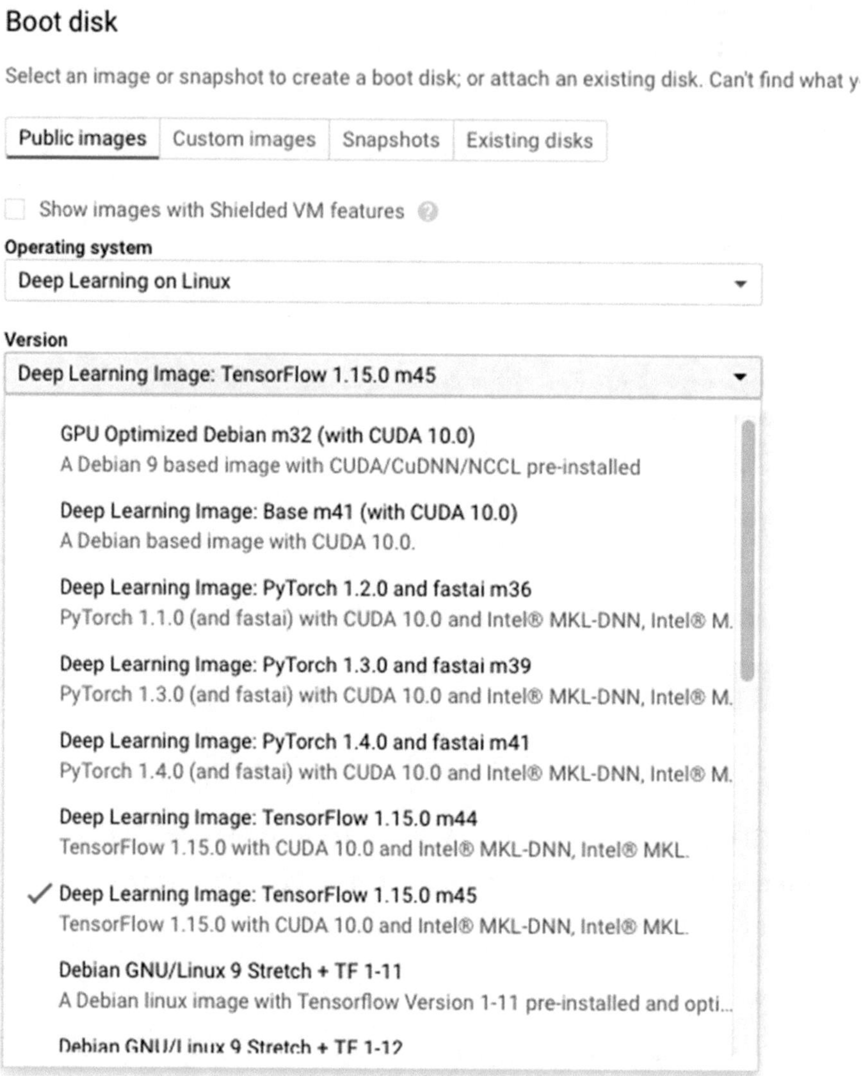

Boot disk

Select an image or snapshot to create a boot disk; or attach an existing disk. Can't find what y

| **Public images** | Custom images | Snapshots | Existing disks |

☐ Show images with Shielded VM features ❓

Operating system

Deep Learning on Linux   ▼

Version

Deep Learning Image: TensorFlow 1.15.0 m45   ▼

GPU Optimized Debian m32 (with CUDA 10.0)
A Debian 9 based image with CUDA/CuDNN/NCCL pre-installed

Deep Learning Image: Base m41 (with CUDA 10.0)
A Debian based image with CUDA 10.0.

Deep Learning Image: PyTorch 1.2.0 and fastai m36
PyTorch 1.1.0 (and fastai) with CUDA 10.0 and Intel® MKL-DNN, Intel® M.

Deep Learning Image: PyTorch 1.3.0 and fastai m39
PyTorch 1.3.0 (and fastai) with CUDA 10.0 and Intel® MKL-DNN, Intel® M.

Deep Learning Image: PyTorch 1.4.0 and fastai m41
PyTorch 1.4.0 (and fastai) with CUDA 10.0 and Intel® MKL-DNN, Intel® M.

Deep Learning Image: TensorFlow 1.15.0 m44
TensorFlow 1.15.0 with CUDA 10.0 and Intel® MKL-DNN, Intel® MKL.

✓ **Deep Learning Image: TensorFlow 1.15.0 m45**
TensorFlow 1.15.0 with CUDA 10.0 and Intel® MKL-DNN, Intel® MKL.

Debian GNU/Linux 9 Stretch + TF 1-11
A Debian linux image with Tensorflow Version 1-11 pre-installed and opti...

Debian GNU/Linux 9 Stretch + TF 1-12

Figure 10-13. CUDA 10–based Linux OS with preinstalled TensorFlow 1.15

4. Figure 10-14 shows the screen that lists all four VMs that we created.

Name ^	Zone	Recommendation	In use by	Internal IP	External IP	Connect
☑ chief	us-east4-c			10.150.0.4 (nic0)	35.230.179.56 ⬀	SSH ▾ ⋮
☑ parameter-server	us-east4-c			10.150.0.3 (nic0)	35.245.124.71 ⬀	SSH ▾ ⋮
☑ worker-0	us-east4-c			10.150.0.5 (nic0)	35.221.12.68 ⬀	SSH ▾ ⋮
☑ worker-1	us-east4-c			10.150.0.6 (nic0)	35.245.253.142 ⬀	SSH ▾ ⋮

VM instances   🗗 CREATE INSTANCE   ⬆ IMPORT VM   ⟳ REFRESH   ▶ START   ◼ S

Figure 10-14. *List of all VMs created*

SSH to Log In to Each VMs

We will use the Cloud Shell and gsutil to log in to all four VMs created in the previous section. Activate the Cloud Shell and click the + icon (marked with a red rectangle in Figure 10-15).

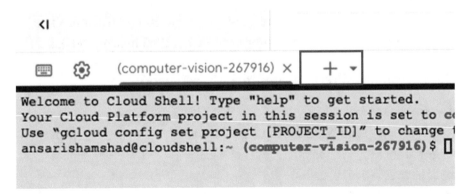

Figure 10-15. *Creating multiple tabs of the Cloud Shell by clicking the + icon*

To log in via SSH, execute the commands (in each of the four Cloud Shell tabs) shown in Listing 10-7.

Listing 10-7. SSH to Log In to All 4VMs Using Cloud Shell

```
SSH to parameter server    gcloud compute ssh parameter-server
SSH to chief               gcloud compute ssh chief
SSH to worker-0            gcloud compute ssh worker-0
SSH to worker-1            gcloud compute ssh worker-1
```

Uploading the Code for Distributed Training or Cloning the GitHub Repository

While logged in via SSH, execute the following command to clone the GitHub repository that contains the distributed model training code (as shown in Listing 10-8). This needs to be done on all machines.

Listing 10-8. Command to Clone the GitHub Repository

```
git clone https://github.com/ansarisam/dist-tf-modeling.git
```

If the git command does not work, install git using the command sudo apt-get install git.

Installing Prerequisites and TensorFlow

The image "Deep Learning on Linux" has all the prerequisites and TensorFlow preinstalled. However, if we want to configure our environment, execute all the commands of Listing 10-9 (review Chapter 1 for the detailed instructions).

Listing 10-9. Installing Prerequisites Including TensorFlow

```
sudo apt-get update
sudo apt-get -y upgrade && sudo apt-get install -y python-pip python-dev
sudo apt-get install python3-dev python3-pip
sudo pip3 install -U virtualenv
mkdir cv
```

```
virtualenv --system-site-packages -p python3 ./cv
source ./cv/bin/activate
pip install tensorflow==1.15
```

Running Distributed Training

Make sure you have cloned the GitHub repository (as shown in Listing 10-8) on all the machines. Also, ensure you are logged in to each of the VMs via SSH (using the Cloud Shell). Execute the following commands on each of the VMs to launch the distributed training.

Here's the command for the parameter server:

```
cd dist_tf_modeling
export TF_CONFIG='{"task": {"index": 0, "type": "ps"},
"cluster": {"chief":["chief:8900"],"worker": ["worker-0:8900",
"worker-1:8900"],  "ps":["parameter-server:8900"]}}';python distributed_
training_ps.py --output_path gs://cv_model_v1
```

Here's the command for the chief node:

```
cd dist_tf_modeling
export TF_CONFIG='{"task": {"index": 0, "type": "chief"},
"cluster": {"chief":["chief:8900"],"worker": ["worker-0:8900",
"worker-1:8900"],  "ps":["parameter-server:8900"]}}';python distributed_
training_ps.py --output_path gs://cv_model_v1
```

Here's the command for the worker-0 node:

```
cd dist_tf_modeling
export TF_CONFIG='{"task": {"index": 0, "type": "worker"},
"cluster": {"chief":["chief:8900"],"worker": ["worker-0:8900",
"worker-1:8900"],  "ps":["parameter-server:8900"]}}';python distributed_
training_ps.py --output_path gs://cv_model_v1
```

Here's the command for the worker-1 node:

```
cd dist_tf_modeling
export TF_CONFIG='{"task": {"index": 1, "type": "worker"},
"cluster": {"chief":["chief:8900"],"worker": ["worker-0:8900",
"worker-1:8900"],  "ps":["parameter-server:8900"]}}';python distributed_
training_ps.py --output_path gs://cv_model_v1
```

Note that all participating nodes must be able to communicate with the parameter servers via the port configured in TF_CONFIG. Also, the nodes must have the necessary read and write permissions to the GCS bucket.

The model checkpoints are saved in GCS at the path gs://cv_model_v1. The trained model is saved as model.h5 in gs://cv_model_v1.

GCP instances with GPUs are expensive. To avoid any charges, you should terminate them if they are no longer used.

Distributed Training on Azure

Microsoft Azure is a cloud computing service used for building, testing, deploying, and managing applications and services through Microsoft-managed data centers.

The distributed training with ParameterServerStrategy in Listing 10-3 will also work on Azure in almost the same way it worked on GCP. The difference between GCP and Azure is the way we create VMs nodes. Instead of repeating the process of distributing the parameter server–based training on an Azure cluster, we will explore a different strategy for distributed training.

We will distribute the training using MirroredStrategy on a single node that has multiple GPUs. In this section, you will learn the following:

- How to create a multi-GPU-based VM on Azure using the web interface

- How to set up TensorFlow to run on GPUs

- What changes are needed to make the code in Listing 10-3 work on multiple GPUs

- How to execute the training and monitor it

Note that the GPU support for TensorFlow is available for Ubuntu and Windows with CUDA-enabled cards. In this exercise, we will create an Ubuntu 20.4–based VM with two GPUs.

Creating a VM with Multiple GPUs on Azure

You need to first sign up at `https://azure.microsoft.com/` to create a free account. Then go to `https://portal.azure.com/` and log in to your account. The free account allows you to create a VM with only one GPU. To create a VM with multiple GPUs, you must activate billing. To activate it, follow these instructions:

1. Click the main navigation (expand the burger icon located in the top-left corner).

2. Select Cost Management + Billing and click Azure Subscription.

3. Click Add.

4. Follow the on-screen instructions.

To create the virtual machine, do the following:

1. On the home page, click the icon for Virtual Machines.

2. Click the button Create Virtual Machine located at the bottom of the page or click the + Add icon located in the top-left corner.

3. Fill in the form to configure the VM. Figure 10-16 shows the top portion of the basic configuration. For the Image field, select Ubuntu Server 20.04 LTS.

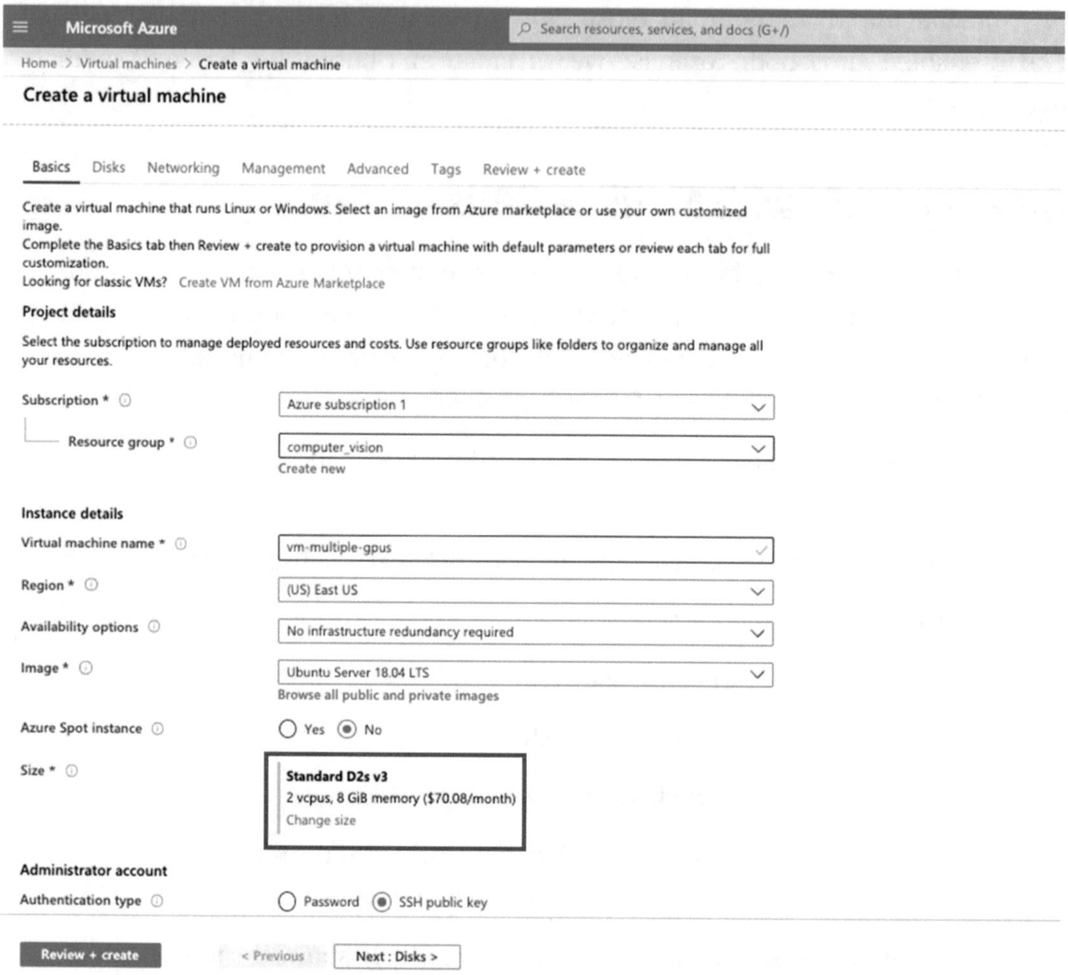

Figure 10-16. *Azure configuration page to create a VM*

4. We will add GPUs to the VM. Click the link Change Size, which is shown enclosed within the red rectangle in Figure 10-16. This will launch the page that shows a list of all the available devices within the region that you selected for the Region field in Figure 10-16.

    As shown in Figure 10-17, first clear all the filters and search for *NC* to find the NC series of GPUs. We will select the NC12_Promo VM size, which gives us two GPUs, 12 vCPUs, and 112GB of memory. Highlight the row corresponding to the size NC12_Promo and click the Select button located at the bottom of the screen.

Visit https://docs.microsoft.com/en-us/azure/virtual-
machines/linux/sizes-gpu for more information about other
VM sizes.

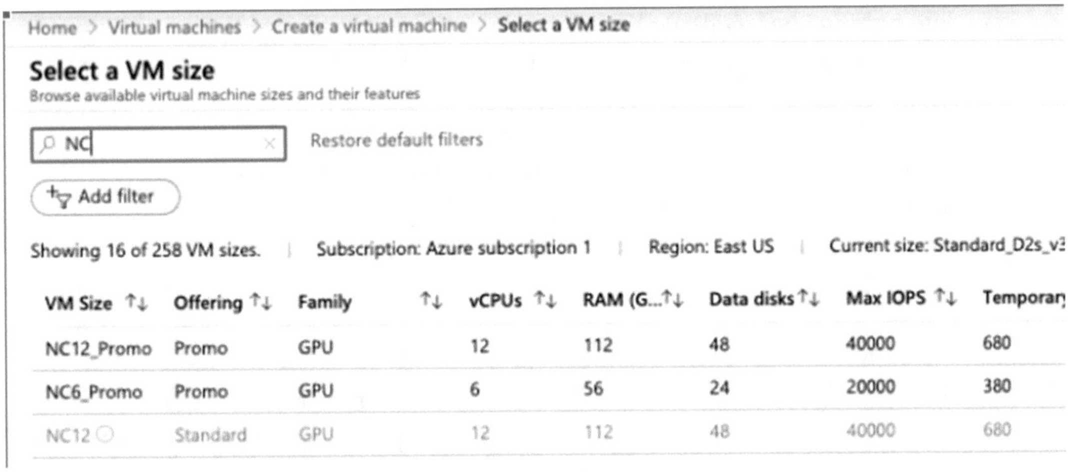

Figure 10-17. *Device size (GPU) selection screen*

If the row corresponding to the GPU you want to use is grayed out,
that means either you have not upgraded your subscription or you
do not have sufficient quota to use that VM.

You can ask Microsoft to increase your quota. Visit https://
docs.microsoft.com/en-us/azure/azure-resource-manager/
templates/error-resource-quota for more information on how
to request a quota increase.

On the basic configuration screen (Figure 10-16, bottom), you can
select either of the following (depending on your security policy)
for the authentication type:

- *SSH public key*: Paste the SSH public key that you will use to
  access this VM.

- *Password*: Create a username and password that you will need to
  supply while connecting via SSH. We will use this option for our
  exercise.

5. Leave everything else as the default and click the Review +
   Create button at the bottom-left corner of the screen. On the next
   page, we will review our configuration to make sure everything
   is selected correctly and then finally click the Create button. If
   everything goes well, the VM with two GPUs will be created. It
   may take a few minutes for our VM to be ready.

   In this case, we did not create any disk because the VM comes
   with a large enough disk size to run our training. This is not
   a persistent disk and will be deleted if the VM is terminated.
   Therefore, in production, you must add a persistent disk to avoid
   losing the data.

6. After our VM is ready, we will see an alert indicating that the VM
   is ready to use, if we have not left the page we were last on. We can
   also go back to the home page and click the Virtual Machines icon
   to see a list of VMs we have created. Click the VM name to open
   the details page, as shown in Figure 10-18.

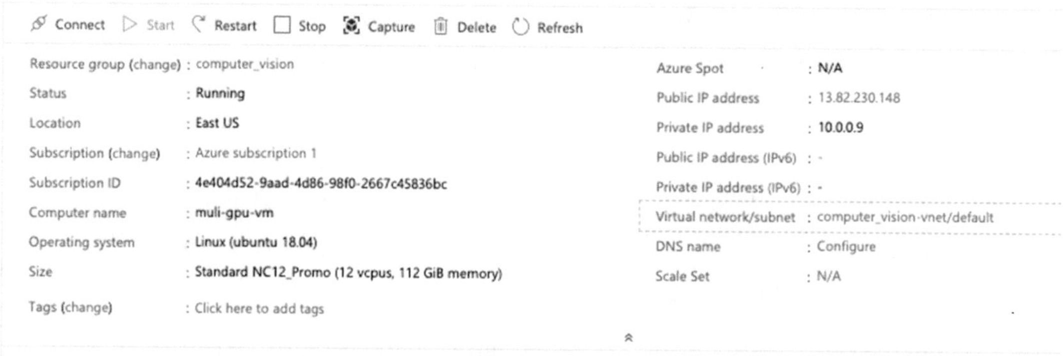

Figure 10-18. *VM detail page showing the public IP address*

7. Note the public IP address or copy it, as you will need it to SSH
   to your VM. Using an SSH client, such as Putty for Windows
   or the Shell terminal in Mac or Linux, log on to the VM using
   the authentication method you selected before. Here are the
   commands to SSH via the two methods of authentication:

- Password-based authentication:

  ```
  $ ssh username@13.82.230.148
  username@13.82.230.148's password:
  ```

- SSH public key–based authentication:

  ```
  $ ssh -i ~/sshkey.pem 13.82.230.148
  ```

If successfully authenticated, you will be logged in to the VM.

Installing GPU Drivers and Libraries

To run TensorFlow on a GPU-based machine, we need to install the GPU driver and a few libraries. Perform the following steps:

1. Execute all the commands of Listing 10-10 on the terminal (make sure you are logged in via SSH).

Listing 10-10. Commands to Add NVIDIA Package Repositories

```
# Add NVIDIA package repositories
$ wget https://developer.download.nvidia.com/compute/cuda/repos/ubuntu1804/
x86_64/cuda-repo-ubuntu1804_10.1.243-1_amd64.deb
$ sudo dpkg -i cuda-repo-ubuntu1804_10.1.243-1_amd64.deb
sudo apt-key adv --fetch-keys $ https://developer.download.nvidia.com/
compute/cuda/repos/ubuntu1804/x86_64/7fa2af80.pub
$ sudo apt-get update
$ wget http://developer.download.nvidia.com/compute/machine-learning/repos/
ubuntu1804/x86_64/nvidia-machine-learning-repo-ubuntu1804_1.0.0-1_amd64.deb
$ sudo apt install ./nvidia-machine-learning-repo-ubuntu1804_1.0.0-1_
amd64.deb
sudo apt-get update
```

2. If the NVIDIA package repositories are successfully added, install the NVIDIA driver using the command from Listing 10-11.

Listing 10-11. Installing the NVIDIA Driver

```
$ sudo apt-get install --no-install-recommends nvidia-driver-418
```

3. You will need to reboot the VM for the previous installation to take effect. On the SSH terminal shell, execute the command sudo reboot.

4. SSH to the VM again.

5. To test whether the NVIDIA driver was successfully installed, execute the following command:

```
$ nvidia-smi
```

This command should display something like Figure 10-19.

```
ansarisam@muli-gpu-vm:~$ nvidia-smi
Wed Feb 19 08:03:17 2020
+-----------------------------------------------------------------------------+
| NVIDIA-SMI 430.50       Driver Version: 430.50       CUDA Version: 10.1     |
|-------------------------------+----------------------+----------------------+
| GPU  Name        Persistence-M| Bus-Id        Disp.A | Volatile Uncorr. ECC |
| Fan  Temp  Perf  Pwr:Usage/Cap|         Memory-Usage | GPU-Util  Compute M. |
|===============================+======================+======================|
|   0  Tesla K80           Off  | 0000C894:00:00.0 Off |                    0 |
| N/A   49C    P0    59W / 149W |      0MiB / 11441MiB |      0%      Default |
+-------------------------------+----------------------+----------------------+
|   1  Tesla K80           Off  | 0000ED6E:00:00.0 Off |                    0 |
| N/A   40C    P0    71W / 149W |      0MiB / 11441MiB |     97%      Default |
+-------------------------------+----------------------+----------------------+

+-----------------------------------------------------------------------------+
| Processes:                                                       GPU Memory |
|  GPU       PID   Type   Process name                             Usage      |
|=============================================================================|
|  No running processes found                                                 |
+-----------------------------------------------------------------------------+
```

Figure 10-19. *Output of the command* nvidia-smi

6.   We will now install the development and runtime libraries
     (Listing 10-12). This will be around 4GB in size.

Listing 10-12. Installing Development and Runtime Libraries

```
$ sudo apt-get install --no-install-recommends \
    cuda-10-1 \
    libcudnn7=7.6.4.38-1+cuda10.1  \
    libcudnn7-dev=7.6.4.38-1+cuda10.1
```

7.   Install the TensorRT library as shown in Listing 10-13.

Listing 10-13. Installing TensorRT

```
$ sudo apt-get install -y --no-install-recommends
libnvinfer6=6.0.1-1+cuda10.1 \
    libnvinfer-dev=6.0.1-1+cuda10.1 \
    libnvinfer-plugin6=6.0.1-1+cuda10.1
```

Creating Virtual Environment and Installing TensorFlow

Follow the instructions provided in Chapter 1 to install all the libraries and
dependencies you will need. We will execute the commands in Listing 10-14 to install all
the prerequisites that we need for our current exercise.

Listing 10-14. Installing Python, Creating Virtualenv, and Installing TensorFlow

```
$ sudo apt update
$ sudo apt-get install python3-dev python3-pip
$ sudo pip3 install -U virtualenv
$ mkdir cv
$ virtualenv --system-site-packages -p python3 ./cv
$ source ./cv/bin/activate
(cv) $ pip install  tensorflow
(cv) $ pip install tensorflow-gpu
```

Implementing MirroredStrategy

Refer to line 9 of Listing 10-3. Instead of instantiating `ParameterServerStrategy`, we will create an instance of `MirroredStrategy`, as shown here:

```
strategy = tf.distribute.MirroredStrategy()
```

All the other lines of Listing 10-3 will remain the same.

We have committed to the GitHub repository the modified code that has the implementation of `MirroredStrategy` for distributed training. The GitHub repository location is `https://github.com/ansarisam/dist-tf-modeling.git`, and the file name containing the `MirroredStrategy` code is `mirrored_strategy.py`.

Running Distributed Training

Log on via SSH to the VM we created earlier. Then clone the GitHub repository, as shown in Listing 10-15.

Listing 10-15. Cloning GitHub Repository

```
$ git clone https://github.com/ansarisam/dist-tf-modeling.git
```

Execute the Python code shown in Listing 10-16 to train the distributed model.

Listing 10-16. Executing the MirroredStrategy-Based Distributed Model

```
$ python dist-tf-modeling/mirrored_strategy.py
```

If everything goes well, you will see the training progress printed on the terminal console. Figure 10-20 shows some sample output.

```
Train for 313 steps
Epoch 1/100
313/313 [==============================] - 3s 9ms/step - loss: 0.5083 - accuracy: 0.8545
Epoch 2/100
313/313 [==============================] - 0s 1ms/step - loss: 0.2110 - accuracy: 0.9370
Epoch 3/100
313/313 [==============================] - 0s 1ms/step - loss: 0.1392 - accuracy: 0.9579
Epoch 4/100
313/313 [==============================] - 0s 1ms/step - loss: 0.0965 - accuracy: 0.9724
Epoch 5/100
313/313 [==============================] - 0s 1ms/step - loss: 0.0681 - accuracy: 0.9817
Epoch 6/100

Epoch 94/100
313/313 [==============================] - 0s 1ms/step - loss: 5.4511e-09 - accuracy: 1.0000
Epoch 95/100
313/313 [==============================] - 0s 1ms/step - loss: 4.9393e-09 - accuracy: 1.0000
Epoch 96/100
313/313 [==============================] - 0s 1ms/step - loss: 4.4751e-09 - accuracy: 1.0000
Epoch 97/100
313/313 [==============================] - 0s 1ms/step - loss: 4.0348e-09 - accuracy: 1.0000
Epoch 98/100
313/313 [==============================] - 0s 1ms/step - loss: 3.6301e-09 - accuracy: 1.0000
Epoch 99/100
313/313 [==============================] - 0s 1ms/step - loss: 3.3683e-09 - accuracy: 1.0000
Epoch 100/100
313/313 [==============================] - 0s 1ms/step - loss: 3.2373e-09 - accuracy: 1.0000
10000/10000 [==============================] - 1s 135us/sample - loss: 4.8399e-09 - accuracy: 1.0000
Evaluation [4.3511385577232885e-09, 1.0]
Predicted [[2.00329108e-27 4.47499693e-28 3.57979841e-23 ... 1.00000000e+00
  2.54690850e-28 5.45738153e-28]
 [1.17271654e-26 7.68435632e-26 1.00000000e+00 ... 0.00000000e+00
  4.65196148e-22 0.00000000e+00]
 [1.35216691e-22 1.00000000e+00 6.96082825e-18 ... 4.59888577e-15
  3.85402886e-19 3.36883012e-22]
 ...
 [1.42788530e-28 3.89840506e-33 4.69148616e-38 ... 1.98623204e-26
  2.19390131e-25 2.64972213e-19]
 [1.18806643e-26 1.40197781e-28 1.63681088e-27 ... 1.11720601e-25
  6.12110026e-13 8.08174249e-37]
 [1.77768293e-28 1.78035542e-37 2.56305317e-32 ... 0.00000000e+00
  3.01141678e-35 0.00000000e+00]]
```

Figure 10-20. *Sample training progress and evaluation outputs*

To check whether the GPUs are being utilized for the distributed training, SSH to the VM from a different terminal and execute the command shown in Listing 10-17.

Listing 10-17. Checking the GPU Status

```
$ nvidia-smi
```

Figures 10-21 and 10-22 show the outputs of this command.

```
Wed Feb 19 00:25:15 2020
+-----------------------------------------------------------------------------+
| NVIDIA-SMI 430.50       Driver Version: 430.50       CUDA Version: 10.1     |
|-------------------------------+----------------------+----------------------+
| GPU  Name        Persistence-M| Bus-Id        Disp.A | Volatile Uncorr. ECC |
| Fan  Temp  Perf  Pwr:Usage/Cap|         Memory-Usage | GPU-Util  Compute M. |
|===============================+======================+======================|
|   0  Tesla K80           Off  | 0000B4A0:00:00.0 Off |                    0 |
| N/A   47C    P0    59W / 149W |      0MiB / 11441MiB |      0%      Default |
+-------------------------------+----------------------+----------------------+
|   1  Tesla K80           Off  | 0000E42F:00:00.0 Off |                    0 |
| N/A   38C    P0    71W / 149W |      0MiB / 11441MiB |      1%      Default |
+-------------------------------+----------------------+----------------------+

+-----------------------------------------------------------------------------+
| Processes:                                                       GPU Memory |
|  GPU       PID   Type   Process name                             Usage      |
|=============================================================================|
|  No running processes found                                                 |
+-----------------------------------------------------------------------------+
```

Figure 10-21. *GPU status before the training starts*

```
[ansarisam@cv:~$ nvidia-smi
Wed Feb 19 00:35:30 2020
+-----------------------------------------------------------------------------+
| NVIDIA-SMI 430.50       Driver Version: 430.50       CUDA Version: 10.1     |
|-------------------------------+----------------------+----------------------+
| GPU  Name        Persistence-M| Bus-Id        Disp.A | Volatile Uncorr. ECC |
| Fan  Temp  Perf  Pwr:Usage/Cap|         Memory-Usage | GPU-Util  Compute M. |
|===============================+======================+======================|
|   0  Tesla K80           Off  | 0000B4A0:00:00.0 Off |                    0 |
| N/A   56C    P0    60W / 149W |     67MiB / 11441MiB |      0%      Default |
+-------------------------------+----------------------+----------------------+
|   1  Tesla K80           Off  | 0000E42F:00:00.0 Off |                    0 |
| N/A   44C    P0    71W / 149W |     67MiB / 11441MiB |      0%      Default |
+-------------------------------+----------------------+----------------------+

+-----------------------------------------------------------------------------+
| Processes:                                                       GPU Memory |
|  GPU       PID   Type   Process name                             Usage      |
|=============================================================================|
|   0      7387     C    python                                        56MiB |
|   1      7387     C    python                                        56MiB |
+-----------------------------------------------------------------------------+
```

Figure 10-22. *GPU status while training is in progress*

If you no longer need the VM, you should terminate it to avoid any costs, as these GPU-based VMs are very expensive. Before terminating the VM, make sure you download and store the trained model and checkpoints to permanent storage.

Distributed Training on AWS

Amazon Web Services (AWS) is a subsidiary of Amazon that provides on-demand cloud computing platforms and APIs to individuals, companies, and governments, on a metered pay-as-you-go basis. In this section, we will explore how to train a distributed model on AWS.

The distributed training of Listing 10-3 will also work on AWS. All we need to do is to create VMs and follow the steps that we did for training the model of GCP.

Similarly, we can train the `MirroredStrategy`-based model on AWS VMs that have multiple GPUs. All the instructions for training on Azure will be the same for AWS, except the method of creating multi-GPU-based VMs.

Here we will explore yet another technique for training a scalable model on the cloud. We will use Horovod to distribute the training on AWS. Let's first understand what the Horovod framework is and how to use it in distributed model training.

Horovod

The official document describes Horovod as a distributed deep learning training framework for TensorFlow, Keras, PyTorch, and Apache MXNet. Horovod aims to make distributed deep learning fast and easy to use. Horovod was developed at Uber and is hosted by Linux Foundation AI.

The source code with documentation is maintained at the GitHub repository at `https://github.com/horovod/horovod`. The official documentation is at `https://horovod.readthedocs.io/en/latest/summary_include.html`.

To use Horovod, we will need to make a few minor changes in the TensorFlow code for model training. We will use the same example code from Listing 5-2 and make changes to make it Horovod compatible.

How to Use Horovod

When we define a neural network, we specify the optimization algorithm, such as AdaGrad, that we want our network to use to optimize the gradients. In distributed learning, the gradients are calculated in multiple nodes, averaged using an all-reduce or all-gather algorithm, and further optimized using the optimization algorithm. Horovod provides a wrapper function to distribute the optimization to all participating nodes and delegates the gradient optimization task to the original optimization algorithm that we wrap in Horovod.

We will use Horovod with TensorFlow to distribute the model training to multiple nodes, each having one or more GPUs. We will work on the same code example from Listing 5-2, make a few minor changes to it to make it Horovod compatible, and execute the training on AWS. To use Horovod, we need to make the following changes in the code of Listing 5-2:

1. Import `horovod.tensorflow` as `hvd`.

2. Initialize Horovod using `hvd.init()`.

3. Pin the GPU that will process gradients (one GPU per process) using this:

   ```
   config = tf.ConfigProto()
   config.gpu_options.visible_device_list = str(hvd.local_rank())
   ```

4. Build the model as we normally do in TensorFlow. Define the loss function.

5. Define the TensorFlow optimization function, as follows:

   ```
   opt = tf.train.AdagradOptimizer(0.01 * hvd.size())
   ```

6. Call the Horovod distributed optimization function and pass the original TensorFlow optimizer from step 5. This is the core of Horovod.

   ```
   opt = hvd.DistributedOptimizer(opt)
   ```

7. Create a Horovod hook to broadcast training variables to all processors:

   ```
   hooks = [hvd.BroadcastGlobalVariablesHook(0)]
   ```

0 means all processors with rank zero (e.g., the first GPU) to all
processors.

8. Finally, train the model using this:

train_op = opt.minimize(loss)

Let's put all these steps together and convert our code from Listing 5-2 into Horovod-
compatible code that can be distributed across multiple nodes with multiple GPUs.
Listing 10-18 shows the complete code that we can execute on a Horovod cluster with
TensorFlow as an execution engine. Code taken from the examples directory of the official
source code of Horovod is maintained at https://github.com/horovod/horovod.git.

Listing 10-18. Distributed Training with Horovod

```
File name: horovod_tensorflow_mnist.py
1    import tensorflow as tf
2    import horovod.tensorflow.keras as hvd
3
4    # Horovod: initialize Horovod.
5    hvd.init()
6
7    # Horovod: pin GPU to be used to process local rank (one GPU per process)
8    gpus = tf.config.experimental.list_physical_devices('GPU')
9    for gpu in gpus:
10       tf.config.experimental.set_memory_growth(gpu, True)
11   if gpus:
12       tf.config.experimental.set_visible_devices(gpus[hvd.local_
         rank()], 'GPU')
13
14   # Load MNIST data using built-in datasets download function
15   mnist = tf.keras.datasets.mnist
16   (x_train, y_train), (x_test, y_test) = mnist.load_data()
17
18   # Normalize the pixel values by dividing each pixel by 255
19   x_train, x_test = x_train / 255.0, x_test / 255.0
20
21   BUFFER_SIZE = len(x_train)
```

```
22   BATCH_SIZE_PER_REPLICA = 16
23   GLOBAL_BATCH_SIZE = BATCH_SIZE_PER_REPLICA * 2
24   EPOCHS = 100
25   STEPS_PER_EPOCH = int(BUFFER_SIZE/EPOCHS)
26
27   train_dataset = tf.data.Dataset.from_tensor_slices((x_train, y_
     train)).repeat().shuffle(BUFFER_SIZE).batch(GLOBAL_BATCH_SIZE,drop_
     remainder=True)
28   test_dataset = tf.data.Dataset.from_tensor_slices((x_test, y_test)).
     batch(GLOBAL_BATCH_SIZE)
29
30
31   mnist_model = tf.keras.Sequential([
32       tf.keras.layers.Conv2D(32, [3, 3], activation='relu'),
33       tf.keras.layers.Conv2D(64, [3, 3], activation='relu'),
34       tf.keras.layers.MaxPooling2D(pool_size=(2, 2)),
35       tf.keras.layers.Dropout(0.25),
36       tf.keras.layers.Flatten(),
37       tf.keras.layers.Dense(128, activation='relu'),
38       tf.keras.layers.Dropout(0.5),
39       tf.keras.layers.Dense(10, activation='softmax')
40   ])
41
42   # Horovod: adjust learning rate based on number of GPUs.
43   opt = tf.optimizers.Adam(0.001 * hvd.size())
44
45   # Horovod: add Horovod DistributedOptimizer.
46   opt = hvd.DistributedOptimizer(opt)
47
48   # Horovod: Specify `experimental_run_tf_function=False` to ensure
     TensorFlow
49   # uses hvd.DistributedOptimizer() to compute gradients.
50   mnist_model.compile(loss=tf.losses.SparseCategoricalCrossentropy(),
51                       optimizer=opt,
52                       metrics=['accuracy'],
```

```
53                        experimental_run_tf_function=False)
54
55   callbacks = [
56       # Horovod: broadcast initial variable states from rank 0 to all
         other processes.
57       # This is necessary to ensure consistent initialization of all
         workers when
58       # training is started with random weights or restored from a
         checkpoint.
59       hvd.callbacks.BroadcastGlobalVariablesCallback(0),
60
61       # Horovod: average metrics among workers at the end of
         every epoch.
62       #
63       # Note: This callback must be in the list before the
         ReduceLROnPlateau,
64       # TensorBoard or other metrics-based callbacks.
65       hvd.callbacks.MetricAverageCallback(),
66
67       # Horovod: using `lr = 1.0 * hvd.size()` from the very beginning
         leads to worse final
68       # accuracy. Scale the learning rate `lr = 1.0` ---> `lr = 1.0 *
         hvd.size()` during
69       # the first three epochs. See https://arxiv.org/abs/1706.02677 for
         details.
70       hvd.callbacks.LearningRateWarmupCallback(warmup_epochs=3, verbose=1),
71   ]
72
73   # Horovod: save checkpoints only on worker 0 to prevent other workers
     from corrupting them.
74   if hvd.rank() == 0:
75       callbacks.append(tf.keras.callbacks.ModelCheckpoint('./checkpoint-
         {epoch}.h5'))
76
77   # Horovod: write logs on worker 0.
```

```
78   verbose = 1 if hvd.rank() == 0 else 0
79
80   # Train the model.
81   # Horovod: adjust number of steps based on number of GPUs.
82   mnist_model.fit(train_dataset, steps_per_epoch=500 // hvd.size(),
     callbacks=callbacks, epochs=24, verbose=verbose)
```

The code sections that use the Horovod APIs are marked in the comments with the label Horovod:. The code is properly commented to help you understand how to use Horovod. All other lines of code were already explained in Chapter 5.

Creating a Horovod Cluster on AWS

You must have an AWS account and be able to log in to your AWS web console. If you do not have an account, create one at https://aws.amazon.com. AWS offers certain types of resources for free for a year. But the types of resources that we need in order to train our model on a Horovod cluster may require you to enable billing. Your account may be charged for the resources you will use to run the distributed training. You may also need to request to increase quotas for certain resources such as vCPU and GPUs. The instructions to increase quotas are available at https://aws.amazon.com/about-aws/whats-new/2019/06/introducing-service-quotas-view-and-manage-quotas-for-aws-services-from-one-location/.

Horovod Cluster

AWS provides a convenient way to create a massively scalable Horovod cluster with just a few clicks. For the purpose of our exercise in this section, we will create a cluster of two nodes, each having only one GPU. Perform the following:

1. Log on to your AWS account to access the AWS management console; see https://console.aws.amazon.com.

2. Click Services, then EC2, then Instances, and then Launch Instance (as shown in Figure 10-23).

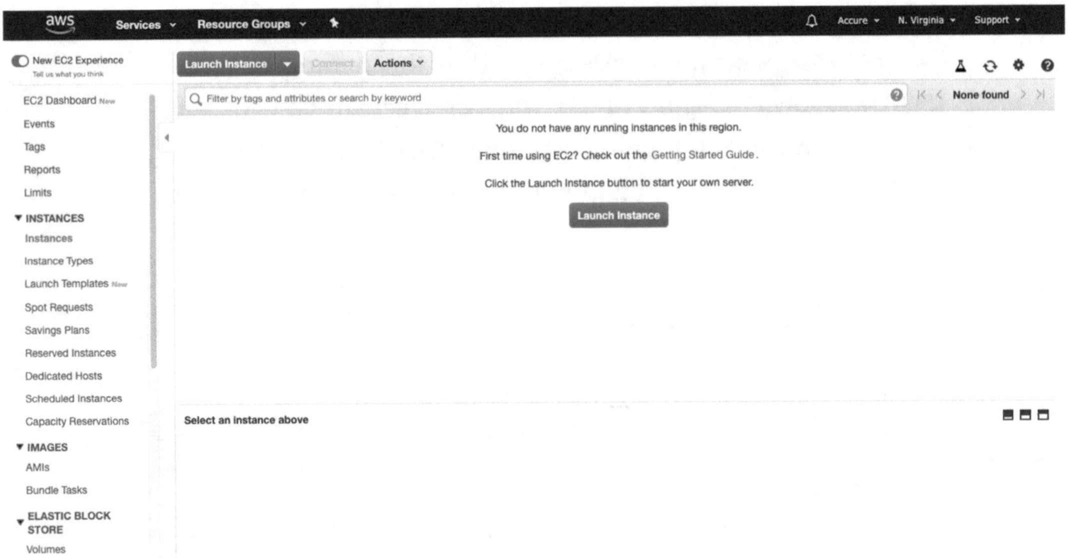

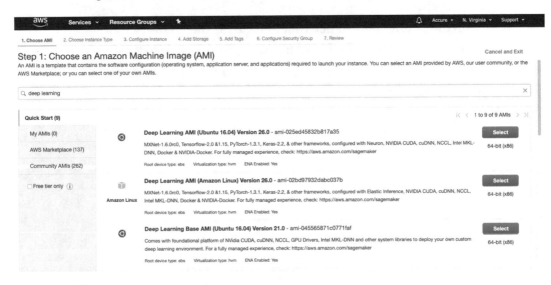

Figure 10-23. *AWS instance launch screen*

3. On the next screen, search for *deep learning* and select
    "Deep Learning AMI (Amazon Linux) Version 26.0 -
    ami-02bd97932dabc037b)" from the list of Amazon machine
    images (AMIs). See Figure 10-24.

Figure 10-24. *AMI selection screen*

4. On the Choose an Instance Type page, select the GPU instances,
    type *g2.2xlarge*, set the vCPUs to 8, and set the memory to
    15GB (as shown in Figure 10-25). You can select any GPU-based
    instance to meet your training requirements. Click the Next:
    Configuration Instance Details button at the bottom of the screen.

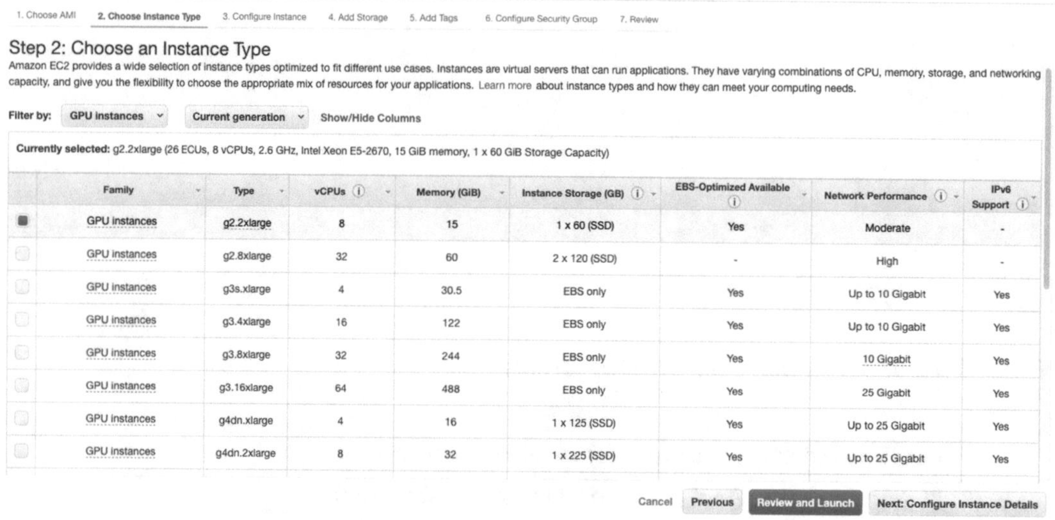

	Family	Type	vCPUs ⓘ	Memory (GiB)	Instance Storage (GB) ⓘ	EBS-Optimized Available ⓘ	Network Performance ⓘ	IPv6 Support ⓘ
■	GPU instances	g2.2xlarge	8	15	1 x 60 (SSD)	Yes	Moderate	-
☐	GPU instances	g2.8xlarge	32	60	2 x 120 (SSD)	-	High	-
☐	GPU instances	g3s.xlarge	4	30.5	EBS only	Yes	Up to 10 Gigabit	Yes
☐	GPU instances	g3.4xlarge	16	122	EBS only	Yes	Up to 10 Gigabit	Yes
☐	GPU instances	g3.8xlarge	32	244	EBS only	Yes	10 Gigabit	Yes
☐	GPU instances	g3.16xlarge	64	488	EBS only	Yes	25 Gigabit	Yes
☐	GPU instances	g4dn.xlarge	4	16	1 x 125 (SSD)	Yes	Up to 25 Gigabit	Yes
☐	GPU instances	g4dn.2xlarge	8	32	1 x 225 (SSD)	Yes	Up to 25 Gigabit	Yes

Figure 10-25. *Choose an Instance Type selection screen*

5. Fill in the Configure Instance Details page (as shown in
    Figure 10-26). In the Number of Instances field, I entered **2** to
    create two nodes in the cluster. You can create as many nodes as
    you need to scale your training.

Step 3: Configure Instance Details

Number of instances ⓘ	2	Launch into Auto Scaling Group ⓘ
	You may want to consider launching these instances into an Auto Scaling Group to help you maintain a help your application stay healthy and cost effective.	
Purchasing option ⓘ	☐ Request Spot instances	
Network ⓘ	vpc-871726fd (default) ⬍	C Create new VPC
Subnet ⓘ	No preference (default subnet in any Availability Zon ⬍	Create new subnet
Auto-assign Public IP ⓘ	Use subnet setting (Enable) ⬍	
Placement group ⓘ	☑ Add instance to placement group	
Placement group name ⓘ	○ Add to existing placement group.	
	● Add to a new placement group.	
	comp_viz_cluster	
Placement group strategy ⓘ	cluster ⬍	
Capacity Reservation ⓘ	Open ⬍	C Create new Capacity Reservation
IAM role ⓘ	None ⬍	C Create new IAM role
CPU options ⓘ	☐ Specify CPU options	

Figure 10-26. *Configuring the instance details*

For the Placement Group setting, check the box "Add Instance to placement group" and choose either to create a new placement group (as shown in Figure 10-26) or add to an existing one. For the Placement Group Strategy field, select cluster.

Leave everything else at the default settings on this page and click the Next: Add Storage button.

6. On the Add Storage page (as shown in Figure 10-27), provide the numbers for the disk size as per your needs. In this example, we will leave everything as is. Click the Next: Add Tags button and then the Next: Configure Security Groups button.

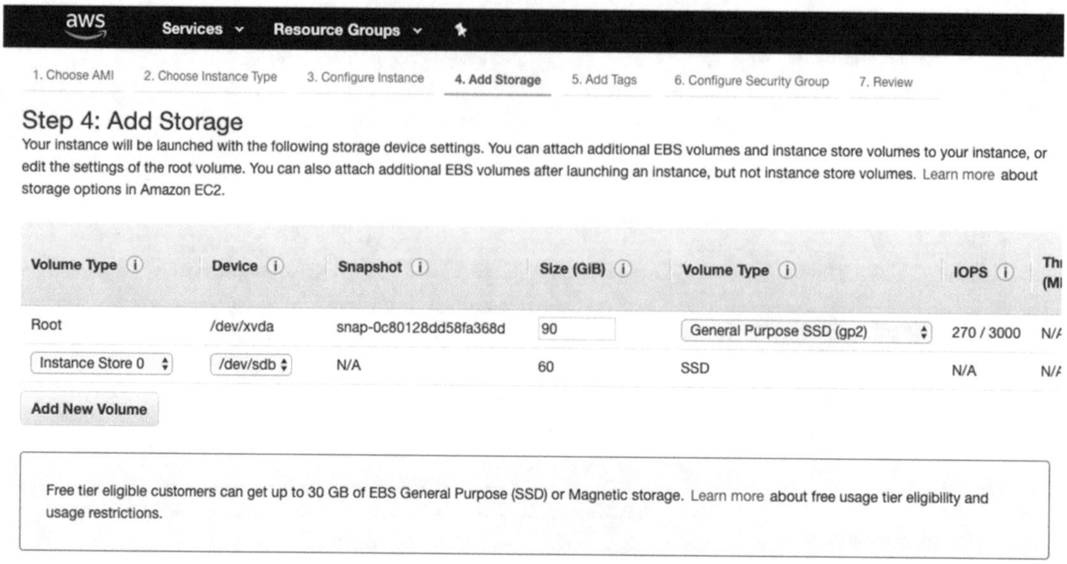

Figure 10-27. *Add Storage page*

7. Either create a new security group or use "Select an existing security group" if you want an existing security group (see Figure 10-28). Click Review and Launch followed by the Launch buttons. This will display a pop-up screen to either create or select a key pair. This key pair is used to log on to the VM using SSH. Follow the on-screen instructions (as shown in Figure 10-29).

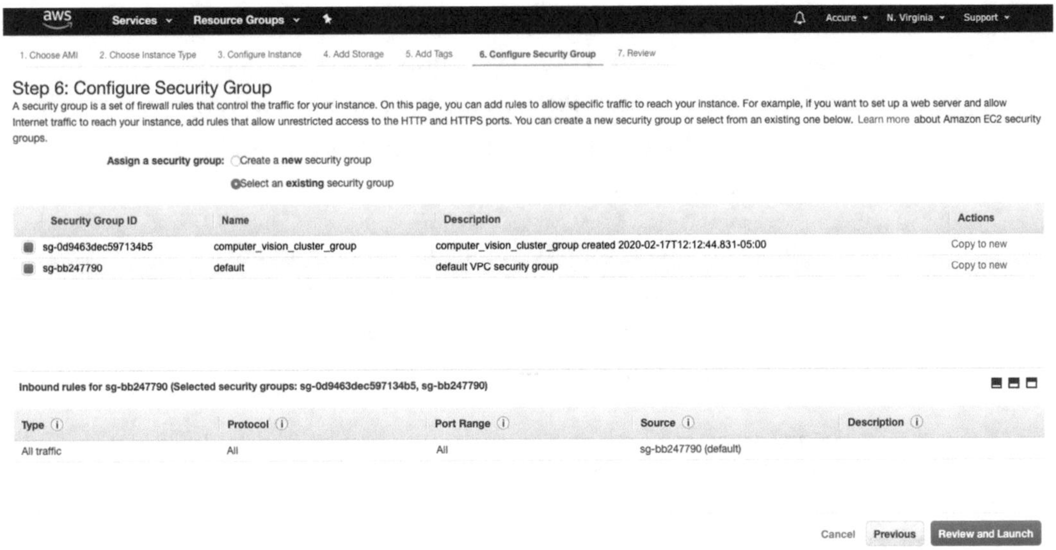

Figure 10-28. *Page to create or select security groups*

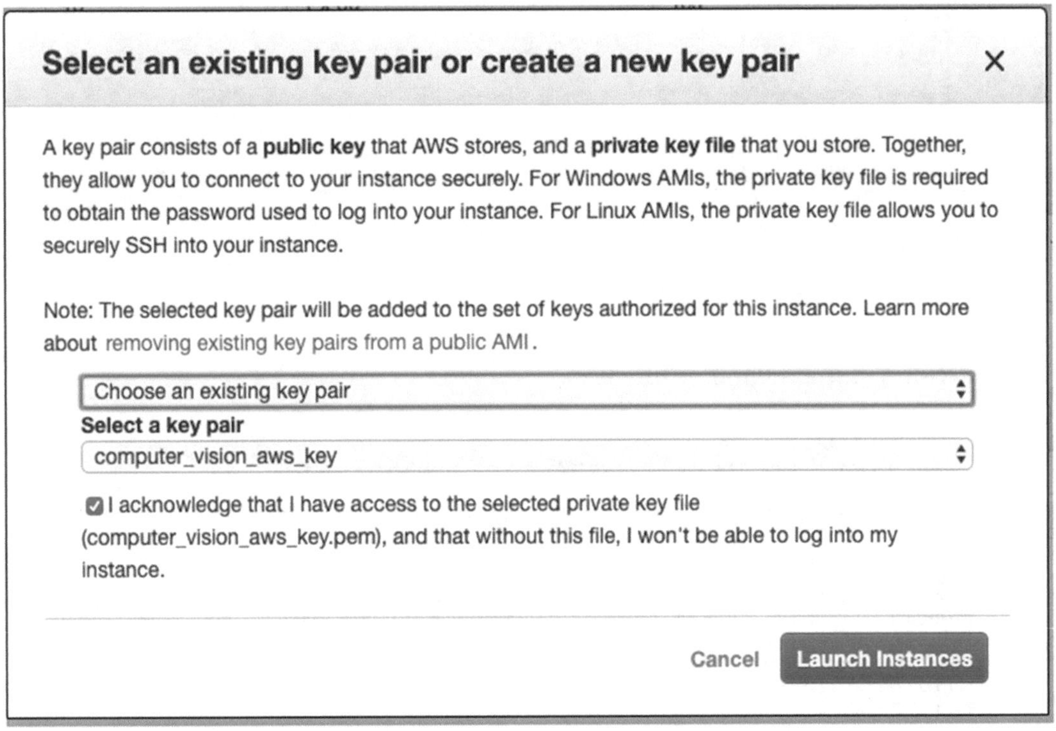

Figure 10-29. *Pop-up screen to create or select a key pair*

8. After the instances are successfully launched, we will need to create passwordless SSH to enable every node to communicate with each other. We create an RSA key on one machine and copy the public key from the rsa_id.pub file to all nodes' authorized_ keys file. Here are the steps:

   a. SSH to machine 1, and from its home directory, execute the command ssh-keygen. Press Enter for every single prompt until you see the fingerprint printed on the screen. The terminal output should look like Figure 10-30.

```
[[ec2-user@ip-172-31-23-129 ~]$ ssh-keygen
Generating public/private rsa key pair.
[Enter file in which to save the key (/home/ec2-user/.ssh/id_rsa):
[Enter passphrase (empty for no passphrase):
[Enter same passphrase again:
Your identification has been saved in /home/ec2-user/.ssh/id_rsa.
Your public key has been saved in /home/ec2-user/.ssh/id_rsa.pub.
The key fingerprint is:
SHA256:874jdAh5vnSgV+gkvCJ85lrxVtHWaXPSZsZFS27IjDs ec2-user@ip-172-31-23-129
The key's randomart image is:
+---[RSA 2048]----+
|              oo|
|         . .+++..|
|    . ...o.*+B+ |
|     = =o...B.  |
| .  .  XS+ E    |
|   o +oo.Boo .  |
|    =..o+ +.    |
|    ... o..     |
|   ..     .oo   |
+----[SHA256]-----+
[ec2-user@ip-172-31-23-129 ~]$ 
```

Figure 10-30. `ssh-keygen` *output*

   b. Copy the content of ~/.ssh/id_rsa.pub to ~/.ssh/ authorized_keys, as shown in Figure 10-31 and Figure 10-32.

```
[ec2-user@ip-172-31-23-129 ~]$ cat ~/.ssh/id_rsa.pub
ssh-rsa AAAAB3NzaC1yc2EAAAADAQABAAABAQC3UrShdR+B0RGOa91nW6zgb6NL2vvaN2pRQALInuvQiYQtn9Oz7P+hq/sBuItz95JRv2
hT61Hg0ntmhRX7onFgVQ00Zht/IAj+WoVKlOS2ozPiypgwiW9ORdiNG5BXwxwBGvhkjsMBh7IXKG31U92+sSxBoAZkfHTGuRWd3m9gzsb6
1KTxpohB2fhbr9MzXnSINV72jgzkAaqZDrgoruh6/0rDdp6Q5C81FBsDfG6dBKSXk2zcBjITYz7joJgMmXXA2tJmLyhBGMEFOFqfIdaBZy
YQahCNmTcnhV/0T6lauGzjjOAYyNfRpiBYbi2MNJnWTm8Cvz8jlb1brlljo+/H ec2-user@ip-172-31-23-129
[ec2-user@ip-172-31-23-129 ~]$ vi ~/.ssh/authorized_keys
[ec2-user@ip-172-31-23-129 ~]$ 
```

Figure 10-31. `cat ~/.ssh/id_ras.pub` *output; copy the entire text starting from* ssh-rsa

```
ssh-rsa AAAAB3NzaC1yc2EAAAADAQABAAAABAQCWJdz9xJiP2c9N2CydvCowo1Pua3pC+M5/Vjpl44YhRwMpIpi6WQYbtjDyFPQCUPQedp
UDtPrOSejHKovY/Ewr9zcH20h1bkq4iLHjkqPTDM56M66jfp00pvNQZHJdgPYolrx5dW8mH85HvNeDBiBvkBERDmUKzdMV6VkXf7aYEU+V
yJnnprTciysfTnHplUTrXVnJRAzv0UVyFnODwbk1wSCKmIrd0CaJSh0RtcFOWb9FOPWmrN7LBAWoIsGxyXlUCEZi/QEyO3pOwtMKH6kuW6
M7ruYDeO/iE0UzT4JJKsrwFOVLr7Yx1puH1eKLSSmZk1N1fDE+dZd5SIJJLcIN computer_vision_aws_key
ssh-rsa AAAAB3NzaC1yc2EAAAADAQABAAAABAQC3UrShdR+B0RGOa91nW6zgb6NL2vvaN2pRQALInuvQiYQtn9Oz7P+hq/sBuItz95JRv2
hT61Hg0ntmhRX7onFgVQ00Zht/IAj+WoVKlOS2ozPiypgwiW9ORdiNG5BXwxwBGvhkjsMBh7IXKG31U92+sSxBoAZkfHTGuRWd3m9gzsb6
1KTxpohB2fhbr9MzXnSINV72jgzkAaqZDrgoruh6/0rDdp6Q5C81FBsDfG6dBKSXk2zcBjITYz7joJgMmXXA2tJmLyhBGMEFOFqfIdaBZy
YQahCNmTcnhV/0T6lauGzjjOAYyNfRpiBYbi2MNJnWTm8Cvz8jlb1brlljo+/H ec2-user@ip-172-31-23-129
~
~
~
~
~
```

Figure 10-32. *Paste the* `id_rsa.pub` *content to the end of the* `authorized_`
keys file

    c.   Copy the `id_rsa.pub` content of one machine to the end of the
authorized_keys files of all nodes.

    d.   Repeat the process to create `ssh-keygen` on the rest of the
machines and copy the contents of `id_rsa.pub` to the end of
authorized_keys of each of the nodes.

    e.   You should verify by logging in via SSH from one machine to
another. It should allow you to log on without any password.
If SSH prompts for a password, that means you do not have
passwordless communication from one machine to the
other. For Horovod to work, all machines must be able to
communicate without a password to other machines.

Running Distributed Training

The AMI we used in this example contains scripts to launch the training in distributed
mode. There is a `train_synthetic.sh` shell script located at `/home/ec2-user/examples/`
`horovod/tensorflow`. You can modify this script to point to your code and launch the
training.

This example script launches a RestNet-based training on the Horovod cluster we
just created. Simply execute it as follows:

sh /home/ec2-user/examples/horovod/tensorflow/train_synthetic.sh 2

The 2 arguments indicate the number of GPUs in the cluster.

If everything goes well, you will have a trained model that you can download to the
machine where you will host the application that uses this model to predict outcomes.

The AMI we used has Horovod already installed. If you want to use a VM that does not have Horovod, follow the installation instructions in the next section.

Installing Horovod

Horovod depends on OpenMPI to run. First we need to install OpenMPI using the commands shown in Listing 10-19.

Listing 10-19. Installing OpenMPI

```
# Download Open MPI
$ wget https://download.open-mpi.org/release/open-mpi/v4.0/
openmpi-4.0.2.tar.gz
# Uncompress
$ gunzip -c openmpi-4.0.2.tar.gz | tar xf -
$ cd openmpi-4.0.2
$ ./configure --prefix=/usr/local
$ make all install
```

It will take several minutes to install OpenMPI.

After OpenMPI is successfully installed, install Horovod using the pip command, as shown in Listing 10-20.

Listing 10-20. Installing Horovod

```
$ pip install horovod
```

Listings 10-19 and 10-20 must be executed on all machines of the cluster.

Running Horovod to Execute Distributed Training

To run on a machine with four GPUs, use this:

```
$ horovodrun -np 4 -H localhost:4 python horovod_tensorflow_mnist.py
```

To run on four machines with four GPUs each, run this:

```
$ horovodrun -np 16 -H host1:4,host2:4,host3:4,host4:4 python horovod_
tensorflow_mnist.py
```

You can also specify host nodes in a host file. Here's an example:

```
$ cat horovod_cluster.conf
host1 slots=2
host2 slots=2
host3 slots=2
```

This example lists the hostnames (host1, host2, and host3) and how many "slots" there are for each. Slots indicate how many GPUs the training can potentially execute on a node.

To run on hosts specified in a file called horovod_cluster.conf, run this:

```
$ horovodrun -np 6 -hostfile horovod_cluster.conf python horovod_
tensorflow_mnist.py
```

VMs with GPUs are costly. Therefore, it is advised to terminate the VMs if they are no longer used. Figure 10-33 shows how to terminate your instances.

Figure 10-33. *Terminating AWS VMs*

Summary

The chapter started with the introduction of distributed training of computer vision models. We explored various distribution strategies supported in TensorFlow and analyzed how to write code for distributed training.

We trained our handwriting recognition model based on the MNIST dataset on the GCP, Azure, and AWS cloud infrastructures. We explored three different techniques of training models on the three cloud platforms: GCP, AWS and Azure. Our example training was based on TensorFlow-supported distribution strategies: `ParameterServerStrategy` and `MirroredStrategy`. You also learned how to use Horovod for large-scale training of computer vision models.

Index

A

Activation functions, 179
    computer vision model, 212
    Leaky ReLU, 184
    linear activation function, 180, 181
    neuron, 179
    ReLU, 183
    SELU, 184, 185
    sigmoid activation function, 181, 182
    softmax function, 186, 187
    softplus activation function, 186
    TanH, 182
Adaptive Gradient (AdaGrad), 195, 199, 498
Adaptive Moment (Adam), 195, 200
AlexNet, 257, 258, 268
Amazon Simple Storage Service
    (Amazon S3), 167
Amazon Web Services (AWS), 458, 497
Anchor boxes, 275, 278, 289
Application-specific integrated circuits
    (ASICs), 465
Arithmetic operations
    addition
        add() function, 52, 54
        addWeighted() function, 52, 55
        images, 51
        OpenCV, 51
        original image, 53
    subtraction, 55, 56
        cats images, 57, 58
        constant subtraction, 60
        cv2.subtract() function, 57

NumPy, 57, 59
    pixel values, 55
    resizing images, 57
Artificial intelligence (AI), 125, 305
Artificial neural network (ANN),
    164, 167, 170
    artificial neuron, 173
    human brain function, 171
    human eye functions, 171
    information processing, human
        neurons, 172
    interpreting device, 173
    manual feature engineering, 236
    MLP (see Multilayer perceptron (MLP))
    neurons, 172
    object recognition, 170
    perceptron, 174
    sensing device, 172, 173
Aspect ratios, 277, 278, 283
Asynchronous data parallelism, 460
Azure Blob Storage, 167, 450

B

Backpropagation method, 201
best.onnx, 365
Bilateral blurring, 81–83
Binarization
    adaptive thresholding
        adaptiveThreshold() function, 88
        binarized image, 89
        definition, 86

513

Printed in the United States
by Baker & Taylor Publisher Services